Contents

Contents

Preface

This is the third edition of *Understanding Housing Defects*. In this new edition, besides reviewing, revising and updating the text, we have also added:
- a chapter on the Investigation of Structural Defects
- additional chapters on services
- sections explaining common problems in basements, chimneys and garden walls
- many more photographs.

But perhaps the most fundamental difference is the move from black and white to full colour. We hope you share our opinion that colour photos and graphics are more realistic, they show more detail, and they are easier to understand (and maybe remember).

Our aim in writing this edition has not changed – we want to provide an interesting, coherent and comprehensive introduction to the causes, investigation and diagnosis of housing defects. There are a number of excellent texts on building failure but these tend to be either highly technical or aimed at specialist practitioners. We have always felt there was a need for a more general text, aimed at those students and practitioners who require a broad understanding of housing defects as part of a wider sphere of academic or professional activity. This group includes Building Surveyors, General Practice Surveyors, Architects, Maintenance Inspectors, Estate Agents, Housing Managers, Environmental Health Officers and Builders.

The revised text, like the original, has three specific objectives:
- to explain why, and how, defects occur
- to enable the reader to recognise and identify building defects
- to provide, where appropriate, guidance on their correct diagnosis.

Additionally, we still hold the view that a broad knowledge of building principles and good building practice are prerequisites for a genuine understanding of building defects. To this end we have, where necessary, introduced each chapter with a brief summary setting out construction principles and describing the evolution of current practice.

The text does not include aspects of building repair and we make no apology for this. Many repairs are obvious, others require input from specialists. In addition, there is often a very wide range of repair methods and materials; to do them all justice would require several volumes.

With regard to Building Services we have extended the text and split it into three discrete chapters. Some defects in building services are generic but many are unique to specific products. We have, however, tried to include guidance on the most common problems.

Duncan Marshall, Derek Worthing and Roger Heath, UWE, Bristol 2009

Building Movement
– Foundations

Subsidence of a Solicitor's Offices at Northwich

Both these problems have been caused by mining. Dudley (The Crooked House pub) used to be a coal mining area, Northwich is famous for its salt mines.

Ben Locke

Cheshire Record Office

SOME KEY TERMINOLOGY

There are some basic definitions that need to be understood before considering foundation failure.

Cohesive soils

Soils that have the natural tendency to 'ball' together when squeezed firmly in the palm of the hand.

Consolidation

The act of increasing the density and strength of the soil by the expulsion of water and air under self-weight and constant external loading.

Desiccation

The process of the drying out of a soil or sub-soil.

Differential settlement

This occurs where settlement (see below) takes place at differing rates under different parts of a building or structure as a result of variations in loading and/or ground conditions.

Heave

The upward or lateral expansion of the sub-soil or fill.

Non-cohesive soils

These are granular soils which range from compact gravel and sand to loose sand.

Settlement

The downward movement of the ground, or any structure on it, which is due to the load applied by the structure. (Note: this definition is the one accepted by surveyors and engineers but the legal definition differs.)

Subsidence

Vertical and downward movement of the ground which is not caused by the imposition of building/foundation loads, ie loss of support from the sub-soil. The two main causes are trees and defective drains; less common are other causes such as mining subsidence and made ground/land fill.

INTRODUCTION

All buildings move. In the majority of cases this will be for quite acceptable reasons and the amount of movement will not adversely affect the building's performance.

Initially, when a building is constructed, movement downwards will be caused by the new loads imposed on the sub-soil beneath. Unless that sub-soil is particularly dense and strong (eg a rock such as granite), a newly constructed building will suffer from initial settlement as the sub-soil beneath is consolidated. Normally, the downward movement of the sub-soil and any resultant cracking of the building are both of a minor nature. With non-cohesive granular sub-soils, such as sand, this initial settlement occurs quickly and has often ceased by the time the construction has been completed. However, in cohesive clay sub-soils, settlement may occur over quite a long period, perhaps lasting 20 years or more from construction until the sub-soil is fully consolidated.

Simplified diagram illustrating loading

LOAD

LOAD

Loads transferred downwards from roof and floors into walls and ultimately foundations.

Upper levels of sub-soil may be affected by seasonal change.

Ground floors can bear directly on the ground or be carried by walls.

Weaker sub-soils initially tend to consolidate under the newly applied loads of the building.

Buildings are also subject to seasonal movement and this will continue throughout the building's life. This is a result of certain sub-soils, in particular clay, being affected by climatic changes as summer changes to winter and back again. Expansion of these sub-soils occurs in winter as water content increases and, conversely, contraction occurs as they dry out in summer.

It should be obvious that such cracks will open and close seasonally but in normal British weather conditions such movement should be within acceptable limits. Unfortunately, for a number of reasons, seasonal movement in water-susceptible sub-soils, such as shrinkable clay, can become excessive and lead to an unacceptable level of cracking and movement.

While accepting that much building movement will be of a minor nature in both extent and effect, there will still be occasions when an inspection of a building will reveal evidence of movement that causes concern. Often, the movement will manifest itself along lines of 'natural' weakness within the above-ground structure, eg windows, doors or junctions between separate elements of construction. It is then necessary to determine the cause(s) of the problem and its implications. This may involve assessing a wider range of defects other than just cracking and movement, eg cracking of an external wall due to foundation failure may also allow water penetration.

Left: The movement in this stone wall has taken place along a line of structural weakness, ie around an opening.
Right: On the other hand, many older buildings are capable of withstanding substantial movement without cracking because they are built in brickwork (ie smaller structural units which can better distribute load) laid in a lime mortar.

FOUNDATIONS – CONSTRUCTION SUMMARY

TRANSFERANCE OF LOAD

The purpose of the foundation is to fully and efficiently transfer the load from the building onto the sub-soil beneath. The exact type and performance requirements of a foundation will depend upon a number of factors including building type, use, load, etc and/or sub-soil type, formation, condition.

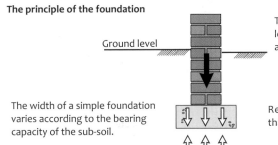

The principle of the foundation

Ground level

The width of a simple foundation varies according to the bearing capacity of the sub-soil.

The building load including the structure and all imposed loads (furniture, snow etc.) exerts downward pressure on any footing/foundation.

Resistance of the sub-soil needs to be equal or greater than the total applied load from the foundation.

The foundation is an essential element of construction in any modern building although many older buildings have no formal foundations as such. It should be designed specifically for a particular building and the sub-soil beneath. For many buildings, especially low-rise housing, its design will be relatively simple although due attention must be given to any peculiarities of the structure and sub-soil.

For other buildings which, perhaps, involve unusual construction, layout or use, or where sub-soil conditions indicate poor bearing capacity, a more sophisticated approach to both design and construction of the foundation will be required.

The pier of the brickwork in between the windows has created excess loading on the foundation/sub-soil. Cracking has been avoided because the walls are built in lime mortar.

SUB-SOILS – A BRIEF OUTLINE

Consideration must be given to the type of sub-soil being built upon, both when dealing with original design and, subsequently, when trying to understand why a building is suffering movement.

Broadly speaking, there are five main types of sub-soil. These are set out below in descending order of bearing capacity.

Rock Strictly speaking, this is not a soil. This should bear loads without problems. However, rock may contain faults or fissures that can collapse under load and result in settlement.

Granular soils such as sands and gravels These should bear loads without problems but sands can be affected by flooding and loose sand is inherently unstable unless contained.

Cohesive soils such as clays These include firm/stiff clays, sandy clays, soft silt, and soft clay. This group includes shrinkable clay. They frequently suffer problems, normally due to the presence of water or changes in water content. Approximately 75% of UK building insurance claims are for subsidence in clay sub-soils.

Organic soils such as peat and topsoil These should never be built upon (unless piles are used) because they contain organic material and a high proportion of air and water.

Made ground Made ground (normally referred to as landfill) is a separate category. In theory, if it is formed with non-organic material it is possible to build directly on it once it has fully settled and consolidated. In practice, it is difficult to guarantee its content and it may contain a high proportion of vegetable and other deleterious materials such as tin cans, old refrigerators, etc which consolidate in an unpredictable manner.

Sub-soils will originally have been laid down in a series of layers or beds of varying thickness. Weaker sub-soils may lie over, or under, stronger ones, or there may be a multi-layered mixture. Good design must ensure that the foundations of a building are taken down to a suitable depth to avoid the effects of adverse climatic factors and, also, to bear on a sub-soil capable of providing adequate support. This may involve building down through weaker layers such as peat or made ground, until a firmer, more stable sub-soil is reached. Alternatively, the design may involve a different approach and make use of a raft foundation that floats on or just below the surface of the sub-soil. Unfortunately, building movement often occurs because insufficient thought has been given to the type of sub-soil beneath the proposed building.

FOUNDATIONS – A BRIEF HISTORY

Historically, buildings were constructed directly on the surface of the ground until it was realised that removal of at least the topsoil normally provided a firmer base. This was considered sufficient for cheaper brick and stone buildings although timber-framed structures were often built on a stone plinth. In the 19th century a thin layer of fire ash or furnace clinker was commonly spread over the exposed sub-soil to provide a level surface on which to build.

This load-bearing wall has a crude foundation comprising a shallow row of boulders rammed into the earth. The circled brush gives some idea of scale.

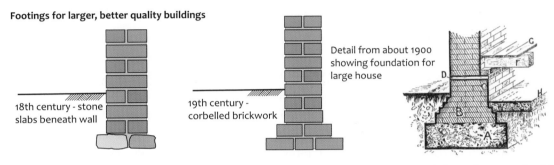

Footings in poorer quality buildings up to late 19th century

Ground level

Depth could be as little as depth of topsoil

Topsoil removed and subsoil surface levelled

25-50mm of ash or clinker

By 1900 new houses were normally required to have concrete or brick foundations (unless on rock). However, they were often less substantial than those used today. In addition Building Control was far from rigorous.

Footings for larger, better quality buildings

18th century - stone slabs beneath wall

19th century - corbelled brickwork

Detail from about 1900 showing foundation for large house

Larger buildings were found to require a more sophisticated approach and this often resulted in slightly deeper foundations and an attempt to spread the greater loads involved over a larger area of sub-soil. Originally this was by the use of stone slabs, but as brickwork became more frequently used in the 19th century, corbelled brick footings were developed. These were often used in better quality housing, especially in the latter half of the Victorian period.

The 19th century saw great advances in the knowledge of how buildings and materials perform and at the same time there was a realisation of the need for minimum construction standards. Although there had been previous attempts to legislate for building standards, these were neither of great detail nor necessarily applied uniformly in all parts of the country. The Public Health Act of 1875 was the first of a number of pieces of legislation that attempted to introduce minimum building standards on a national basis in England and Wales (except for London, which had its own rules and regulatory structure). Subsequent Public Health Acts, Building Acts and, since 1965, Building Regulations, have set increasingly higher standards of construction which, in part, reflect greater understanding of how buildings perform. London has, quite separately, followed the same raising of standards through its own building legislation although since 1987 the Building Regulations have applied here as well.

Strict application of minimum standards by local authorities under the powers given to them under

Left: A section through a proposed house of the 1890s showing shallow strip foundations.
Right: Although of a later date (1920s), this design shows strip foundations under stepped brick footings.

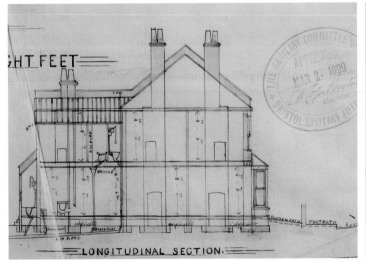

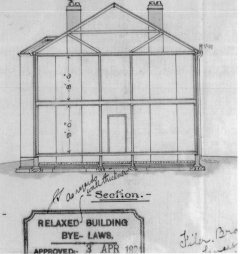

legislation resulted in both deeper and more sophisticated footings in the late 19th and early 20th centuries. In particular, corbelled brick footings began to be used for all types of construction. Eventually, these were superseded by concrete strip foundations in the early part of the 20th century.

Concrete strip foundations

It is unusual to find concrete strip foundations in 19th century buildings and they only became common from the 1920s onwards, at the same time as external cavity walls became the accepted standard. Strip foundations are designed to spread building load in a similar, but cheaper, way to corbelled brickwork. By this time it was also recognised that seasonal movement in certain types of sub-soil, in particular shrinkable clay, required a foundation depth of at least 750mm deep below ground level. Subsequent knowledge, in part as a result of investigations into building movement caused by a series of extremely hot and long dry periods in the mid-1970s and the 1980s, has resulted in recommended foundation depth in climatically affected sub-soils (eg shrinkable clays) being increased to a minimum of 1000mm. This is increased to up to 2.5m or greater where located close to or above removed trees. However, in more stable sub-soils, such as gravel or firm sand, a shallower foundation can still be constructed, eg at least 450mm deep in firm sand, although the risk of frost attack must be recognised.

There have been great advances over the years in the development of new forms of construction, materials and components. Pressure caused by increased demands on land use allied to our greater knowledge has resulted in the development of a variety of alternative foundation techniques. Some have been introduced because they are more economic, others because they allow inherently more difficult sub-soils to be built over.

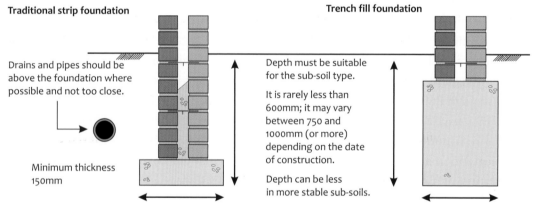

Traditional strip foundation

Drains and pipes should be above the foundation where possible and not too close.

Minimum thickness 150mm

Trench fill foundation

Depth must be suitable for the sub-soil type.

It is rarely less than 600mm; it may vary between 750 and 1000mm (or more) depending on the date of construction.

Depth can be less in more stable sub-soils.

Width of foundation depends on bearing capacity of ground.
Strip foundations need additional width to provide working space.

Trench-fill foundations

This is a modern alternative to a strip foundation that is suitable for use down to depths of about 1000–1500mm below ground level. Below this level it becomes uneconomic. Its advantages are its simplicity and relative cheapness because its use of mass concrete avoids the need for block-work or brick-work below ground level. It can be found in buildings constructed from the 1970s.

Trench-fill foundations on a modern housing estate.

Raft foundations

A raft foundation is basically a slab of reinforced concrete at, or just below, ground level. It acts as both a ground-bearing slab over the floor of the building and a footing to the walls. It enables the loads of the building to be spread over a large sub-soil area and allows areas of either known or suspected weak sub-soil to be bridged over. Often used above underground mining areas, it requires careful design by structural calculation. A 20th century approach to a variety of sub-soil problems.

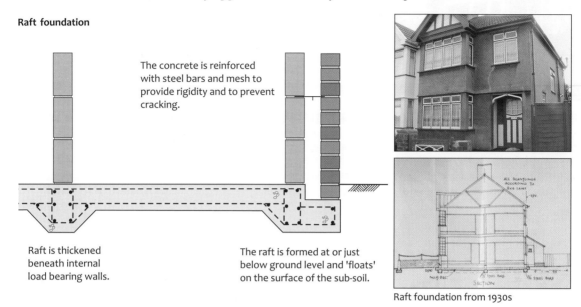

Raft foundation

The concrete is reinforced with steel bars and mesh to provide rigidity and to prevent cracking.

Raft is thickened beneath internal load bearing walls.

The raft is formed at or just below ground level and 'floats' on the surface of the sub-soil.

Raft foundation from 1930s

Piled foundations

Filled ground, such as former quarries, and ground with a weak sub-soil profile, such as alternating layers of soft clay and other similarly soft sub-soils, may require piling in order to properly carry the building load down to a sub-soil layer capable of supporting the load. The piles support ground beams which act as bridges beneath the walls above. In these circumstances the ground floor might well be a reinforced concrete slab or pot and beam construction spanning the full distance between structural walls to avoid any problems of inadequate support. There are a variety of different types of concrete pile ranging from mini-piles, suitable for depths down to about 2m below ground level, to driven or bored piles which can be used down to much greater depths – 8m or more. This form of foundation is a late 20th century technique as far as low-rise domestic construction is concerned and is rarely found in buildings constructed before 1970, except as remedial underpinning.

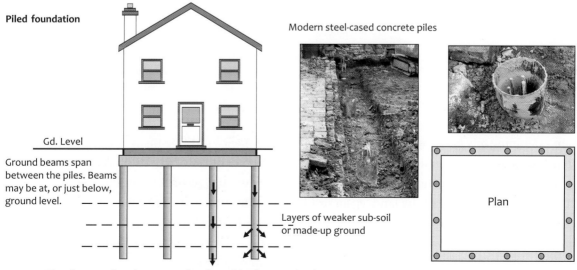

Piled foundation

Modern steel-cased concrete piles

Gd. Level

Ground beams span between the piles. Beams may be at, or just below, ground level.

Layers of weaker sub-soil or made-up ground

Plan

The piles are taken down to a sub-soil capable of supporting the load of the building. They can be end-bearing and/or friction bearing.

In modern construction piles are usually formed in concrete or steel. In Amsterdam timber piles were common. Rot in the piles has caused extensive differential settlement in this building.

COMMON STRUCTURAL DEFECTS OF FOUNDATIONS AT CONSTRUCTION STAGE (NEW BUILD)

POTENTIAL PROBLEM AREAS	IMPLICATIONS	TYPICAL DEFECTS
Insufficient depth of foundation (especially in shrinkable clay sub-soils)	Seasonal movement due to changes in water content of sub-soil. Movement due to trees and shrubs	Cracks in walls especially at 'natural' lines of structural weakness, eg windows, doors, junctions with extensions and bays
Insufficient width of foundation	Load insufficiently spread over sub-soil. Tendency for settlement to occur	Cracking in walls above ground level
Soft spots in sub-soils	Foundation is inadequately supported by the sub-soil leading to settlement	Cracking of walls above the affected section of foundation
Insufficient steps in strip foundation on sloping ground	Tendency for building to slide or creep especially if there is soil erosion	Displacement movement of building/ section of building. Cracking in walls
Trees and large shrubs in close proximity (in shrinkable clay sub-soils)	Sub-soil subject to changes in water content due to desiccation (lack of moisture) and heave (excess moisture)	Cracking in walls especially along natural lines of structural weakness
Recent removal of trees or large shrubs (especially in shrinkable clay sub-soils)	Sub-soil subject to a period of heave (over several years)	Cracking in walls especially along natural lines of structural weakness
No provision of compressible material to piles and ground beams in shrinkable clay sub-soils (to allow for heave)	Any expansion (or heave) of the sub-soil will lead to structural movement	Upward movement with cracks along lines of weakness in structure
Clay sub-soil with high sulfate content	Sulfates attack concrete foundations and below-ground cement mortar jointing	Movement in above-ground walls. Expansion and deterioration of below-ground cement mortar and concrete
Inadequacy of structural support from land fill (made-up ground)	Excessive settlement over a prolonged (often indeterminate) period of time	Cracking of walls. Complete collapse of building
Close proximity of drain runs which are deeper than foundations and/or lacking concrete cover	Inadequate support to the foundations	Cracking of walls above ground level. Walls out of plumb due to lateral displacement

CONCLUSION

When examining any building for movement problems, reference must be made to the date of construction of each part. Consideration will need to be given to the implications of dealing with different ages, types and styles of construction within the same property. Initially, records should be examined to determine previous use(s) of the land, eg an in-filled pit. Unfortunately, it may be difficult to ascertain these facts without some element of detailed and, possibly, destructive investigation, eg the removal of render/plaster or the excavation of trial holes. Foundations and their defects, by their very nature, are invariably difficult to precisely identify without some form of excavation.

FOUNDATION DEFECTS – THE CAUSES

Not all defects are as dramatic as these. The Temple Church in Bristol was started in 1398. The tower soon leant because the alluvial soil was too weak to support the heavy load of the tower. The upper stage (built at a different angle in an attempt to straighten it) was built in 1460. According to local history the Americans, stationed in Bristol at the end of the Second World War, intended to demolish the tower, incorrectly assuming the lean to be the result of bomb damage.

The 'defect' on the right was caused by an explosion at Bridgnorth Castle during the English Civil War.

Peter Smith

INTRODUCTION

There is a wide range of reasons for foundation failure but often it is because of inadequate design and/or poor construction. Historically, this was due to lack of technical knowledge of construction and poor understanding of how sub-soils perform when under load. With more recently constructed houses, it has to be accepted that such failure is often due to incompetence, eg lack of proper site investigation, poor design and/or bad site practice.

The investigation of any suspected foundation failure will generally involve assessment of movement that is initially apparent in other elements of the structure, eg walls, floors, roof. It is only after evidence of above ground movement has been analysed and its wider implications considered that physical investigation of the underground parts of a building's structure will be undertaken, (probably by means of trial excavations). The examination of the foundation, or lack of it, will include investigation of the sub-soil beneath the building. At the same time, reference will need to be made to topographical factors such as hills, valleys, water courses, trees and vegetation, in order that the effects of such external influences upon the structural performance of a particular building are fully recognised. Similarly, other factors of a man-made nature will need to be considered with regard to their effect upon the building, eg proximity of underground drains and services, proximity of recent construction, traffic movement, and industrial activities. *(See Chapter 3 for a more detailed discussion of site investigation.)*

The extent of foundation problems can be determined from the number of insurance claims made annually for subsidence. In 2003, they amounted to a total of over 54,000, with a total estimated cost of approximately £390 million, although this was after a drought year.

INITIAL SETTLEMENT

Certain sub-soils, such as clays and similar materials, may be subject to varying amounts of compression due to the pressure applied by a new building. Often, this is due to water contained within the sub-soil being squeezed out as the sub-soil consolidates under the newly applied load. Such settlement will be of a variable degree depending upon the type of sub-soil. A shrinkable clay sub-soil with a high water content will consolidate further and over a longer period of time than a firmer clay with less water. Movement of this nature will be at its greatest in the first few years after construction and will normally cease within 10–20 years of erection. Most buildings will suffer from it to a greater or lesser degree.

Normally, the effects of such movement are within acceptable levels of damage to the structure and finishes of a building, with cracks up to 5mm across. These can be filled as required for decorative purposes. The cracks can occur anywhere within floors, walls and ceilings but are commonly experienced at the junction between two or more elements, ie at the weakest points in the construction. Sometimes initial settlement is also linked to seasonal movement *(see later section of this chapter)* and the cracks can open and close accordingly.

Occasionally, the damage caused by initial settlement can be greater than 5mm or be spread over a longer period of time. If the movement is significant and proved to be progressive, then remedial under-pinning and crack repair may be required.

Originally constructed in the 17th century on a shrinkable clay sub-soil, this building experienced extensive initial settlement.

It also needs to be recognised that even apparently firm sub-soils such as gravels and softer rock formations may compact slightly under the initial load of a building. In these circumstances, any cracking of the structure and finishes will probably be of minor nature.

SUBSIDENCE DUE TO SUB-SOILS CONTAINING ORGANIC MATTER

The construction process requires the removal of all topsoil incorporating organic or vegetable matter. This is usually to a depth of 150–225mm across the whole area of the building. Normally, such removal will have disposed of all soil containing vegetable matter. However, certain deeper sub-soils, in particular peat that contains high levels of organic matter and water, will also be unacceptable.

The effects of compressible sub-soils

Cracking pattern will be unpredictable. Cracks tend to be wider at the bottom.

Ground level

Settlement due to a compressible sub-soil can be extensive and uneven

Shallow foundation on organic material (peat)

Any applied load will compress such a material and result in movement of the building above. These types of sub-soil should be completely removed or, if of sufficient depth, they will require the use of deep foundations, such as piles, bearing through the weak layer onto a firmer stratum beneath.

The effect of building on easily compressible sub-soil can be cracking within the walls and other elements of the structure as the inadequately supported foundations settle. Such settlement can be extensively and unevenly distributed around a building resulting in significant cracking with no discernible pattern. The pressure from the weight of the building will continue until the organic sub-soil is fully squeezed and compressed. If the sub-soil is deep, catastrophic failure of the building will occur. Remedial under-pinning may be carried out but often with great difficulty. Frequently, the only choice when dealing with this problem is demolition of the building.

SUBSIDENCE DUE TO SEASONAL MOVEMENT

Introduction

A number of sub-soils are affected by seasonal change. This is due to seasonal variation between wet/cold and hot/dry weather. The extent to which a sub-soil is affected depends mainly on its water content. When a great amount of water is present, normally in winter, the sub-soil will expand and, conversely, it will contract when there is a reduction in water content, normally in summer. The presence of trees and vegetation, or their removal, can accelerate or exacerbate the effects of such changes.

Frost heave

This is a problem that can occur when the ground has a high water table or after a period of high rainfall. Water in the soil expands as it turns to ice in cold weather. The effect upon the sub-soil is known as frost heave and it results in the ground expanding, primarily upwards. Sub-soils that can be adversely affected in this way include silts, fine sands and chalk.

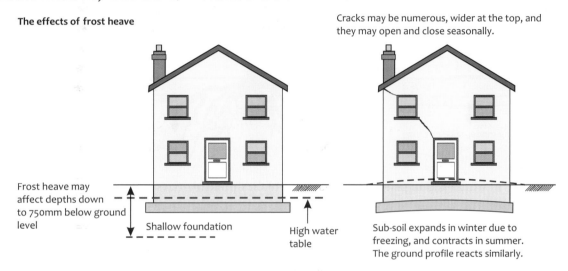

The effects of frost heave

Cracks may be numerous, wider at the top, and they may open and close seasonally.

Frost heave may affect depths down to 750mm below ground level

Shallow foundation

High water table

Sub-soil expands in winter due to freezing, and contracts in summer. The ground profile reacts similarly.

In the UK it is unusual for frost to affect a greater depth than 700mm below ground level. When frost heave does occur it will have a very noticeable seasonal effect with cracks in a building opening and closing between winter and summer. It can sometimes be severe enough to require under-pinning down to a non-frost susceptible level below 700mm to avoid its effects.

Shrinkable clays

Clay soils are capable of absorbing and releasing large amounts of water but not all clays do so. They range from firm clays with relatively low water content to shrinkable clays that contain high levels of water. The higher the water content of the clay the more susceptible it is to climatic change and seasonal movement. Shrinkable clays can expand and/or contract by between as much as 50mm and 75mm seasonally. They are also subject to greater initial settlement as water being squeezed out under a new building's load will cause the sub-soil to consolidate. A look at the geological map of England and Wales reveals that the clays that are subject to shrinkage and swelling are to be found mostly south and east of a line

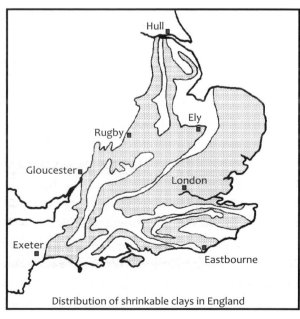

Distribution of shrinkable clays in England

drawn from Gloucester to Hull. Those north and west of this line tend to contain a higher proportion of sand and are less prone to shrinkage.

In summer, shrinkable clay sub-soils will normally tend to contract due to reduction of their water content. This reduction of the soil moisture content is known as 'desiccation'. It can be made worse by the presence of trees and shrubs that take up moisture through their roots. The resultant contraction of the sub-soil can lead to 'subsidence' – the building above will sag/move downwards. In such circumstances cracking will tend to be wider at the bottom rather than the top. However, the presence and effect of trees may confuse the damage pattern.

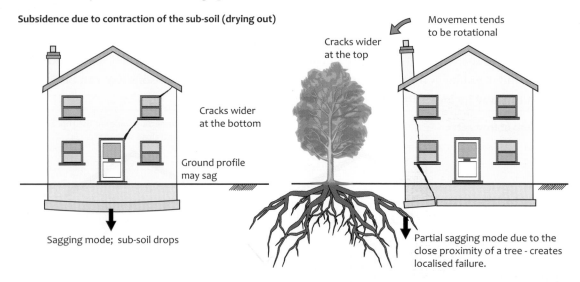

Subsidence due to contraction of the sub-soil (drying out)

Movement tends to be rotational

Cracks wider at the top

Cracks wider at the bottom

Ground profile may sag

Sagging mode; sub-soil drops

Partial sagging mode due to the close proximity of a tree - creates localised failure.

Conversely, in winter and in periods of excessive rainfall, shrinkable clays will tend to expand as water content increases. The increase in volume of the clay leads to an effect known as 'heave' – the building above will rise upwards (known as 'hogging'). The effect is rather similar to frost heave (*as described previously*) in which the cracks tend to be wider at the top and narrower at the bottom.

The effects of seasonal changes in shrinkable clays can generally be avoided by ensuring that foundations are formed at a sufficient depth to avoid the problems of subsidence and heave. This was understood although not well practised in the late 19th century. At this time and for much of the 20th century a sufficient depth was thought to be 750mm but research in the last 30 years indicates that a depth of at least 1200mm is needed (*more near trees – see later*). The research was originally instigated following the very long hot and dry period of 1975/76

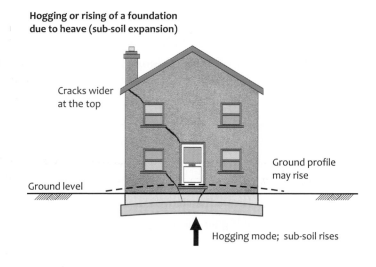

Hogging or rising of a foundation due to heave (sub-soil expansion)

Cracks wider at the top

Ground profile may rise

Ground level

Hogging mode; sub-soil rises

when a considerable number of buildings were affected by subsidence in shrinkable clays. The increased knowledge gained from that experience and subsequent similar hot, dry periods in the 1980s and early 1990s has led to a requirement for deeper foundations in these clays. The research has also indicated that in many cases the movement sustained by a building in such excessively dry spells can be matched by that movement closing back up as sub-soil conditions return to their norm in subsequent seasons. Unfortunately, this closing-up process does not always return a building to its original state of stability or condition. It can also be delayed by further dry weather conditions after the original hot spell.

When subsidence due to a shrinkable clay sub-soil is suspected, the best approach to any investigation is to proceed slowly and methodically unless there is the likelihood of collapse. Initial investigation should indicate whether the movement is due to subsidence or heave but it is usually more difficult to confirm that the movement is progressive, ie still continuing. Often, this can only be found out by monitoring the building over a period of time. This could be for weeks, months or even years. Monitoring may reveal a variety of conditions – the movement has stabilised and is not getting worse; it is progressing slowly/quickly and is worsening; it has changed direction and the cracking is closing. Once monitoring has confirmed what is happening, decisions can be made on any remedial work needed.

This trench in a shrinkable clay sub-soil has been excavated to a depth of at least 1200 to avoid the effects of seasonal change.

These cracks (left) appeared after a long hot summer. The house is built on shrinkable clay which is susceptible to changes in moisture content. Investigation of the building (centre) revealed a shallow foundation under the single storey extension resulting in differential settlement. The cracking is both vertical and horizontal – running along the line of the DPC to behind the water pipe and along the stone cill.

SUBSIDENCE DUE TO TREES AND SHRUBS – PROXIMITY TO BUILDINGS

Movement due to the close proximity of trees to a building is a common problem. This is due to the effect of the tree on the water content of any sub-soil that it is growing in. Trees require water to survive and grow. Much of this is taken up through the root system. A mature poplar can take up to 50,000 litres of water from the sub-soil each year. Other trees including oak can have a similar effect and the more mature a tree the greater the amount of water it requires.

The root radius of a tree is often equal to or greater than its height above ground level. In some cases, eg willow, poplar, elm, the radius can be up to twice the height. The older the tree, the greater the radius and the greater its effect upon the sub-soil.

TABLE OF POPULAR SPECIES OF TREES KNOWN TO CAUSE SUB-SOIL PROBLEMS		
Species	**Mature height (metres)**	**Likelihood of damage to close buildings**
Oak	25–30	1 (Most likely)
Poplar	30–35	2
Ash	25–30	3
Plane	25–30	4
Willow	25–40	5
Elm	25–30	6 (Least likely of the six shown)

Note: It should not be assumed that the above list of trees is comprehensive. There are a great number of other species that are known to cause problems, including many so-called small ornamental species such as hawthorn. Even shrubs and climbers, such as pyracantha, wisteria and rambler roses, have been identified as the causes of building movement. Also some leading authorities differ with regard to likelihood position.

Left: Local authorities are often liable for subsidence in adjoining properties caused by trees planted in streets.
Right: Even a supposedly safe species such as silver birch can pose problems, in this case affecting the adjoining pavement.

Sub-soils containing trees are affected to a varying degree by the type, age and size of the trees and, in some cases, by the number of trees growing. Gravels, firm clays and, to some extent, sandy sub-soils are less affected than shrinkable clays and other water sensitive sub-soils. The effect of any tree growing in one of the latter types of sub-soil is to reduce the volume of the sub-soil as it takes up ground water. This is on top of any drying out due to seasonal change *(see previous sub-section)* and leads to contraction of the clay and the downward movement of any foundation supported. The result can be cracking of the walls and other structural elements of any building above due to subsidence. At its worst, it can affect the complete building.

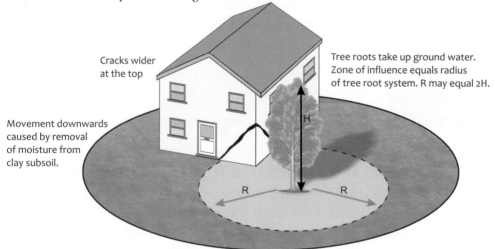

Cracks wider at the top

Tree roots take up ground water. Zone of influence equals radius of tree root system. R may equal 2H.

Movement downwards caused by removal of moisture from clay subsoil.

H

R   R

Seasonal changes can influence the effect of trees upon sub-soils. In particular, prolonged warmer and dryer spells can accelerate sub-soil shrinkage because moisture taken up by the tree root system is not replaced quickly enough to prevent drying up and desiccation of the sub-soil. In such periods the root system may expand in its search for moisture and thus increase the ground area adversely affected by subsidence.

Generally, the further north and west a property is situated in the British Isles the higher the rainfall that it will experience. Any moisture deficit caused by trees and shrubs will be more readily replaced the greater the distance that the building is located from London and the south-east of England. The National House Building Council (NHBC) recognises this by allowing a progressive reduction in the depth of a foundation to a building that is close to trees the further north and west from London that it is situated.

Further problems can be caused by tree roots due to their physical presence and strength. As the tree grows, the radius of the root system increases and the individual roots grow in size. This, in time, can lead to the displacement of the surrounding sub-soil, upwards and outwards pressure on foundations and walls in the vicinity of the root and even growth through underground walls of softer/weaker materials.

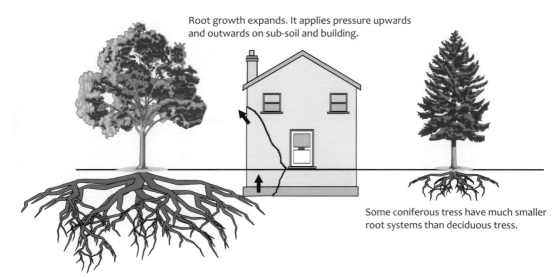

Root growth expands. It applies pressure upwards and outwards on sub-soil and building.

Some coniferous tress have much smaller root systems than deciduous tress.

The Building Research Establishment (BRE) has undertaken considerable research into the effect of trees, subsidence and heave in clay sub-soils. This is wide ranging and continuing. Analysis of its database indicates that oak trees are by far the most damaging species of tree. However, the recommendation by the NHBC of up to 2.5m (or greater where appropriate) deep foundations for all new housing near trees should be adequate to prevent damage by seasonal movements. Unfortunately, there are a great number of houses and other buildings with much shallower foundations than 2.5m (some of quite recent construction) that were built before stricter guidelines were issued for buildings erected close to trees. Furthermore, building owners and occupiers generally do not understand the problems caused by the close proximity of trees to buildings and frequently exacerbate the potential problems by new planting, often of totally inappropriate species. Similarly, trees are often removed by owners, occupiers and developers, sometimes leading to problems associated with heave.

Another recent contributory factor is the modern practice of householders paving over front gardens or having extensive patios. These impervious surfaces increase the zone of influence of any nearby trees, lead to a greater potential for the drying out of the sub-soil and the effects of subsidence, and also contribute to flash flooding.

Left: The close proximity of this protected oak meant that deep foundations were required to the adjoining house. Right: This pyracantha is not a problem – yet!

Trees – the effect of their removal

The removal of trees often causes a process of water re-absorption in a susceptible sub-soil such as shrinkable clay. This is known as ground heave or ground swell because the sub-soil has a tendency to expand as it takes up groundwater that would previously have been absorbed by the roots of the removed tree. The affected area will extend to the limit of the root system of the former tree and the process of ground heave may occur over a period of several winter seasons until the sub-soil has a water content approaching that normally found when unaffected by trees.

The effect upon any building within the radius of the former root system is an upward movement of the foundation or footing. Cracking in walls above will tend to be wider at the bottom than the top. The movement may affect part or all of a building and may continue over several years.

If a tree is causing concern, the advice of an experienced arbiculturalist should be sought to consider effective tree management procedures, eg alternative pruning remedies such as crown thinning, crown reduction, pollarding or root severance or a root barrier can all be considered.

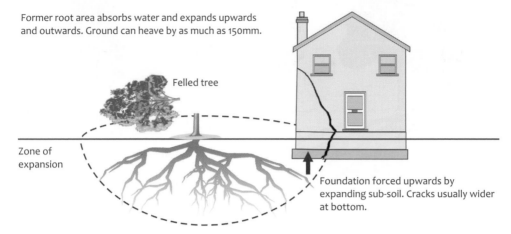

Former root area absorbs water and expands upwards and outwards. Ground can heave by as much as 150mm.

Felled tree

Zone of expansion

Foundation forced upwards by expanding sub-soil. Cracks usually wider at bottom.

Shrubs, hedges, etc

The close proximity of shrubs to a building, especially if mature or in large numbers, can have the same effect as trees – sub-soil shrinkage and building movement. Conversely, the removal of plants, particularly dense hedges and mature species, can lead to ground heave problems.

Greenfield developments on shrinkable sub-soils can be adversely affected by ground heave because they often involve the removal of hedgerows as well as associated ditches.

DIFFERENTIAL SETTLEMENT

This is the term used to describe the circumstances in which separate parts or elements of a building are subject to quite independent movement, eg a bay window or rear annexe of a house settles downwards while the main part of the building remains stable. This is often due to the fabric of the separate areas being quite distinct in form and/or construction, or the foundations being formed at different depths. Any part of a building formed in lighter construction will impose lower loads on its foundations and settlement may be less than for the main structure. Alternatively, shallower foundations will be more susceptible to climatic change, sub-soil shrinkage and the effect of trees or their removal.

The effect is that the two parts of the building settle or move at different rates or even in opposite directions. Cracking will tend to occur at the interface between the parts and will often indicate that they are tending to slip or slide past each other. Sometimes, the effect of differential movement will be movement of one part away from the next, possibly linked with settlement as well. Differential settlement or movement can occur in buildings of all styles, ages and periods. It is commonly found in older buildings where attention to foundation depth and detailed construction was less strictly controlled than in more recent years.

The lean-to extension (right) was built on inadequate foundations. The rotational movement is most evident at the top. The extension shown left also has a shallow foundation – in this case, the movement has been exacerbated by a fractured drain pipe.

Typical late Victorian terrace

Shallower foundations to bay

Differential movement may be vertical (up/down). It may also be rotational.

Shallower foundations to annexe

Cracking may occur in walls, ceilings and floors along movement line

Plan

Dotted line indicates movement plane, ie interface of different forms/areas of construction where cracking can occur.

SUBSIDENCE CAUSED BY DRAINS

A surprising amount of building movement is the result of underground drainage problems which themselves may have been caused by settlement of the building. Such a circumstance is where a drain has been cracked or crushed due to the effect of building load. This can occur where the drain has been laid too close to the building and at a depth lower than the foundation or footing. The broken drain will tend to leak water into the surrounding sub-soil and can affect it in a number of ways that result in movement problems. It may, for example, wash a sandy sub-soil away from under a foundation (the cause of up to 20% of insurance claims for subsidence) or increase the water content of a shrinkable clay and make it more sensitive to building load and thus lead to settlement or subsidence. Alternatively, the drain pipe may not fail but the trench in which it lies may cause problems. Generally, any back-filled trench will not be as capable of providing as good support as the original subsoil. Any pressure on the back-fill will tend to consolidate it, leading to settlement of any close building.

These exposed rigid mortar joints have cracked and are leaking into the soil.

Drainage problems

Corners of buildings are often subject to the effects of drain problems as builders take the shortest route to save time and cost.

Plan

Corner of building subject to settlement

IC

Inspection chamber

Section

GL

Angle of load from the foundation

Softer/less compact soil in backfilled trench can cause instability. Drain cracks under building pressure. Displacement/crushing of drain alone may be sufficient to cause subsidence. If drain cracks or leaks, water adversely affects sub-soil and causes subsidence.

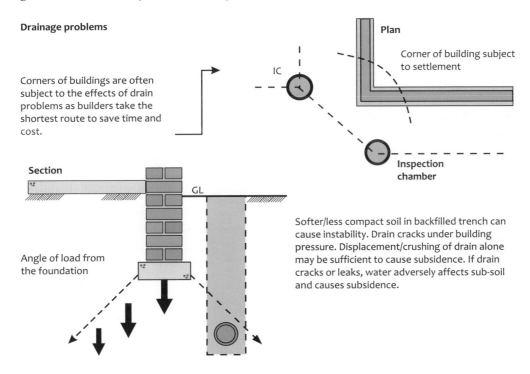

Drain-related problems can sometimes be exacerbated by trees, particularly poplars and willows, which can be extremely aggressive in their search for water. It is not uncommon for such tree roots to break into drain runs at weeping or defective joints to obtain moisture, especially in dry climatic conditions. Tree roots may also exploit the situation where a drain has already been damaged by building settlement. The leaking drain results in increased root growth of nearby trees and this can lead to more extensive drain problems and soil, especially if of a sandy nature, being washed away, creating increased building movement.

Other causes of drain failure include damage by traffic, ground movement (eg landslip), rigid pipe joints or inadequate movement joints where pipes pass under buildings or join inspection chambers.

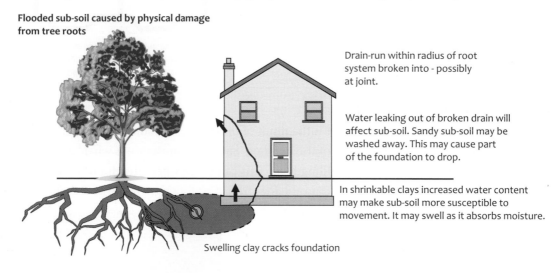

Flooded sub-soil caused by physical damage from tree roots

Drain-run within radius of root system broken into - possibly at joint.

Water leaking out of broken drain will affect sub-soil. Sandy sub-soil may be washed away. This may cause part of the foundation to drop.

In shrinkable clays increased water content may make sub-soil more susceptible to movement. It may swell as it absorbs moisture.

Swelling clay cracks foundation

SUBSIDENCE DUE TO LAND FILL

The NHBC has stated that construction over poor land fill is the largest single cause of foundation failure for new houses – over 50% of all failures. It must be recognised that any land fill site will be inherently unstable over a long period of time and that, if built-over, appropriate foundation design must be undertaken. This means the use of piled foundations to avoid the effect of the filled ground consolidating. Alternatively, a ground improvement process, such as vibro-compaction, may be undertaken in conjunction with reinforced concrete raft or beam foundations.

Recent pressures upon land use have resulted in the increasing development of poorer quality land including reclaimed land and former fill sites. However, this is not a new approach to land use. The

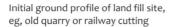

Initial ground profile of land fill site, eg, old quarry or railway cutting

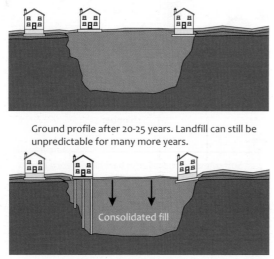

Ground profile after 20-25 years. Landfill can still be unpredictable for many more years.

Consolidated fill

Any new building will be subject to settlement unless suitable piled foundations are used.

The quarry on the left was closed down some 50 years ago. Since then it has been filled in and built on. Look at the two men in the bottom right to get some idea of scale.

Victorians would often raise the ground level on river flood plains to overcome flooding and subsequent generations over the last 100 years or so have constructed on these filled areas as towns have expanded.

Filled ground, however well formed, will need time for the rubble, waste and other matter to consolidate. In normal circumstances a minimum period of at least 25 years, and perhaps longer, will be needed before the ground is sufficiently stable to build on. Unfortunately, the quality control of the fill material has, in the past, tended to be very poor or lacking altogether. Instead of inert building and waste materials being deposited, this has allowed large amounts of organic and other inappropriate materials to be used for fill. Apart from the adverse effects of obnoxious gases and combustibility, the

rotting of organic matter and rusting of metals can create voids in the fill that may take decades to consolidate properly. This means that even apparently stable filled ground may still be subject to unacceptable levels of settlement.

Most land fill is associated with former quarries, gravel and clay pits, open-cast mines and railway cuttings. However, the unusual must also be considered if a building is suffering from settlement, eg raised flood plains, back-filled former streams and deep culverts in town centres.

Generally, the problem caused by land fill is settlement of ground and building. Occasionally, the opposite may occur – in both Teeside and Glasgow there have been instances of expansion of the filled site caused by the swelling of pyritic shales used as fill material. Pyritic shale contains iron sulphide that oxidises and expands.

This 1950s house was built on a reclaimed industrial site, the fill included shale (a mining waste). This expands when wet and has led to heave affecting the floors but not, fortunately, the foundations.

It is wise to undertake at least a desk study when developing land or buildings where there is the possibility of fill having occurred. However, only a physical survey involving trial holes and/or bore-holes will provide the fullest evidence of sub-soil conditions.

This end of terrace house (top left) is on a sloping site and partly built on filled ground (behind the retaining wall). Viewed from the rear (bottom left), the movement is self evident. There is extensive cracking in the end wall (see photo on right). The adjacent property (bottom right) seems to be crack free.

SUBSIDENCE DUE TO MINING-RELATED PROBLEMS

Many areas in England and Wales have been extensively mined for coal, lead and copper, and stone for building. Some areas are well known for mining, eg Yorkshire, Nottinghamshire and South Wales (coal); Cornwall (lead and copper); Cheshire (salt). Others may not be so well known because mining finished many years ago, eg Bristol, North Somerset and South Gloucestershire (coal mining), Woolwich and Norwich (chalk), Dudley and Bath (limestone), Droitwich (rock salt).

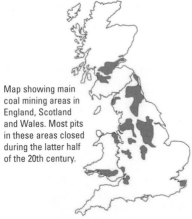

Map showing main coal mining areas in England, Scotland and Wales. Most pits in these areas closed during the latter half of the 20th century.

This map of mainland Britain gives an indication of how extensively one type of mining – coal mining – has affected many areas while the photograph shows a street in Droitwich, a town that has suffered subsidence because of brine extraction in the past. This street used to be level.

The effect of mining upon any buildings above depends upon the type of sub-soil over the mineral-bearing stratum. Strong, dense rock such as granite tends to result in little, if any, adverse effect upon any building. However, where the sub-soil above is of a weaker or softer nature there is a tendency for the land above the mined area to be subject to subsidence. Often, the miners will have left pillars of solid material to support the ground above the mine, the intention being that there would be a bridge effect between each pillar. More often than not, this did not work as over time the bridged areas collapsed causing the ground above to subside. This resulted in low spots above the mined areas and a lunar-like landscape. Sometimes, the thickness between the mine roof and the buildings above is very shallow, eg only 2 metres in some cases above the limestone mines of Bath.

mine-explorer.co.uk

Pillar and chamber mines can be found throughout the UK.

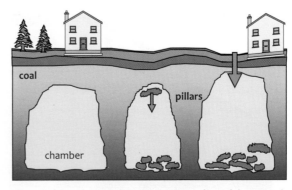

As roof of chamber slowly collapses the surface of the ground develops a 'lunar' landscape.

B & NES Council

Near left: This Victorian terrace was affected by mining subsidence some 100 years ago. The buildings were constructed with flexible lime mortars so cracking was avoided. Far left: The limestone mines below Bath are the subject of a Government-sponsored stabilisation scheme.

Such subsidence might take place rapidly or over a longer period of time. This means that any location where mining has taken place in the past must be viewed with caution even if there is no evidence as yet of subsidence occurring.

Older buildings would almost certainly not have been designed or constructed to take account of movement of this nature. Similarly, modern buildings would not have taken mining problems into account unless direct need had been proved or anticipated. All of these buildings would probably be affected by ground movement that is the result of mining having taken place beneath. Such movement will tend to be stealthy rather than suddenly catastrophic. Many older buildings so affected have a greater tolerance of movement because they are constructed with lime mortar. This results in a more flexible structure. On the other hand, modern cement-jointed cavity walls, while being stronger than comparable monolithic construction, are also more brittle with the result that they are less tolerant of movement. With both older and modern buildings, massive subsidence may result in eventual collapse.

Original ridge line

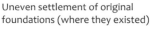

Uneven settlement of original foundations (where they existed)

Terrace of houses subject to mining settlement. Cracking will relate to degree of settlement. Doors, window heads etc. will lean, tilt and stick. Subsidence may lead to collapse.

The modern approach is to build on raft foundations and provide movement joints between units to allow settlement without damage.

Shallow mine workings should also be stabilised.

Modern buildings in mining areas that are subject to subsidence should be constructed with raft foundations so that they can 'float' on the surface of the ground. Longer buildings and terraces of houses should have movement joints incorporated to allow differential settlement to occur without causing structural problems.

Azadour Guzelain

The catastrophic failure of the garage (left) was due to unknown deposits of gypsum being washed away.
The failure in the photo on the right was caused by the collapse of a coal mining tunnel.

A further problem can be that caused by the covering, if any, to shafts of former mines where no record has been kept of their position. Often, the cover consisted of timber baulks or sleepers with a metre or two of over-burden placed on top. It is not unusual for these to have been subsequently built over and the first indication of their presence may occur when the timber sleepers rot. The overburden gives way and the structure above is suddenly left unsupported.

The Coal Authority was established by the UK Government in 1994 and keeps records of all known coal mine shafts but many are unrecorded. It will provide information where it has records on any environmental and stability risk from any past, current or proposed underground and surface coal mining activity that affects property, eg it will advise on a shaft's recorded position (not necessarily its actual position) if it is within 20m of a property.

Sometimes, a shaft can open up that is not caused by mining. Many old wells were capped by similar means and occasionally one of these collapses. There will almost certainly be no record but any property constructed over the site of a former building may be subject to this problem.

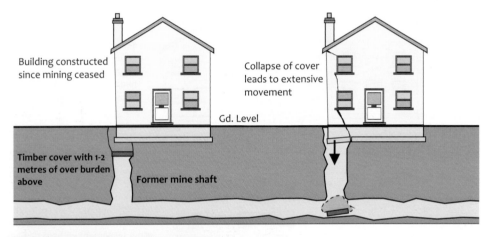

Building constructed since mining ceased

Collapse of cover leads to extensive movement

Gd. Level

Timber cover with 1-2 metres of over burden above

Former mine shaft

Department of Environmental Protection, Pennsylvania

Subsidence due to coal mining affects many countries. In Pennsylvania (USA) the mines are quite shallow – subsidence is a common problem, as a consequence it is often excluded from insurance policies.

LANDSLIP

Landslips, or landslides, are mass gravitational movements of soil or rock. The landslip may be slow, quick, intermittent or continuous. Many sections of the British coast suffer from landslips which cause coastal erosion and instability, eg Devon, Dorset, Kent, Yorkshire and the Isle of Wight. However, it is not just a coastal problem. Recent inland landslips have occurred in the Wye Valley in Gloucestershire, at Daventry in Northamptonshire and at Halifax in Yorkshire. Railway and highway cuttings and embankments throughout the country are frequently affected by landslips and adjoining properties can be adversely affected. Waste tips are notoriously prone, the worst disaster being that at Aberfan in South Wales in 1966 when the school and adjoining houses were engulfed, killing 145 people. In the winter of 2000–2001 there were more than 500 instances of landslip reported after the wettest winter since records began in 1727.

They occur for a variety of reasons, eg on sloping ground with clay soils there is a danger that the upper layers can slowly move downhill when they slide over lower layers; in many cases susceptible soils or waste heaps can slip as a result of a prolonged period of rain. It can often be recognised at the site investigation stage by the terraced appearance of the ground. Soil creep can occur in gradients as shallow as 1:10 and buildings with shallow foundations can obviously be damaged. A good guide for further information on landslips is *Planning Policy Guidance 14: Development on Unstable Land* published by the Department of Communities & Local Government, which refers to landslips as landslides.

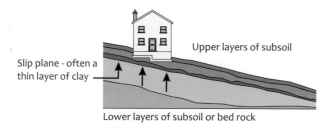

Slip plane - often a thin layer of clay

Upper layers of subsoil

Lower layers of subsoil or bed rock

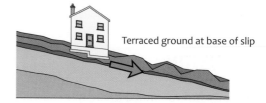

Terraced ground at base of slip

Red arrow indicates direction of land slip (often a result of a rise in ground water pressure after intense rain).

Sealand Aerial Photography

Lyme Regis Museum

The aerial photograph shows a massive coastal landslip near Lyme Regis while that on the right shows building damage following a landslip at Lyme Regis in the mid 1960s.

SULFATE ATTACK

This is a problem that can occur where there is a high water table linked to moving ground water and a sub-soil (often ancient sedimentary clay) that contains calcium, magnesium, and/or sodium sulfate. Such clays can be found in England anywhere below a line drawn between Lancaster and Middlesborough. Sulfates in the presence of moisture will attack any cement-based material (eg mortar in brickwork/block-work, concrete in foundations) due to a chemical reaction between the sulfate and one of the constituents of cement – either calcium hydroxide or calcium aluminium hydrate. The result is a rapid expansion of the compound formed by the reaction.

The chemical process is discussed in more detail in Chapter 4 as the sulfates found in the clays used for the manufacture of some types of bricks can lead to similar problems. Sulfate attack often occurs in burnt colliery shales, which were used for hardcore for solid ground floor slabs in the first half of the 20th century *(refer to Chapter 5)*.

The effect on that part of the building below ground level is to expand and crack concrete foundations. The mortar joints in the walls beneath damp-proof course level will suffer similarly. The expansion resulting from the sulfate attack may lead to upward movement of the wall and may also be seen as cracked and broken down mortar joints below DPC level. Foundations subject to such an attack may, if left long enough, disintegrate. The process will be accelerated if the concrete, or cement mortar, is also subject to frost attack. In any case, as the foundation deteriorates progressively, it will become less capable of supporting the imposed building loads, leading to settlement. Where such sub-soils are identified, any construction below damp proof course level should incorporate sulfate-resisting cement in floors, walls and foundations. If the problem is discovered in an existing building, it may result in the need to completely remove and reconstruct the affected elements.

A separate type of sulfate attack has been recognised since the late 1980s – thaumasite sulfate attack (TSA). This attacks concrete rather than mortar, requires low temperatures (generally less than 15 °C) and the presence of carbonate, eg limestone – provided by the course and fine aggregates in the concrete. It does not depend on the presence of calcium aluminium hydrate and is becoming an increasing problem.

CONCLUSION

This chapter has examined the major causes of foundation failure in domestic buildings. It has also described typical patterns of cracking associated with problems of settlement, heave and subsidence.

In the future, the implications of climate change, with increasing temperatures, wetter winters and dryer summers, will have to be taken into account when assessing building performance and failure.

In practice, the pattern of cracking often provides the first clues in what can be a lengthy diagnosis. However, be prepared for the unusual. The cracking in the single storey extension, shown on the right, was caused by impact damage. Brake failure, not foundation failure, accounts for this apparent defect.

Finally, it should be noted that buildings affected by foundation failure may not require drastic or expensive remedy. Remedial action involving tree removal or pruning or relining drains is more often the solution than under-pinning.

CHAPTER 2

Building Movement – Walls

INTRODUCTION

A building will suffer some degree of movement throughout its life. Often, it may not have its performance affected in any significant manner. On the other hand, the movement may be so great, or happen in such a way, that performance is adversely affected.

The walls of newly constructed buildings commonly suffer cracking as a result of the initial drying out of the materials used in the building process. This is usually caused by the evaporation of the water used in the process and can take up to 12–18 months after construction has finished before it ceases. Normally, the severity of the cracking is within acceptable limits and is insufficient to require more than minimal repair. Usually, filling in of the cracks alone will suffice.

Similarly, a building's structure will be subject to thermal and moisture movement caused by changes in climate and/or internal living conditions affecting building materials. This will result in expansion and contraction of both the materials and the elements of structure that they form. Mostly, the amount

These four photographs show buildings that have suffered extensive wall movement. Older buildings often can sustain considerable movement because they incorporate timber framing or lime mortar. However, the integrity of the arched opening is depending wholly on the head of the timber doorframe.

A detail of one of the buildings in bottom photo which emphasises how much movement brickwork in lime mortar can accommodate.

of movement will be insufficient to adversely affect the overall performance of either the material involved, the element concerned or the building as a whole, especially if expansion in one season is offset by contraction in the next. Modern design and construction should incorporate expansion details to accommodate such movement but, in practice, both older and modern buildings can suffer structural failure from these problems, the latter because, often, insufficient allowance was made for movement.

CONSTRUCTION SUMMARY

FUNCTION OF THE WALL

There are two primary functions of the wall. They are listed below:

1. **To support floor and roof loads and transfer them to the foundations.** This may be undertaken by either external or internal walls.
2. **To give environmental protection to the internal areas of the building.** This is performed by external walls.

In addition, a wall may be required to perform a further function:

3. **To define separate areas of use.** This is normally undertaken by internal walls or partitions.

Walls are divided into **load-bearing** and **non load-bearing** depending upon whether or not they perform function 1 above. The following diagrams indicate the principles of each.

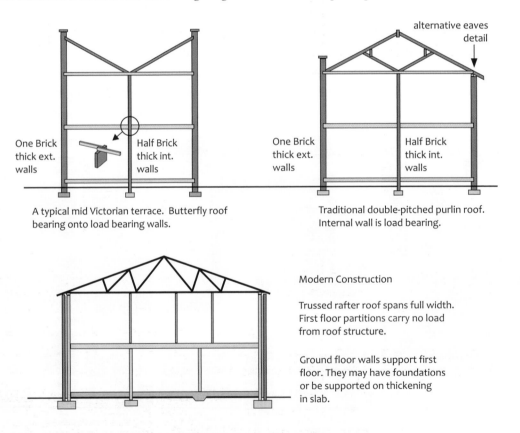

A typical mid Victorian terrace. Butterfly roof bearing onto load bearing walls.

Traditional double-pitched purlin roof. Internal wall is load bearing.

Modern Construction

Trussed rafter roof spans full width. First floor partitions carry no load from roof structure.

Ground floor walls support first floor. They may have foundations or be supported on thickening in slab.

Whichever functions a wall is undertaking it must have suitable strength, stability and durability to perform properly. These factors must be carefully assessed at the design stage and each will influence the final choice of form and construction.

WALLS – BRIEF HISTORY

The first walls were probably constructed from branches and other vegetable matter before heavier and larger pieces of timber were used. Bricks of sun-baked clay have been used to form buildings for at least 4000 years and stone for probably longer. Generally, people used whatever material was readily available locally and this can be seen in the localisation of construction forms and types of materials used. This state of affairs continued until the 19th century and the Industrial Revolution, which transformed transportation and industrial production. During the 19th century, canals and then railways enabled materials to be transported more or less anywhere, although local style and appearance still tended to be followed.

In England and Wales the oldest domestic buildings still standing were originally constructed of stone, brick, or timber frame with infill panels – probably of wattle and daub, often replaced with brick. Few are older than the 14th century and from the 18th century onwards they are generally of brick or stone monolithic (solid wall) construction. Brick walls were initially constructed using a simple bond, eg garden wall bond incorporating alternate courses of stretchers and headers, but more sophisticated bonds were developed for thicker walls to enable greater loads and heights to be sustained, eg English and Flemish bonds. *(Refer to Chapter 4: Brickwork and Stonework for details.)*

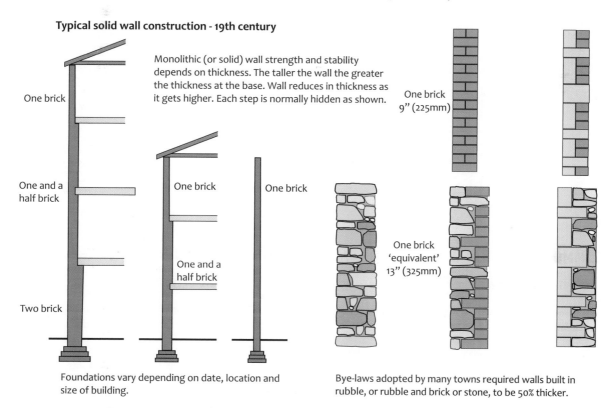

Typical solid wall construction - 19th century

Monolithic (or solid) wall strength and stability depends on thickness. The taller the wall the greater the thickness at the base. Wall reduces in thickness as it gets higher. Each step is normally hidden as shown.

One brick

One and a half brick

Two brick

One brick

One and a half brick

One brick

One brick 9" (225mm)

One brick 'equivalent' 13" (325mm)

Foundations vary depending on date, location and size of building.

Bye-laws adopted by many towns required walls built in rubble, or rubble and brick or stone, to be 50% thicker.

External cavity walls started to appear in the 1830s but were very uncommon until the twentieth century. From the 1930s, such walls are the normal form of construction for housing and it is extremely unusual to find solid external walls in buildings constructed since that date. Modern cavity walls may have brick external skins but are mostly constructed with lightweight concrete skins and incorporate high levels of insulation.

Material and labour shortages after each of the World Wars led to alternative construction approaches being sought, each time resulting in the use of innovative framed construction, mostly concrete or steel and, less frequently, timber. The numbers of these houses built in the inter-war period were relatively small (approximately 50,000) but between 1945 and 1960 approximately 500,000 were erected. Subsequently, since the late 1970s timber frame houses have become a significant percentage of new construction totals. *(Refer to Chapter 15: System Building for a summary of their defects.)*

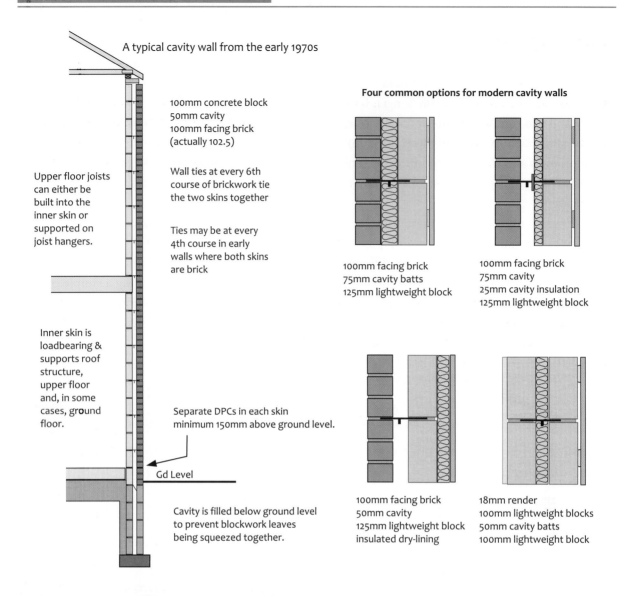

A typical cavity wall from the early 1970s

Four common options for modern cavity walls

100mm concrete block
50mm cavity
100mm facing brick
(actually 102.5)

Wall ties at every 6th course of brickwork tie the two skins together

Ties may be at every 4th course in early walls where both skins are brick

Upper floor joists can either be built into the inner skin or supported on joist hangers.

Inner skin is loadbearing & supports roof structure, upper floor and, in some cases, ground floor.

Separate DPCs in each skin minimum 150mm above ground level.

Gd Level

Cavity is filled below ground level to prevent blockwork leaves being squeezed together.

100mm facing brick
75mm cavity batts
125mm lightweight block

100mm facing brick
75mm cavity
25mm cavity insulation
125mm lightweight block

100mm facing brick
50mm cavity
125mm lightweight block
insulated dry-lining

18mm render
100mm lightweight blocks
50mm cavity batts
100mm lightweight block

COMMON STRUCTURAL DEFECTS OF WALLS AT CONSTRUCTION STAGE (NEW BUILD)

POTENTIAL PROBLEM AREAS	IMPLICATIONS	TYPICAL DEFECTS
Below-ground brickwork or blockwork in sulfate-bearing clay sub-soil	Sulfates in sub-soil will attack cement mortar	Expansion and cracking of mortar below ground plus movement in wall below and above ground
Lack of adequate movement joints	Excessive thermal expansion in long lengths of wall	Vertical cracking at regular intervals in long sections of wall and at short returns. Horizontal movement above the line of the DPC
Inadequate restraint provided to walls from floors and roof	Walls will tend to move horizontally	Bulging and bowing of walls
Lack of lintels over the openings – small and large (both above and below ground level)	Unsupported masonry	Movement cracking above opening Dropped head of opening
Lintels with insufficient end-bearing	Crushing of supporting brickwork or blockwork due to excessive point-loading	Lintel drops at one end Vertical cracking above end of lintel
Insufficient wall-ties	Lack of restraint on individual leaves of a cavity wall	Bulging of wall externally and/or internally
Incorrect mortar mix	Inherent weakness in the structural capability of the wall	Disintegration of mortar jointing Bulging and bowing and/or buckling of walls

WALL MOVEMENT – THE CAUSES

INTRODUCTION

Reference has already been made in Chapter 1 to movement as a result of failure of, or defects in, the foundations or the sub-soil beneath a building. There is also the need to assess whether the apparent movement under investigation is as a result of a failure or problem in that part of the building above ground level. Normally the investigation will be triggered by evidence of movement in the latter and the assessment will involve finding whether the cause(s) are above and/or below ground related.

As with foundation failure, there is a wide range of reasons for wall failure, many due to either inadequate design and/or poor construction. The underlying factors are also similar – historically, lack of technical knowledge; more recently, incompetence leading to poor design and bad site practice.

DISTORTION OF WALLS (BOWING, BULGING AND BUCKLING)

Walls distort for a number of reasons. Some are poorly built which results in distortion from the start; some acquire distortion due to failure of the wall and/or adjoining elements; others have distortion thrust upon them when unforeseen loads are imposed. It is a problem that can affect any solid masonry or cavity wall and may occur for a number of reasons. In approximate terms, it is only when a solid wall is more than one-sixth of its thickness out of plumb in any storey height, or its overall height, that distortion needs serious consideration. For a cavity wall, the relevant thickness is that of the thinner leaf.

Distortion

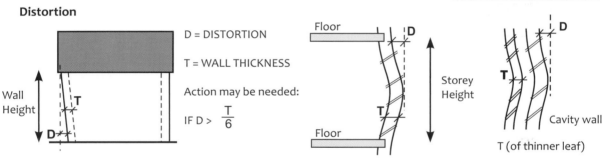

D = DISTORTION

T = WALL THICKNESS

Action may be needed:

IF $D > \dfrac{T}{6}$

Wall Height · T · D

Floor · D · Floor · T · Storey Height

T · D · Cavity wall · T (of thinner leaf)

The causes of distortion leading to bowing, bulging and buckling include the following:

Poor building practice

Walls are sometimes constructed so poorly that they are inherently distorted. This can often be observed in older buildings and is, unfortunately, found in some of recent construction where walls can be seen to be out of plumb. Even though the wall may be distorted, it can still be stable and unless there is evidence of current progressive movement it is often acceptable to take no further action apart from monitoring at regular intervals.

Examples of bowed walls. The bowing on the front wall in the left-hand photograph was so great as to require its reconstruction. At the party line, the junction of the new and adjacent walls shows that the bowing is obviously a problem along the line of the terrace.

Subsequent distortion

This can be caused by a number of factors including:

Inadequate thickness of the wall The relationship between a wall's height and its thickness is known as the 'slenderness ratio'. If a wall is tall and thin it will have a greater tendency to buckle or bow under applied load than a short and thick wall. Historically, this has resulted in walls that are very thick at ground level and are stepped in thickness as they get taller. Modern understanding of how walls perform, and the use of structural calculations in the last 50 years or so, has meant that in recent years taller, thinner walls have been successfully constructed.

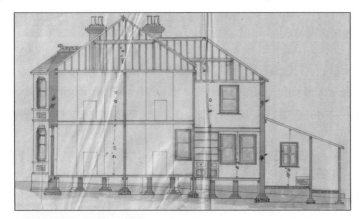

External walls in solid construction should be at least one-brick thick (225mm, 9"). In some houses, usually 3 to 4 storeys or more, the wall thickens towards its base. Problems of buckling should not, therefore, occur as a result of an inadequate slenderness ratio. Floor joists and roof timbers built into wall also offer some restraint. The house above has an 18" front wall in stone (450mm), 4" internal walls (100mm), and a rear wall 9" (225mm) at the top and 14" (350mm) at the base.

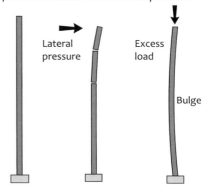

Tall thin walls will tend to fall over under lateral pressure or buckle under vertical pressure.

Lateral pressure

Excess load

Bulge

If wall is too slender (ratio of width to height), or if there is inadequate restraint, or loading is too high, buckling and cracking may occur.

Lack of lateral restraint or support A large unrestrained or unsupported area of wall will tend to lean or fall over especially if it is subject to external load or pressure (thrust). Walls are normally restrained by being tied to adjoining walls, floors and roofs. Alternatively, walls can be supported by the provision of buttresses. Either approach can provide the necessary support. Current statutory requirements and good building practice should result in all elements of construction being properly and fully tied

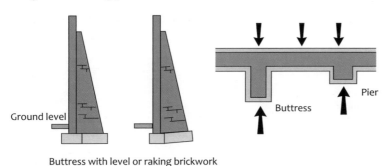

Ground level

Buttress with level or raking brickwork

Buttress

Pier

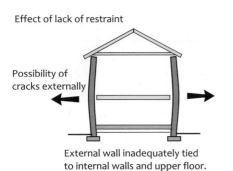

Effect of lack of restraint

Possibility of cracks externally

External wall inadequately tied to internal walls and upper floor.

Tie rods added to provide restraint.

together. Lower standards were applied formerly. The older the building, the more likely it is to suffer from problems associated with a lack of restraint, eg walls were often butted up to each other, bonding was either non-existent or minimal. Often, only one or two courses of bricks per storey were toothed into each other; floor joists running parallel to a wall would have no lateral ties to support that wall. Where it is possible to do so, restraint can be provided by the addition of tie bars or rods.

Subsequent thoughtless alterations to a building can result in restraining elements being removed without the provision of alternatives.

The effect of alterations

Chimney breast removed to increase floor area (removes buttress effect). May be full or single storey height.

Internal wall removed to change layout - restraint reduced. Even a new door opening in an existing wall can have the same effect.

Staircase installed - floor joists no longer restrain external wall.

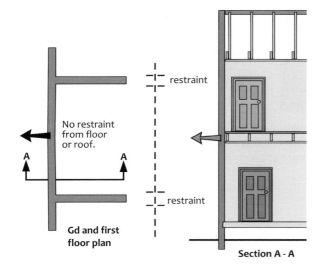

No restraint from floor or roof.

A — A

Gd and first floor plan

restraint

restraint

Section A - A

If a wall is not adequately restrained it may move. This cracking is slowly getting wider.

Thrust from other elements This is often associated with the problem of roof spread where a roof construction which is inadequately tied together results in an eccentric thrusting force being applied at the top of a wall. The affected area of wall is pushed outwards. The appearance may be a bulge or bow, although often brickwork may become stepped as the individual courses slide across each other.

Thrust can also be exerted laterally by the thermal expansion of roofs and floors, especially when they are of concrete construction (eg concrete flat roof). *(Refer to later section in this chapter on thermal movement.)*

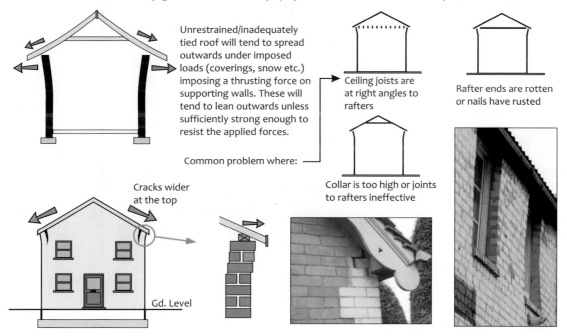

Unrestrained/inadequately tied roof will tend to spread outwards under imposed loads (coverings, snow etc.) imposing a thrusting force on supporting walls. These will tend to lean outwards unless sufficiently strong enough to resist the applied forces.

Ceiling joists are at right angles to rafters

Rafter ends are rotten or nails have rusted

Common problem where:

Collar is too high or joints to rafters ineffective

Cracks wider at the top

Gd. Level

Excess loading Buildings are not always used as originally intended and this may result in over-loading of the structure. An example of this is the greatly increased amount of furniture and personal possessions commonly to be found in the modern family home. Cheaper Victorian and earlier housing generally has thinner/lighter forms of construction for upper floors and walls than would be acceptable now. Apart from any over-stressing of the floors, the higher-than-designed-for loads may result in unforeseen pressures causing bulging/bowing of adjoining walls.

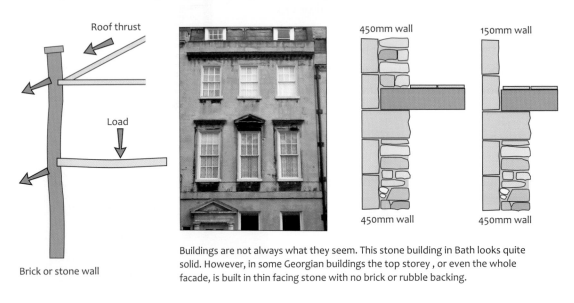

Buildings are not always what they seem. This stone building in Bath looks quite solid. However, in some Georgian buildings the top storey , or even the whole facade, is built in thin facing stone with no brick or rubble backing.

Loss of bonding between materials The bulging, bowing or apparent buckling of a wall may not be due to structural failure. Older, solid masonry walls are often formed of several different materials bound together in a series of layers by a mortar (probably of lime). Walls of all ages and types may be finished with a thin layer of facing material bonded to the structural element behind. Historically, this was frequently achieved by the use of iron ties, the modern equivalent being metal cavity ties. The bond between the surface material and the underlying material(s) may fail.

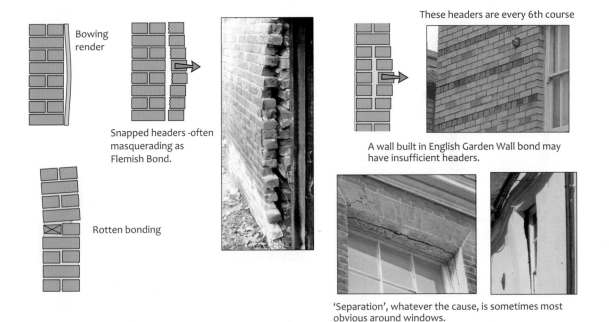

'Separation', whatever the cause, is sometimes most obvious around windows.

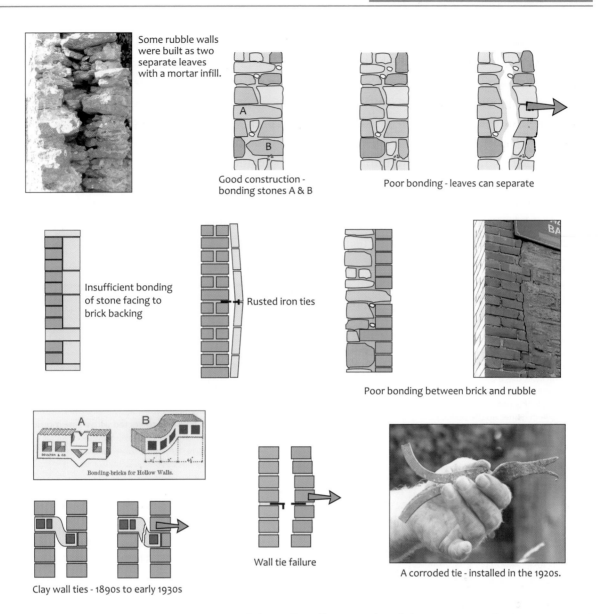

Some rubble walls were built as two separate leaves with a mortar infill.

Good construction - bonding stones A & B

Poor bonding - leaves can separate

Insufficient bonding of stone facing to brick backing

Rusted iron ties

Poor bonding between brick and rubble

Bonding-bricks for Hollow Walls.

Clay wall ties - 1890s to early 1930s

Wall tie failure

A corroded tie - installed in the 1920s.

In some circumstances the appearance of the wall suffering such a failure may give the impression of a more serious problem involving building movement. A check should be made on whether movement identified on one face of a wall is repeated on the other face – lack of such evidence may indicate a failure of bonding rather than a loading problem.

The bulging between the windows in the central photographs is caused by the thin external facing of ashlar losing its bond with the brick backing. This is often caused by rusting of the tie irons. Traditionally these were made from wrought iron which, although less brittle than cast iron and therefore more tolerant of minor movement, is prone to rust.

THERMAL MOVEMENT

The external walls of a building are, by their nature, exposed to the elements and are, therefore, subject to the effect of temperature changes. This often results in expansion and contraction of the wall.

The exact amount of expansion experienced will depend upon the materials used in the construction of the wall. In some cases, the excessive length of the wall or its detailing may influence the amount of movement. Usually, any expansion due to the effect of solar heat gain will be followed by contraction as the affected element cools down during subsequent climatic change. Unfortunately, the amount of contraction does not always equal the original expansion and the net result is permanent cracking. The cracking actually occurs during the cooling period of the cycle. It will also be found that the affected wall expands/contracts above any DPC – over which it tends to slide. This factor provides supporting evidence that the problem is caused by thermal movement.

The terrace expands as it gets hot. If there is a DPC it may slide over it and over-sail the substructure. As the terrace contracts during a cooler period it is unable to pull the building back to its original shape; cracks therefore occur. Over the years the movement can get worse as dust and debris fill the cracks. If there is no DPC, lateral movement is sometimes less prevalent at ground level - this can lead to a slight outwards lean higher up the building.

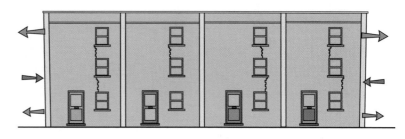

Cracking may be less obvious in buildings constructed with a lime mortar.

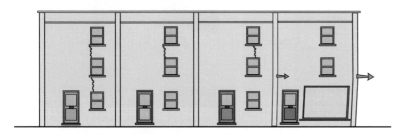

If shop windows are at the end of a terrace, long-term thermal movement can lead to distortion. The photo (below right) shows the flank wall of the terrace. At some point in the past the wall has been anchored back to the first party wall to try and limit further movement.

Modern design should incorporate the use of expansion joints to accommodate this type of movement although they are sometimes omitted because of poor design or construction. Older buildings are unlikely to have any such provision. However, those built with a lime-based mortar will be more likely to avoid any serious problems as, being a more flexible material, it will accommodate thermal movement more successfully than the cement-based mortar used in buildings during the 20th century. (An important point to note is that the materials used in movement joints suffer deterioration due to age and fatigue and, unless periodically checked and renewed on a regular basis, may themselves become a defect.)

Walls can also be affected by thermal movement in adjoining elements or at weak details in the wall. Thermal movement of other elements of construction can also cause problems due to the lateral

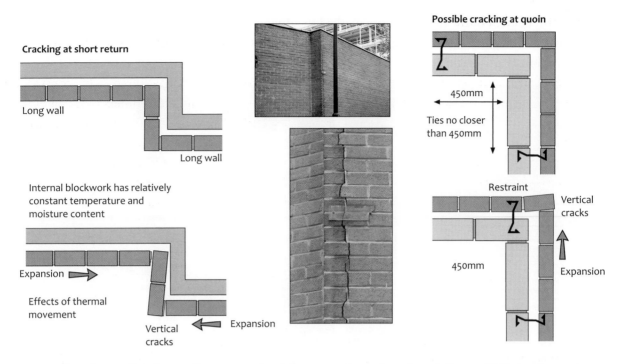

pressures placed on adjoining walls as a result of the expansion of roofs and floors. *(Refer to the previous sub-section on 'Thrust from other elements')*. The building below is suffering from two quite separate problems. Expansion of a concrete roof has caused the horizontal displacement beneath the parapet wall while expansion due to moisture movement in the long length of the flank wall has led to the vertical cracking *(see next section)*.

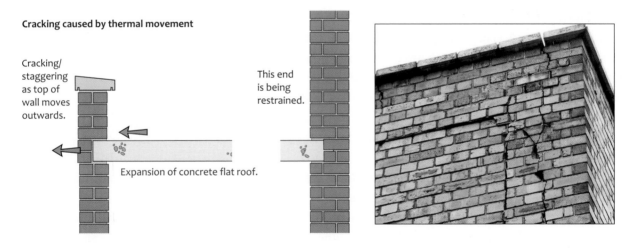

MOISTURE MOVEMENT

Expansion and contraction of the fabric of a building can also occur because it is affected by moisture or subsequent drying out. The cause of the moisture may be weather-induced (rain, snow), a defect (leak), condensation, or as the result of the building process.

Contraction of a material due to initial drying

Brief reference has already been made in this chapter to the effect of the cracking caused by initial drying out of a building once the construction process has been completed. Shrinkage cracking of this nature is quite normal. It is the most common form of cracking found in newly constructed buildings. Portland cement products, lightweight concrete, sand-lime bricks, calcium silicate bricks, some plasters and timber are all subject to initial drying out and cracking. With cement-based products, the greater the cement content of the mix, the more marked the drying shrinkage. While such cracking can be extensive and may occur over a long period of time (up to 20/25 years), it rarely results in serious problems but it is irreversible.

Two examples of cracking caused by the initial drying out of sand-lime bricks. These bricks were frequently used for construction in the 1960s and 1970s.

Expansion of a material due to moisture absorption

On the other hand, clay products, such as bricks, undergo initial expansion as moisture is re-absorbed after firing. This is only slight – between 0.1 and 1.0mm/m in unrestrained brickwork – but the effect is increased in long/large areas, where the overall movement can be considerable and results in vertical cracking that is similar in appearance to that of thermal movement. Brick panels in concrete-framed buildings can also be affected – the brickwork is constrained by the frame/floors/roof and expansion results in bowing and horizontal/vertical cracking of the panel *(see last photograph in previous Thermal Movement sub-section)*.

The majority of moisture induced expansion occurs in the first few months but it can continue for up to 20 years after construction and is irreversible.

Not all moisture-induced movement will be of such great extent that it adversely affects a building's performance. Failure may more often lead to weather ingress problems rather than serious structural failure. However, the modern designer should have anticipated moisture (and thermal) movement by incorporating movement joints with appropriate filler material capable of expanding or contracting as necessary.

FAILURE OF ARCHES AND LINTELS

Often, movement cracks in a wall will be observed above window and door openings. They may be indicative of failure of the wall itself or its foundations. However, the problem may be as a result of the failure of the arch or lintel above the opening.

In traditional masonry walls, openings were normally formed with brick flat or segmental rough arches above. If the mortar in the joints of the arch weathers excessively, or the wall around the opening is subject to some movement, the result can be slumping or movement within the arch itself. Ultimately, this may lead to its collapse together with any area of supported wall above.

A range of lintel problems can be seen here. The top right photograph shows a timber lintel which had only render as weather 'protection'; in the top left, the window head is collapsing as a result of the inner timber lintel deflecting because it is over-loaded; in the bottom photograph, the stone lintel has been affected by foundation settlement.

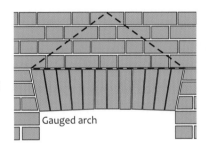
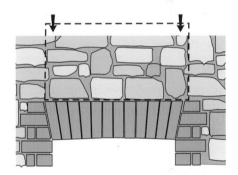

Area of brickwork supported by lintel is quite small due to brick bonding. In rubble walls the area (and load) supported can be much higher.

Gauged arch

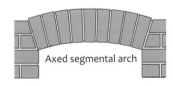

Axed segmental arch

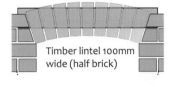

Timber lintel 100mm wide (half brick)

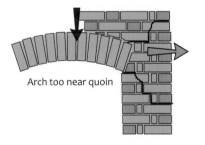

Arch too near quoin

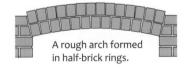

A rough arch formed in half-brick rings.

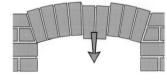

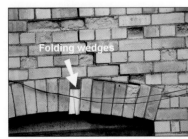

Folding wedges

Arch failure can be a consequence of mortar failure, rot in the backing lintel or building movement. In many cases the timber lintel, if sound, will prevent total collapse. Many rendered buildings hide simple arches.

43

In the mid-20th century the concrete boot lintel was popular for a time. The problem with this particular type of lintel is that it is not built into the external skin of the cavity wall into which it is incorporated and this often results in rotation of the lintel. Even simple reinforced concrete beam or plank lintels can fail if the ends are inadequately supported where built in.

Lintel section, plan and elevation

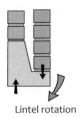

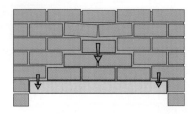

Lintel rotation

Concrete boot lintels (photo - below left) were popular from the 1930s to the 1950s. Some of them were only supported by the internal leaf. The weight of the external leaf could cause rotation in the lintel, leading to cracking in the brickwork.

The lintel in the middle is suffering form carbonation. This is explained in more detail in the chapter on System Building. The cracking is caused by rusting of the reinforcement. The lintel on the right has insufficient bearing - it should be at least 150mm either side. To make matters worse the lintel is bearing on a brick off-cut.

Victorian and older buildings often had window and door openings formed with one or more timber lintels. In external walls the timber lintel was commonly positioned behind the external facing material of brick or stone so that it did not show on the elevation. However, the lintel was, frequently, insufficiently far enough behind the face of the wall to be unaffected by damp penetration. This may result in the lintel being affected by wet rot or possibly dry rot *(refer to Chapter 12: Timber Pests)*. Any lintel so affected will eventually lose structural capability and this can lead to collapse of the timber itself and loss of support to the wall above the opening. Timber lintels are also subject to the threat of insect infestation and a bad attack can result in similar failure of the lintel and wall above.

The cause of the movement in each of these photographs was found to be the failure of the timber lintel behind the stone/brick window head.

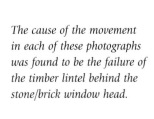

CORROSION OF STEEL AND IRON COMPONENTS

Unprotected steel and iron will rust when in the presence of water. The formation of the rust (or hydrated iron oxide) results in a four-fold increase in the volume of the affected metal. The only way to prevent this is to coat the metal to provide a barrier between it and any water. This is often achieved by the use of hot or cold galvanising. This process involves applying a coat of zinc over the complete surface of the metal. It will offer protection for between 20 and 60 years depending upon the thickness and type of galvanising applied.

The expansion of the rusting metal can cause movement, spalling or cracking of the surrounding fabric of the building. It is a common problem in many houses with earlier, inadequately galvanised wall ties or concrete reinforced with steel and either contaminated with chlorides or carbonated by atmospheric carbon dioxide.

The rusting of the steel reinforcement in the window lintel and in the concrete column has led to the spalling of the concrete cover of each component.

Many buildings have unprotected steel and iron components, such as steel joists supporting floors and above openings, which are built into the structure. The problem of rust can be very serious where the metalwork is in an external solid masonry wall which, unless of some thickness, may be incapable of offering adequate protection from water penetration. Similarly, steelwork within the building may be affected by water leaks from defective services. In modern houses of cavity wall construction any unprotected steel components should only be built into the internal skin of the external wall and should, in normal circumstances, not be affected by water penetration.

A rusting steel RSJ embedded in a brick solid wall is the cause of this problem

At its worst, rust-affected steelwork can completely lose its strength and may lead to total structural failure of any supported part of the building. The actual process of expansion as the rust progresses should be evident in its effect upon any adjoining structure. This may show in the forcing apart of the fabric vertically and/or horizontally with increasing strength as the defect progresses.

45

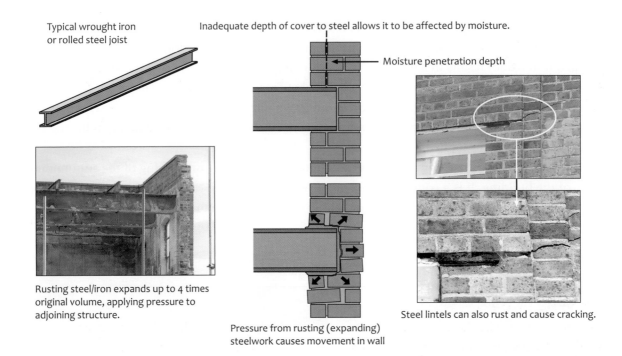

Typical wrought iron or rolled steel joist

Inadequate depth of cover to steel allows it to be affected by moisture.

Moisture penetration depth

Rusting steel/iron expands up to 4 times original volume, applying pressure to adjoining structure.

Pressure from rusting (expanding) steelwork causes movement in wall

Steel lintels can also rust and cause cracking.

RETAINING WALLS AND GARDEN WALLS

RETAINING WALLS

Retaining walls are subjected to a mixture of lateral pressures and they must always be designed and constructed to withstand them. The principal pressure is normally formed by the force exerted by the volume of soil supported but this is often increased by further applied forces. These may include groundwater pressure, loads from adjoining buildings and loads from traffic on nearby roads.

Common problems are set out below:

Slenderness of the wall

A retaining wall that is too slender to withstand the pressures being applied against it will tend to bow or lean outwards. This is a defect that is commonly observed in garden walls that are retaining soil. It is often as a result of too little thought being given to the design implications of any external landscaping. The higher the ground being retained, the thicker the wall, although it can be of stepped width – retained heights up to 900mm should have a 225mm thick brick wall. This is required to become progressively thicker at the base as the height of the wall increases. The addition of piers or buttresses may allow a thinner wall to be formed, but never less than 225mm.

Drainage problems

All retaining walls require the provision of drainage points at frequent intervals in their length and height. Their provision is a common omission and a frequent cause of movement due to the pressure of the groundwater behind the retaining wall. The drain points should be backward sloping to prevent any weeping and the subsequent staining of the wall. Over time the holes can become clogged, reducing their effectiveness. The use of a French drain behind the retaining wall is also highly effective in relieving groundwater pressure.

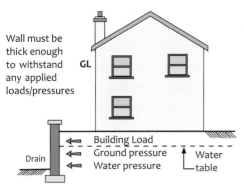

Wall must be thick enough to withstand any applied loads/pressures

GL

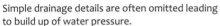

Building Load
Ground pressure — Water table
Water pressure

Drain

Simple drainage details are often omitted leading to build up of water pressure.

Progressive thickening of brick retaining wall

1B

900

Weepholes

A French drain improves drainage and relieves water pressure.

Weepholes should fall backwards slightly to prevent dripping.

Proximity of trees

A further cause can be the close proximity of trees and shrubs. The growth of the root system over a number of years can lead to increasing lateral pressure on any adjoining retaining wall.

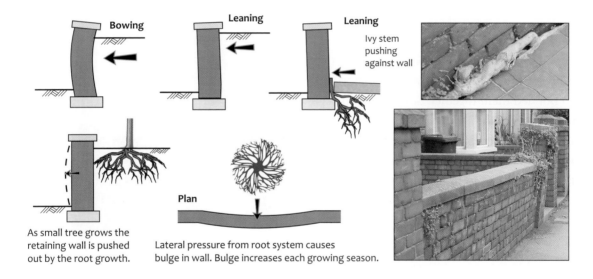

Bowing

Leaning

Leaning

Ivy stem pushing against wall

As small tree grows the retaining wall is pushed out by the root growth.

Plan

Lateral pressure from root system causes bulge in wall. Bulge increases each growing season.

Sulfate attack

Brick retaining walls can suffer from sulfate attack. This can be avoided by the use of sulfate-resistant bricks and mortar or the encasing of the retaining face with water-proof material. Brick walls should also have movement joints at approximately 10 metre intervals. These are frequently omitted and can lead to movement due to moisture ingress or thermal expansion.

Frost damage

Moisture penetration leading to frost damage can be a problem where brick or stonework is susceptible to groundwater penetration. Suitable frost-resistant materials should be used in any retaining wall construction.

GARDEN WALLS

These may act as boundary markers or be dividing space within a garden. Older garden walls were constructed in clay bricks and/or stonework – both stone rubble and freestone were used. Modern walls may be of the same materials but are often constructed with concrete blocks, sometimes rendered. The top should be protected from the weather and a variety of capping materials have been used including stone copings, concrete copings, clay tiles and bricks-on-edge, either laid flat or castellated (which is known as 'cock and hen').

Because they are commonly regarded as of no great aesthetic or structural importance, garden walls are frequently ill-considered, leading to them being poorly designed, constructed, maintained and/or repaired. They are susceptible to a wide range of problems of which the following are the more common:

Shallow foundations

Victorian and older walls were constructed off the shallowest of foundations; frequently they had no foundation *(refer to Chapter 1)*. Modern garden walls should have a foundation that is at least 500mm deep. This is particularly important where a wall is 2 metres high or taller, but often this is not the case. A lack of depth can result in instability, especially if the wall is of any height and/or in an exposed location.

Older stone walls

Older stone walls were commonly constructed with two skins in-filled with rubble and lime mortar which bonded them together. Sometimes through-stones were incorporated. The top may be finished with a cock-and-hen capping or a coping. Over a period of time one or both skins can detach from the backing, leading to partial or total collapse of the wall.

This stone boundary wall was partially demolished to provide a vehicle access. It is still generally intact but will deteriorate if left in this state.

Proximity of trees

The close proximity of a tree and its roots can lead to the cracking and/or lifting of a wall as the tree grows (right).

Tall, thin walls

Half brick walls constructed in long lengths without stiffening piers at regular intervals and tall, thin walls will both have a tendency to be overturned by high (in some cases, relatively low) winds. The exposure of the location will be an important element. These types of walls can be strengthened at initial erection by the inclusion of piers or by staggering or curving their construction, which permits increased height in exposed locations.

Even with the piers and being sheltered behind the hedge, this 100mm thick concrete block wall is too high.

Movement joints

Lack of adequate movement joints, especially in brick walls, can result in bowing and cracking due to moisture or thermal movement. Such joints should be provided at intervals of between 8 and 12 metres depending on the type of brick, eg 8 metres for engineering bricks. A frequent error is to provide the joint but not continue it through the capping.

The expansion joint needs to continue through the capping bricks, as well.

Frost damage

The omission of a proper capping or a defective capping can lead to excessive moisture penetration and frost damage to the face of brickwork or stonework. Similarly, the lack of a DPC can lead to frost damage at the base of a wall. Another cause of frost damage is the use of inappropriately soft bricks or stones. *(For more information on frost damage see Chapter 4: Brickwork and Stonework.)*

These soft London Brick Company facing bricks were totally unsuitable for use in a garden wall.

This Cotswold limestone wall has extensive frost damage

Mortar joints

In older walls, lime mortar joints can decay over time leading to unstable brickwork. Repair work, such as pointing up the face of a joint with a strong and inappropriate cement mortar, can result in the repair taking away the face of the brickwork or stonework. *(See Chapter 4 for more information.)*

CHIMNEYS

CONSTRUCTION

Until the 1960s, most new houses were built with chimneys. After this time, few new houses were built with chimneys. Chimney stacks were usually built in brick (sometimes even in stone buildings) because brick is better able to withstand heat, and the complex flue shapes sometimes required were easier to form in small units (ie bricks). Before the introduction of national Building Regulations for England and Wales in 1965, the construction of fireplaces and chimneys was controlled by generally accepted good practice and local bye-laws. Inevitably, construction details varied slightly from area to area.

A typical Victorian terrace in Stratford-on-Avon.

An example of a stone cottage with a brick chimney and flue.

With regard to chimney stacks, bye-laws usually required that the:
- stack had to be taken at least 3 feet (900mm) above the ridge line
- flues had to be surrounded by at least half-a-brick (102mm)
- stacks could not be more than six times the height of the stack's narrowest dimension.

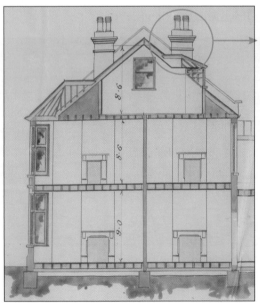

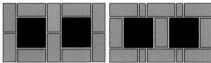

Odd course and even course

In very good quality work the outer chimney wall is one-brick thick. Half-brick thick is much more common.

Odd course only shown

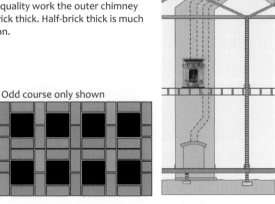

This example (an original drawing from about 1905) shows chimneys front and back, each one serving two flues.

This example shows a gathered stack. In a 'handed' terraced property the chimney is sometimes shared; hence the eight flues.

The Building Regulations 1965 introduced a requirement that all flues should be lined (terra cotta linings are most common) and specified more demanding requirements regarding the position and height of stacks.

DEFECTS

Generally

It is easy to ignore chimney stacks when inspecting a building. Chimney stacks which look sound from the ground may look much worse on closer inspection. Sometimes, a look from an adjoining house window can provide a useful chimney view. A thorough inspection may require ladders; at the very least, it requires a good pair of binoculars.

The following section considers only structural problems associated with the flues and stack. Other problems due to damp penetration or operational defects are not discussed.

Tall, thin stacks

Chimney stacks are obviously in a very exposed position. In high winds it is the taller stack, particularly if it is narrow, which is most at risk. Tall stacks are most likely to be found on houses with hipped roofs. Although bye-laws adopted by most towns permitted a height six times the narrowest width, modern Building Regulations only permit a multiplier of 4.5. Building Regulations are not retrospective although the change reflects health and safety concerns. A stack which has been in position for several years may not be about to topple over but it is still potentially dangerous. If there are any signs of movement, cracking or open mortar joints, a more detailed inspection may be necessary.

Tall, thin stacks require careful examination - they are potentially very dangerous, particularly in high winds. Demolition, strapping or bracing may be required.

Sulfate attack

Cracked mortar joints or a pronounced and permanent lean in a chimney are often the result of sulfate attack. Sulfate attack can occur where clay brickwork, bedded in cement mortar, remains saturated for long periods. The reaction between sulfates, the cement in the mortar and water forms a compound known as tri-calcium sulfoaluminate which causes expansion and cracking of the mortar joints as it crystallises. Sometimes the face of the bricks can spall, most commonly around their edges. The source of the sulfates is often the bricks themselves but they can also be introduced from air pollution and from

Sulfate attack results in expansion and cracking of cement based mortar joints. Cement renders can also be affected.

In some locations, the sulfate attack is more pronounced on one side of the stack than the other; hence the leaning chimney.

51

the exhaust gases of slow-burning fuel appliances. This means that sulfate attack can occur in stacks built from any type of masonry. A common consequence of sulfate attack in chimneys is a pronounced lean caused by the different wetting and drying cycles between elevations, with that elevation facing the prevailing weather more readily suffering the problem.

Chimney stacks constructed in limestone are also susceptible to sulfate attack. This is an example in the Cotswolds.

Failure of the withes

The thin half brick walls (withes) which separate the flues can fail. Failure is usually caused by deterioration in the bedding mortar (the brick becomes loose) or cracking in the bricks themselves. It is not unknown for loose bricks to drop onto the grate. Loose render dropping onto the grate is more likely to be caused by failure of the parging (or rendered lining) of the flue.

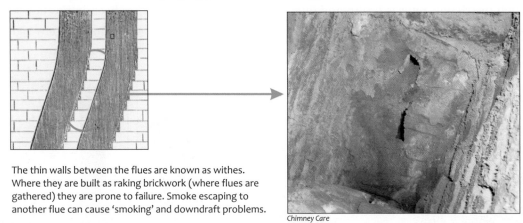

The thin walls between the flues are known as withes. Where they are built as raking brickwork (where flues are gathered) they are prone to failure. Smoke escaping to another flue can cause 'smoking' and downdraft problems.

Chimney Care

Failure due to lack of support

Some structural defects may only be visible from inside the roof space. Where chimney breasts at lower levels have been removed, the brickwork in the roof void (assuming it has not also been removed) should be properly (ie structurally) supported.

The stack on the left looks quite normal but just below the roof line, and in the room below, the breast has been removed. Corbelled brickwork is all that supports the stack.

Timbers should not normally be built into a chimney stack. Sometimes it is unavoidable. There is no risk as long as the timber is at least 225mm clear of any flue.

Other common defects in stacks include:

- general deterioration of the mortar joints, particularly common where the brickwork has been pointed rather than jointed
- vertical or horizontal splits caused by the heat expansion of linings
- frost attack in the bricks or stonework
- potential damage caused by the weight (lever action) of aerials
- broken or loose pots; cracked flaunchings.

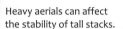
Splitting or cracking may affect stack stability.

Heavy aerials can affect the stability of tall stacks.

This freestone stack is in such poor condition due to age and inappropriate repair with cement mortar that it requires total reconstruction in new stone. Note the hole just above the roofline.

No amount of cement mortar can prevent the deterioration of this brick stack.

Over the last 20 years or so, chimneys have become more common in new construction. It is not unusual, for example, to find a stack containing a single flue from, say, a 'living' gas fire. Modern flues are usually built in pre-formed concrete or clay blocks hidden inside traditional stacks. However, some modern houses have artificial stacks made from glass fibre or stainless steel with brick-slip facings. These are not normally designed to contain flues – they are for appearance only.

An artificial stack ready for lifting into place. These tend to be of lightweight construction so a secure fixing is vital to protect against wind damage.

CONCLUSION

It would be impossible in any textbook to set out a truly comprehensive compendium of all of the potential structural problems that may affect a building. There have always been infinite variations of construction technique, some good and others not so. Similarly, this applies to the many materials used in construction. Many traditional construction techniques would be unacceptable in modern buildings but, even with the advent of regulated standards in the late 19th and 20th centuries, the range of alternative construction solutions available still gives rise to failure.

Sometimes, this means that the problem being experienced may not be truly understood until a full (and destructive) investigation can take place. The reason for the movement in the building below was only identified when the external render was removed to expose the softwood timber plates, known as levelling plates. These had been installed by the builder partly to level up the poor quality brickwork and also to provide a wall plate to the upper floor. This was a common form of construction in the 19th century. Years of water penetration resulting in wet rot had led to disintegration of parts of the plates and movement in both the wall and the upper floor.

Alternatively, the process of destruction may have already taken place. The reason for the spectacular failure of this external skin of brickwork was the force of nature – lightning!

Knightstone Housing Association

CHAPTER 3

The Investigation of Structural Defects

INTRODUCTION

As stated in Chapter 1, it is important to recognise that evidence of movement in one element of the structure may be as a result of failure in a different element, eg cracks in a wall may be caused by failure of the foundation or movement of the roof structure. This means that any investigation should involve a full examination of all structural elements and components that might be causing, or be affected by, the problem under inspection. It is, therefore, both difficult and unwise to consider an affected element in isolation. Due consideration needs to be paid to the various structural elements – roof, walls, floors, foundations – and their effect upon each other. In particular, the close relationship between walls and foundations needs to be recognised.

One of the key principles of defect investigation is that of 'following the trail'. This principle has been stated in numerous legal judgments and should always be born in mind when investigating any building defect. It is especially important when examining structural defects and involves reviewing all of the available explicit and implicit evidence in terms of both cause and effect. A simple example is to examine the outside face of a wall if a crack is found internally. Only when all available evidence has been considered can a judgment be formed as to the cause of the problem and its remedy. The courts have stated that it is essential to 'follow the trail' from the defect originally noted in order to draw the correct conclusions and undertake the proper corrective action. Failure to do so might be considered negligent, especially if one is acting in a professional capacity.

This Victorian house is suffering from the movement identified in the enlarged photograph. 'Following the trail' will involve a full examination of the external and internal areas of the building as well as the surrounding garden area. A walk-over survey and desk survey (see later section of this chapter) are also likely to be undertaken.

Further general factors to consider will be the age of the building and its current and past use. Its age will indicate the original construction in terms of general form of design and materials used so it is important to have knowledge of alternative types of design and materials used in any particular period of construction. The age will also have an important bearing on the defects encountered, those found in a Victorian house of monolithic construction being quite different from a 1930s house with cavity walls. Another important factor to consider is the use of the building since its construction. Many buildings, whether old or modern, will have been originally constructed for quite a different use than that currently noted. Some buildings experience several changes of use over their life. Ill-thought out and executed alterations, inappropriate changes of use, poor maintenance and repair, all need to be considered when undertaking investigation of structural defects as they can act as indicators of the likely extent and degree of the problem being experienced.

This chapter has been divided into a number of sections. However, building failures and their investigation cannot be compartmentalised. A thorough reading of the complete chapter is recommended in order to be able to make the correct decisions as to which procedures will apply for any particular investigation.

PRINCIPLES OF INVESTIGATION

MOVEMENT DUE TO FOUNDATION FAILURE

The failure of the foundations of a building almost inevitably leads to cracking in the walls. However, there are many other causes of cracking in walls, such as thermal movement or stress induced within the wall from floors and roofs. Only careful investigation and consideration of all the facts and implications will lead to the correct diagnosis.

Often, the only physical evidence of building movement due to foundation failure will be cracking or distortion of the walls above ground level. Initial assessment of the defects in the wall will be required, followed by further investigations as the causes are explored and a foundation-related problem becomes suspected.

This 1930s semi-detached house is suffering from movement in the front and flank walls. The causes are unknown and a full investigation is required before any judgement can be made and remedial works commenced.

The investigation may include one or, more usually, several of the following procedures.

'Walk over' survey

A physical review of the surrounding area might be undertaken in order to assess local features having implications for the suspected problem, eg topography such as hills and dales; natural and man-made watercourses; excavations; filled areas of ground; ground erosion or instability; proximity and condition of surrounding buildings; active/recent building works; position and type of trees and shrubs; local place names (eg Fishponds, Clay Hill, Bog Lane, Quarry Way). This village

is a few miles north of Bristol. The sign indicates potential problems linked to coal mining that need to be investigated in any property to be surveyed. Further consideration needs to be given to the type of sub-soil; heath-land is often found to be formed over a sandy sub-soil.

The survey also needs to consider the proximity and type of trees, the effects of the newly constructed tower blocks behind the houses and the effect of the sloping ground.

Desk study

A desk study involves inspection of any of a number of sources of documented information including maps (eg current and old Ordnance Survey maps, geological maps), photographs from aerial surveys, plans deposited with local authority planning and building control departments; and plans from utility companies. It could also include maps and reports from the British Geological Survey, the Coal Authority, the Cheshire Brine Subsidence Compensation Board, and articles from local historical societies. Comparison between older and more recent editions of OS maps etc will reveal changes of land use – see example on next page.

Below-ground investigation

This can involve one or more of a number of approaches. These include the excavation of trial pits by hand or machine to examine the footings of the wall and the sub-soil immediately beneath it, and/or the sinking of small bore holes to obtain deeper sub-soil samples. Alternatively, augered or bored holes, which can be drilled to greater depths, can provide small diameter (up to about 100mm) core samples for laboratory testing. The number of pits/holes will depend on the pattern of cracking and the results of any monitoring *(refer to later section of this chapter)*.

A trial pit can provide much basic information about the sub-soil, the depth and size of foundation, the ground just below the foundation, tree roots, water levels, broken drains, etc. Initial testing of the sub-soil may be carried out on a simple basis, eg squeezing a small sample in the hand to determine water content or driving a metal bar into the sub-soil to indicate its firmness/density. However, such tests can only provide crude/simple information and generally more sophisticated testing needs to be undertaken. Laboratory testing can determine soil type, bearing capacity, water content, organic content, chemical content, plasticity and other relevant information.

Excavating a trial pit beneath an existing foundation in order to determine the type and condition of the sub-soil. On this site cracking was caused by tree roots drying the clay soil. This type of site investigation is often an expensive, lengthy and potentially dangerous operation.

Reading the 1903 Ordnance Survey map of this area of Walsall reveals that it has a highly significant industrial and natural history that is not apparent on the modern aerial survey photograph, eg a colliery, old shafts, a glue works, clay pits, brick works, a stream, The Bog. The implications of the findings will have to be considered as part of the survey.

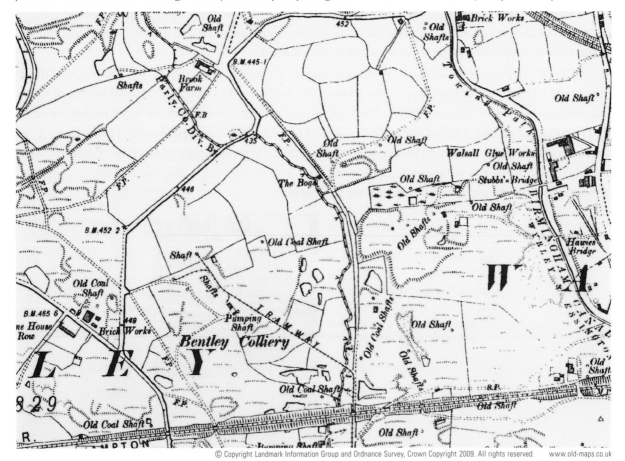

Above-ground investigation

This involves one or more of a number of activities including thorough internal and external inspection of the building to identify where cracking and/or distortion is taking place, recording and assessment of the actual evidence *(refer to next sections)*, the opening up of adjoining walls, floors, ceilings and roof voids to determine whether the element(s) involved are influencing the structural movement and/or the monitoring of suspected movement over a period of time *(refer to later section in this chapter)*. Horizontal and vertical dimensions can be taken by means of tape measures, levelling instruments, plumb-lines and electronic equipment. A visual record needs to be kept; normally this will be by means of record sketches. These can be supported with photographs; nowadays digital cameras make this task relatively easy.

High level inspection requires appropriate inspection apparatus, in this case (right hand photo) a crane together with a cradle. In many investigations, a ladder may be sufficient.

In any case, it may take some time before a final opinion can be reached. Very occasionally the movement discovered is so great, or progressing so rapidly, that judgment is needed straight away. In such circumstances, human safety or the prevention of further damage elsewhere may necessitate immediate emergency works to prevent further movement. The building may even require demolition.

The defect discovered may be due to movement or failure only in the element obviously affected. Alternatively, it may be part of a more complex problem involving structural failure elsewhere in the building, eg overloading of a foundation due to additional structural loads, or it may be found to be due to a cause other than movement, eg failure of a material because of impact damage. In any case, the investigation of both cause and effect should be as full as possible.

Knightstone Housing Association

MOVEMENT ASSOCIATED WITH WALL FAILURE

Visible cracking in a wall

In the previous section, it was noted that wall movement may be indicative of any of a number of structural problems including failure of the wall itself or of another building element such as the foundation, a floor or the roof. If movement is discerned in a wall, a key part of any investigating procedure will be to examine the wall itself.

The investigation should commence with recording basic information on each crack. This will include:

Element – identify the element(s) affected
Location – where in the element(s) affected is the crack? Does it run though window/door openings?
DPC – is it above/below the DPC? Does it pass through it?
Length – including start and finish points
Width – this may vary along its length
Depth – is it through partial or full thickness of the wall?
Pattern – is it vertical, horizontal, diagonal, stepped?
Effect – is there evidence of tension, compression, expansion, rotation or shear?
Alignment – are surfaces on either side level or is one side proud? Is wall out of plumb?
Age – often indicated by the amount of dirt and/or dust in the crack – new cracks tend to be clean, older cracks dirtier/darker
Relationship – is there other evidence of the crack extending into adjoining elements? What construction are they?

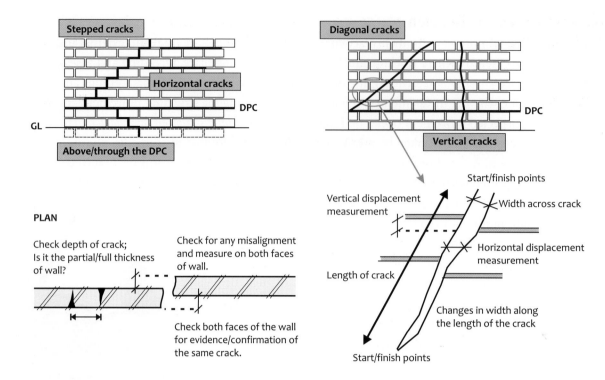

The result is often a fully dimensioned and annotated sketch or drawing. This may detail a single crack or be more comprehensive – perhaps covering a series of adjoining elements and the cracks that they contain.

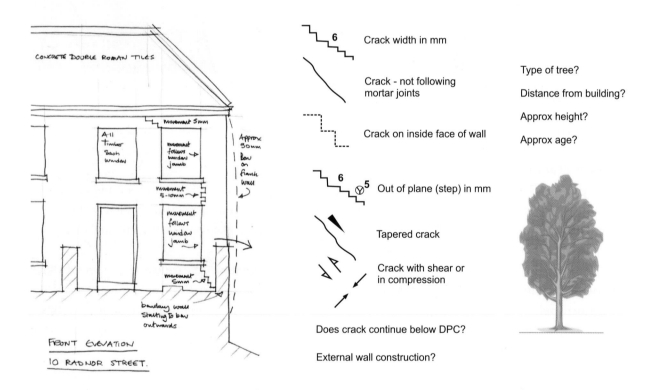

There are a number of conventions for recording cracks – some of these are shown on the diagram above.

Determining when the evidence of movement should cause concern will depend upon the particular factors involved, eg type of building, form of construction, material(s) involved, geological information, topographical features. The amount of movement itself may be the trigger that sets in motion the need for investigation and repair. Considerable research has been undertaken by the Building Research Establishment (BRE) into the overall implications of crack size or width and its guidance has been produced in tabulated form as shown right:

Classification of visible damage to walls with particular reference to ease of repair of plaster and brickwork or masonry (reproduced from BRE Digest 251: Table 1)

Note: Crack width is only one factor in assessing the category of damage and should not be used on its own as a direct measure of it.

CATEGORY OF DAMAGE	DESCRIPTION OF TYPICAL DAMAGE	APPROX. CRACK WIDTH (mm)
0	Hairline cracks of less than about 0.1mm width are classified as negligible. No action required	Up to 0.1
1	Fine cracks which can easily be treated during normal decoration Perhaps isolated slight fracturing in the building Cracks rarely visible in external brickwork	Up to 1.0
2	Cracks easily filled. Redecoration probably required. Recurrent cracks can be masked by suitable linings. Cracks not necessarily visible externally. Some external repointing may be required to ensure weather-tightness. Doors and windows may stick slightly	Up to 5.0
3	The cracks require some opening up and can be patched by a mason. Repointing of external brickwork and possibly a small amount of brickwork to be replaced. Doors and windows sticking. Service pipes may fracture. Weather-tightness often impaired	5–15 (or several of, say, 3.0)
4	Extensive repair work which requires breaking out and replacing sections of walls, especially over doors and windows. Windows and door frames distorted, floor sloping noticeably*. Walls leaning or bulging noticeably*. Some loss of bearing in beams. Service pipes disrupted	15–25 but also depends on the number of cracks
5	This requires a major repair job involving partial or complete rebuilding. Beams lose bearing, walls lean badly and require shoring. Windows broken with distortion. Danger of instability	Usually greater than 25 but depends on the number of cracks

*   * Note: Local deviation of slope from the horizontal or vertical of more than 1:100 will normally be clearly visible. Overall deviations in excess of 1:150 are undesirable.

It should be pointed out that most buildings have cracks in them. The great majority are cosmetic and are usually caused by thermal or moisture movement. Cracking of a structural nature, particularly cracking caused by foundation movement, is not very common. Distinguishing the exact cause of different types of crack is not always easy and careful thought is needed before coming to a judgment. Use of the above table will assist in deciding just how critical the amount of movement that has been identified is.

This crack is about 5–6mm at its widest point.

The cracks in this building range from 5–50mm and there is an obvious danger of instability.

The cracks in this building are of a significant width. Note the shoring which has been put in place because the suspicion is that this movement is progressive. Monitoring is needed to determine the state of the movement.

INVESTIGATION OF DISTORTED WALLS

Apart from cracking, a wall may be subject to distortion, eg bowing or leaning over. When such a defect is identified it will be necessary to measure a wall's thickness, the amount of distortion – both horizontally (with a spirit level or optical level) and vertically (by means of a plumb-line), and the overall extent of wall area affected in order to determine the exact amount of movement.

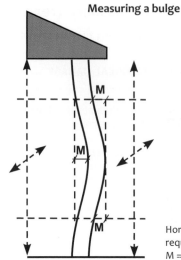

Measuring a bulge

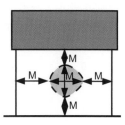

Area affected may be part or whole of wall/storey

Horizontal and vertical measurements required internally and externally.
M = measurement required.

The construction and materials of the wall also need to be ascertained. The investigation should reveal whether the whole thickness of the wall is distorted or only one face is affected. The former would appear to indicate failure of the wall as a whole; the latter only one part – perhaps a loss of bonding between backing and facing materials.

Any examination might include:

Thickness – measurement of wall or skin thickness

Distortion – amount or depth of distortion and the extent of the distortion (height/width, whole or partial thickness of the wall, etc)

Construction of the wall – form, materials, type and size of openings, height of wall, width of wall at all levels, etc

Loading – type(s) and amounts of designed and unforeseen loads imposed on the wall (including lateral and/or eccentric thrusts from other walls, roofs and floors)

Performance of materials – are they decaying? Is adhesion/bonding being lost?

Evidence of lateral restraint – is the wall properly tied to adjoining elements ie walls, roofs and floors?

Continuing movement – is the movement static or progressive?

The investigation may require an internal examination of the wall by means of a boroscope (a type of optical probe that is inserted into a pre-drilled hole in the wall) or by the removal of selected bricks/stones from the affected area. Further examination of adjoining elements (walls, floors and roofs) may also be necessary to identify discernible crack patterns and evaluate the implications for the distorted wall.

The bowing in the wall of this terraced house in North Bristol became significant enough to warrant supporting scaffold until its cause and remedy could be determined.

INVESTIGATION: MOVEMENT LINKED TO ROOFS OR FLOORS

The problems associated with roofs and floors are discussed in detail in Chapters 5 and 6 (floors) and Chapters 7 and 8 (roofs). However, as stated in the Introduction to this chapter, an investigation of structural movement needs to consider all potential causes and this must include the likelihood of roof and or floor movement being involved.

The following factors need to be investigated when considering whether a roof or a floor is linked to building movement:

The effects of vertical and horizontal forces from roofs and floors These may be due to the self-load of the roof or floor but may also be caused by superimposed or applied loads, eg wind, snow and/or rain loads on roofs; furniture and/or stored goods on floors. Problems caused by floor loadings, especially in older buildings, are often linked to a change of use of a building and also to the increased amount of furniture and personal possessions to be found in the modern household. All of these factors need to be taken into consideration when the investigation is undertaken.

The cause of this movement is not apparent from an external inspection. The investigation will need to consider possible causes relating to the foundation, the wall, the adjoining floor and the roof.

Lack of restraint from adjoining elements When a roof or a floor is properly constructed, it will be linked to a wall, especially an external wall, in such a way as to act as a tie and thus provide restraint. Modern Building Regulations require mechanical connection between the roof, floors and external walls. However, this was not the case in Victorian and earlier buildings. Any investigation of building movement will need to include close examination of any adjoining floor and the roof, if only to positively eliminate the other element as a cause of the problem.

GENERAL EXAMINATION OF THE BUILDING

Once the nature and position of any cracks, bulges, etc have been recorded a more detailed inspection of the building and not just the affected area can begin. The purpose of the inspection is to provide further evidence to support an accurate analysis. This process will include an assessment of the crack's significance *(see Classification above)* as well as its cause.

The nature and scope of the inspection will be dictated, to some extent, by the nature of the client brief. Whatever the nature of the brief, the following guidance should be regarded as a minimum procedure for any inspection.

From inside the building:

- look in the roof space – there may be signs of cracking in the walls or at junctions of walls. There may be obvious roof defects such as lack of gable restraint ties or lack of effective ties preventing roof spread
- check the decorations in each room; try and assess their age (left-hand photo overleaf)
- look for cracks in corners of rooms and at vertical junctions with partitions. Do not forget the junctions of walls and ceilings. Pay particular attention to the junction of any annexe to the building
- check ceilings for level – sagging ceilings may indicate undersized joists
- inspect inside built-in cupboards – cracks etc tend to get overlooked inside cupboards and old decorations are often not renewed
- check the operation of windows and doors; look for openings out of square, binding windows, doors that have been planed down and cracks at architrave mitres
- at floor level, look for gaps below skirtings – either suggesting floor has dropped or wall behind skirting has moved

- check whether a floor has a slope. A marble can be used
- try and assess joist depth – a heel drop test may confirm undersized joists or lack of effective strutting. Jumping up and down also works but may cause a great deal of vibration with older floors.

All internal surfaces of a building must be carefully inspected. In the left-hand photograph, the wallpaper is split along the line of movement. Similarly, the right-hand photograph shows the importance of full inspection externally – the wall is over 100mm out of plumb.

From outside the building:

- use a plumb bob on external walls (from a bedroom window) – see right-hand photograph above
- look for signs of dishing etc in the roof (dishing may suggest roof spread)
- check for gaps between window frames and the surrounding wall
- look for vertical movement between openings on different floors
- check whether masonry joints are horizontal using a spirit level, water level or laser level
- identify provision for drainage and positions of drains, soakaways etc. Lift inspection chamber covers to make sure there are no blockages. Flush WC to see how water runs away. Consider drains test to check for leakages
- identify tree and shrub species, age, position and height etc
- consider whether trees and shrubs are still growing or whether trees and shrubs may have been removed
- identify, if possible, sub soil type and consider need for trial pits
- consider the condition and proximity of nearby buildings as well as their current and past use
- note any recent structural alterations or additions such as patio doors, extensions, roof conversions
- note any significant changes to gardens, paths and patios (they may affect drainage). Similarly, consider changes to adjoining property – buildings and land
- check whether cracks been repaired – if so, how long ago?

MONITORING OF MOVEMENT

INTRODUCTION

It is current practice to avoid costly remedial works, such as under-pinning, whenever possible. This is, in part, because research into sub-soil related failures since the severe drought of the mid-1970s has revealed that, in the long term, movement in the majority of severely affected buildings will return (more or less) to a 'safe' or acceptable level of movement distress as subsequent weather patterns return to normal. Therefore, if it is suspected that the movement has not already ceased, consideration should be given to monitoring its progress prior to making any decision on remedial works. However, there is often pressure from householders and some insurance companies to undertake underpinning, in particular, as it is seen as a complete and final solution to any movement problems, whereas no work other than minor repair may actually be necessary in the long-term.

Sometimes the amount of movement requires temporary shoring or scaffolding to arrest progressive movement. In such circumstances, further monitoring may not be necessary if the period of movement can be already determined.

Monitoring involves the regular (usually monthly) inspection and recording of the amount of movement. This is commonly achieved by means of fixed devices such as tell-tales or crack markers (studs) on the face of a wall. It can also be helpful to keep a photographic record as well but only in support of a system of measurement as photographs are not normally accurate enough to calculate the precise amount of movement and its direction.

Normally, monitoring would not be undertaken for a period exceeding 12 months, although occasionally it is carried out for up to 24 months. For movement where a clay sub-soil is involved, the period should cover a full seasonal cycle of summer-winter-summer with readings every 4–6 weeks. At the end of the selected period, there should be evidence available to indicate (1) whether the movement is static or progressive, (2), if the latter, its direction, and (3) whether it is seasonal. This will enable an appropriate decision to be taken on the amount and type of remedial works required.

There are a number of different methods of monitoring. All involve the regular periodic reading of a measuring device. They include the following methods.

VISUAL DEVICES

The simplest method is to measure the crack with a ruler or to measure across a couple of pen or pencil lines either side of the crack. Alternatively, a glass tell-tale can be attached either side of a crack by means of mortar dabs but it provides only limited information. If it falls off, it will confirm that movement is still taking place and if it remains in position that the movement has ceased. Neither simple measurement nor glass tell-tale indicates the direction of any movement. Glass tell-tales are also, unfortunately, commonly subjected to vandalism. Both of these methods are of no use for long-term monitoring.

A better method, externally, is to fix two screws either side of the crack and measure the distance between the screws with an appropriate gauge such as a calliper with a digital read-out. A more sophisticated and accurate version is to use three screws so that both horizontal and vertical movement can be measured. Internally, a system of three pencil dots can be used.

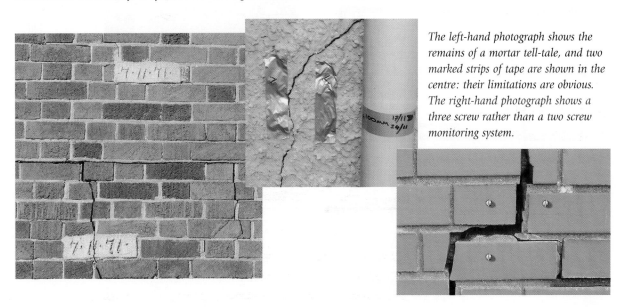

The left-hand photograph shows the remains of a mortar tell-tale, and two marked strips of tape are shown in the centre: their limitations are obvious. The right-hand photograph shows a three screw rather than a two screw monitoring system.

A number of companies produce plastic tell-tales and these are considered to be the most effective of the simpler visual devices. They are manufactured in a range of types, the simplest being fixed straight across a crack and measuring in two dimensions (horizontally and vertically), the more sophisticated being able to measure in three dimensions and/or cracks at the junctions of adjoining walls. The problem with these is that they only really measure to an accuracy of 1mm when an accuracy of ±0.1mm is required. Some surveyors consider the use of three screws and a calliper to be a more accurate method.

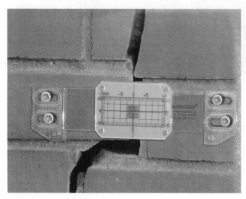 

The plastic tell-tale on the left is manufactured in two halves, each half is fixed either side of the crack. If the crack opens, or closes, the red cross-hairs show the amount of horizontal and vertical movement relative to the graded background. The tell-tale on the right measures movement across a crack and 'out of plane' movement.

OPTICAL LEVELLING

Differential movement or distortion can be measured with an optical level and staff. Variations in level can be recorded and plotted as shown in the graphic below. This is difficult in buildings that are rendered or which are built in stonework because commonly a level line (eg a brick joint) is needed as a starting point. This obviously assumes that the mortar bed is level in the first place. However, any building can be level-monitored by fixing markers (often stainless steel pins) into the face of the wall. Level and verticality monitoring is normally undertaken every 6–8 weeks over a period of 6 months. This period has been shown to be sufficient to provide clear evidence of the pattern and the extent of any movement.

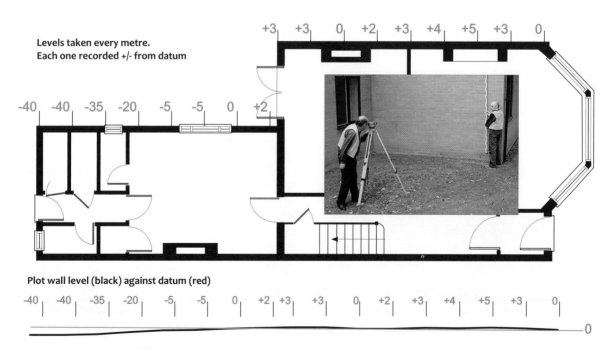

The photograph shows a levelling team at work. Plotting the results as a graph or as an isometric model clearly show the nature and extent of any movement.

ELECTRONIC MONITORING

Remote monitoring by means of electronic equipment is rarely used for investigating movement in smaller domestic properties but can be of benefit in monitoring historic/important buildings or those parts of a larger building that are difficult to access. A mixture of transducers (to measure linear displacement/strain), electrolevels (to measure angular displacement) and piezoelectric and strain gauge accelerometers (to monitor vibration) can be connected to a datalogger. The information collected is transmitted to an office-based computer either by a modem or wirelessly. Alternatively, it is possible to connect the datalogger to a manually-held portable instrument. The data collected can be stored and viewed at any time by the monitoring team.

ANALYSING THE FINDINGS

The findings of the investigation and associated monitoring should enable the surveyor or other construction professional to analyse and then make appropriate judgments about (1) the cause(s) of the movement, (2) whether the condition is passive or progressive, (3) how to remediate its condition and effect(s), and (4) offer advice on financial implications. The key point about any investigation and associated analysis is that they should be as comprehensive as possible to ensure that the decision made (eg to remediate or not) is the most suitable for dealing with the problems being experienced by the building.

The exact method of analysis will depend upon a number of factors including: public safety; the type of construction; the use of the building. One simple method is to consider the following five criteria:

CRITERIA	COMMENTS
1. What is the scale of the damage?	See Table above from BRE Digest 251
2. How does it affect the building now?	Is it purely cosmetic, ie just affecting the building's appearance? Does it affect the use or serviceability of the building, eg are windows difficult to open, do cracks allow rain penetration? Does it affect the stability of the building – is it safe? See Classification table above
3. Is it likely to get any worse?	Are there signs that the movement has ceased or is it progressive? Should it be monitored (see above)?
4. How might it affect the building in the future?	Is there a future risk of collapse?
5. What is the cause?	Is it caused by foundations and or ground conditions? Is it caused by problems in the superstructure?

Providing a detailed list of crack types and possible causes is an almost endless task because there are so many factors and permutations to consider. The brief summary below relates to traditional masonry buildings (solid or cavity construction). A more detailed discussion of causes and effects can be found in Chapters 1 and 2.

A: If the crack appears to be static (ie there are no signs of progressive movement) and the crack is no more than a few millimetres wide, the most likely causes are:
- shrinkage of new materials – this is mostly likely to be evident inside a building; usually at junctions of building elements
- settlement of new buildings (this will happen almost immediately on sandy soils; it can take a few years on clay ones)
- cracking caused by loading – most common around lintels and other beams.

B: If the crack opens and closes over the course of several months, it suggests that the cause may be:
- thermal movement in the superstructure (cracks caused by thermal and moisture movement are normally between 1mm and 5mm wide)
- seasonal shrinkage in a clay subsoil.

Note: Cracks like this tend to 'ratchet' as they accumulate dirt and dust.

C: The following patterns of cracking suggest some form of foundation or ground problem:
- cracks which show on both faces of a solid wall
- cracks which show on both faces of a cavity wall – ie on the outside of the outer leaf and the inside of the inner leaf
- cracks which taper – either wide at the bottom and narrow at the top or vice versa
- distortion in door and window openings
- walls out of plumb and ground floors out of level
- cracks which run across (ie above and below) the DPC
- cracks are often greater than 5mm wide
- broken drains or disrupted services.

Note that in old buildings the use of flexible lime mortars may often limit cracking. Old cracks are also often hidden by 'post-event' renders or filling.

If part of the ground has subsided cracking is likely to be stepped and tapered (wider at the top). If part of the ground has heaved the crack is likely to be wider at the bottom – whether or not the crack is stepped depends on the nature of the masonry: if the masonry shears the cracks may be almost vertical. If drains have cracked, the effect will depend on the nature of the subsoil: sands may be washed away, clays may heave.

The bottom right-hand graphic shows how heave in a central section can be confused with subsidence on either side.

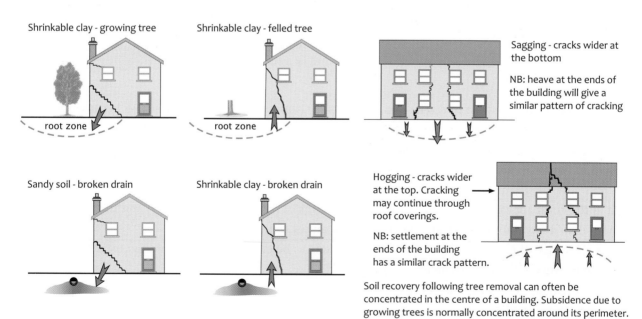

D: If a wall bulges or leans (with or without associated cracking) the cause is likely to be:
- thrust from a badly tied roof
- lack of restraint at the gable (trussed rafters not strapped)
- lack of restraint at upper floors
- lack of restraint from chimney breasts, partitions and return walls
- lateral thrust from expanding solid ground floors.

E: Movement in chimney stacks:
- vertical cracks in chimney stacks are usually caused by cracking of clay linings
- horizontal cracks are more likely to be caused by sulfate attack
- pronounced leans are often caused by sulfate attack.

CONCLUSION

It should be understood from reading this chapter that an investigation into any suspected structural movement must be thorough. The person carrying out the investigation should also be prepared for the unusual. Many would suspect subsidence as the cause of the defects below (photos left and bottom), but how many would think the chimney damage is caused by an earthquake (right)? The key factor in any investigation, however methodical and detailed, is that the investigator must not ignore evidence, however trivial and must not jump to conclusions.

Department of Environmental Protection, Pennsylvania, USA

Mathew Jackson Photography

Cheshire Record Office

CHAPTER **4**

Brickwork and Stonework

SECTION 1

INTRODUCTION

This chapter looks at defects in brickwork and stonework, mainly in relation to their use in external walls. It briefly considers the historical development of these two materials and comments on aspects of construction which might affect their durability.

Historical background

Brick is perhaps the main material which helps define the appearance of a modern house in this country. However, brick has competed with stone in terms of popularity and status for many centuries. There is evidence of brickmaking going back to Roman times and after the 'Dark Ages' it re-emerged in East Anglia – an area with little available building stone but with clay deposits. Generally however, up until the 16th century, timber had been the dominant structural material for most vernacular buildings. Brick and stone started to be adopted more commonly for a number of reasons, including timber supply shortages and, perhaps more significantly, the disastrous experience of fires which led James I, and later, after the Great Fire of London in 1666, Charles II, to issue proclamations insisting on the use of stone or brick.

The dominance of either brick or stone has waxed and waned over the centuries. Partially this was a factor of geography and, therefore, availability of materials. It was also affected by the development of trade skills and by advances in technology which allowed for more efficient quarrying of stone, or excavation of clay, as well as industrialised kilns for brick production which improved the ability to control both quantity and quality. The creation of canals, and later the rail system, had a dramatic effect, particularly on the ability to distribute bricks to areas with no clay deposits. Less pragmatic, but perhaps more significant factors, were the issues of style and status. For example, brick was popular in the Restoration period and the artistic skills practised by the bricklayers in both decorative and precision work during this era was, and still is, admired. The Georgian period is often characterised by the stone buildings of cities such as Bath, but brick was a fashionable material for developments in, for instance, Liverpool, London and Bristol. Because of the links between style and status, the aspiring classes looked to these materials to demonstrate social position. In the Regency and early Victorian periods the appearance achieved by ashlar stone was aspired to, but, because stone was expensive, the look was often achieved by using cheaper bricks covered externally with stucco (a type of render) with lines drawn out to mimic the mathematically precise ashlar blocks. Unfortunately, builders often used substandard

71

bricks in this type of work. This damaged the general reputation of bricks and the practice still causes technical problems in these buildings today. There were, however, surges in the status of brick, including that which arose during the 'Battle of the Styles', one example being the Queen Anne revival in the 1870s. At other times in the Victorian period brick was equal to stone in status, partly because of the increasing precision with which bricks could be made. This precision, allied to the wide range of colours, finishes

and shapes that became available, appealed to the Victorians' sense of pride in technological achievement as well as their liking for the colour and decoration. Clearly though, this appeal was not universal; many late Victorian houses in cities such as Bristol have stone walls at the front with brick being relegated to the side and rear elevation, where it was often rendered. There are, however, geographical areas where brick was used to the front elevation with stone being relegated to the side and rear.

Overview of brick and stone defects

Given the UK climate it is inevitable that external walls may remain wet for long periods. Water penetration is, in fact, the common denominator in most of the defects referred to in this chapter. Although the durability of a wall depends to a large extent on the correct specification of stone/brick and the mortar, the overall design, practical detailing and regular maintenance are all key factors affecting durability. Many of the defects referred to in this chapter can be designed-out in modern buildings by proper attention to such items as dpcs, flashings, roof overhangs, copings and projecting sills. Material selection and detailing should also take into account the potential effect on brick or stone of exposure to frost and the sun.

The exposure of a building will be related to:

• the position of the building – is it particularly exposed by being on the coast or on top of a hill? Is it relatively exposed because of its height or perhaps sheltered by surrounding buildings or trees?

• geographical location and orientation – some elevations will face the prevailing wind and rain, others will be warmed by the sun, while the north elevation will remain cold. Sheltered positions, with little exposure to sunlight or wind, may restrict evaporation, resulting in water remaining in the wall longer.

Particular features which are vulnerable include: chimney stacks, parapets, aprons under windows, garden walls, and retaining walls, ie areas which receive a lot of rain, or are saturated because of contact with the ground. Plain vertical walls are not generally a high risk if they are adequately protected by roof overhangs, correct sill details etc. However, horizontal projections are vulnerable, and may wet the adjacent wall if they shed water inadequately.

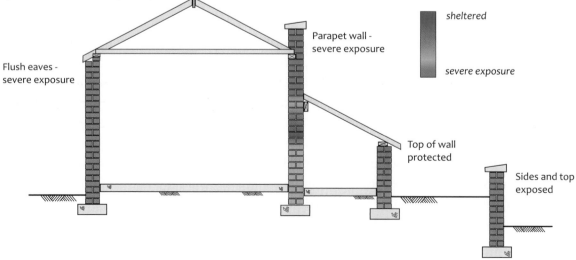

SECTION 2

BRICK

At least 90% of the bricks produced in the UK are made from clay. Concrete bricks, which were introduced in the 1950s, are also produced, as are calcium silicate (sand lime) bricks.

Calcium silicate bricks were first patented in the 1860s in the USA and developed in a number of different countries over the following century. They are produced from hydrated lime and sand: sometimes crushed flint is added (flint lime bricks). Calcium silicate bricks are generally resistant to frost attack and are virtually free from soluble sulfates. Their popularity appears to go in and out of fashion, although they did gain a poor reputation in the 1960s and 1970s. This was to some extent a result of designers' failure to deal adequately with the tendency of the bricks to undergo moisture shrinkage.

Solid and cavity walls

In order to fulfil its structural function a brick or stone wall has to be bonded. Modern walls are built with two single skins, usually with the outer protective skin separated from the internal loadbearing skin by a cavity. The internal skin is nowadays usually constructed in blockwork (unless it is a timber frame house). The two elements are tied together by wall ties. Where the outer skin is in brick this is usually bonded in stretcher bond. Some houses are constructed with blockwork to the inner and outer skin and finished with an external render.

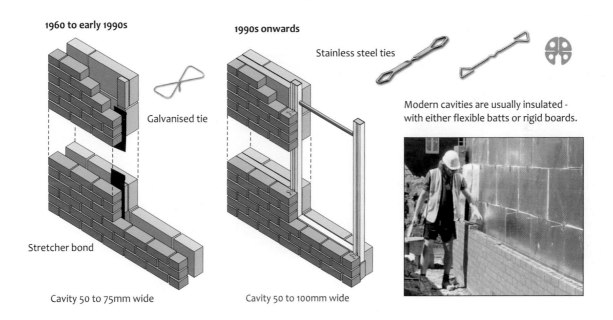

1960 to early 1990s

1990s onwards

Stainless steel ties

Galvanised tie

Modern cavities are usually insulated - with either flexible batts or rigid boards.

Stretcher bond

Cavity 50 to 75mm wide

Cavity 50 to 100mm wide

Before the cavity wall became common solid walls were the norm. Although half brick thick walls were sometimes used for small extensions on poorer quality houses, solid walls were at least 225mm, or one brick thick. Solid brick walls were usually constructed in English or Flemish bond.

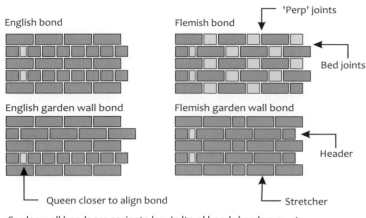

English bond

Flemish bond

'Perp' joints

Bed joints

English garden wall bond

Flemish garden wall bond

Header

Queen closer to align bond

Stretcher

Garden wall bonds are easier to lay. In 'true' bonds headers must be exactly the right length if the wall is to have a fair face both sides. In garden wall bonds, with their fewer headers, waste is reduced.

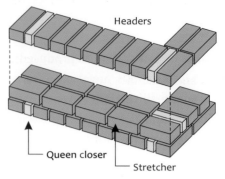

English bond comprises a row of stretchers followed by a row of headers.

Headers

Queen closer

Stretcher

A quarter brick known as a Queen closer is required to line up corners and openings.

English bond produces a stronger wall, although Flemish bond tended to be considered the more aesthetically pleasing. In the Victorian period a number of other bonds were developed including more economical, but less stable, garden wall versions of English and Flemish bond. Despite their name garden wall bonds were often used in houses, although they were usually confined to side or rear walls and were often rendered.

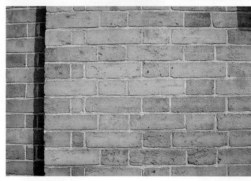

Far left:
English Garden Wall bond is often found on side and rear elevations;

Near left top:
English bond;

Near left bottom:
Flemish bond.

Header bricks were used to tie solid walls together. Unfortunately in some speculative construction, headers were 'snapped'.

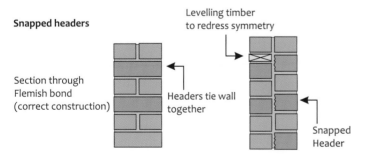

Snapped headers

Levelling timber to redress symmetry

Section through Flemish bond (correct construction)

Headers tie wall together

Snapped Header

Snapped headers were often used where internal and external bricks were of different sizes. They were also used to save money. Facing bricks were much more expensive than commons. 'Snapping' a header created two facing bricks.

Note: these headers might be 'snapped'.

Timbers would sometimes be built into a wall in order to add stability. Although there is some debate as to their exact purpose, they may have been intended to provide lateral constraint and continuity while the lime mortar set. Another reason for burying timbers in the wall was to provide easier fixings for timber panelling and the ends of floor joists.

Snapped headers caused (and still cause) defects because there is a lack of tying-in between the inner and outer face *(see photograph on page 38 as an example)*. Bonding/levelling timbers are vulnerable to rot and beetle attack; if they lose their strength they will no longer resist the compressive loads in the wall and bulging and/or cracking may occur.

Other examples of Victorian bonding included cavities, for example, rat trap bond and Dearnes bond.

'Rat trap' bond was formed by laying bricks on edge. The void in the centre could fill with rainwater.

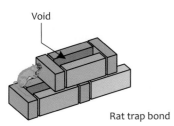

Void

Rat trap bond

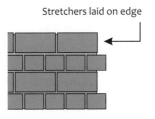

Stretchers laid on edge

Dearne's bond was similar but the headers are laid flat.

Both of these bonds, while economical, were short lived because they allowed water penetration to the interior. They also lacked the stability of English or Flemish bond.

Although there are examples from earlier periods, cavity walls, as we would recognise them today, were essentially introduced in the Victorian period, when it became accepted that having two separate skins helped in preventing penetration of moisture to the interior and when it was technically easier to economically construct thinner individual skins. Cavity walls are referred to in documents from the beginning of the 1800s. However, their introduction was slow and developed at different rates in different parts of the country. They tended, for instance, to be adopted more readily in areas of high exposure where rainwater penetration was a particular problem. By the 1930s cavity walls were common-place in new construction, although they were not necessarily dominant in all parts of the country.

Victorian

inner leaf

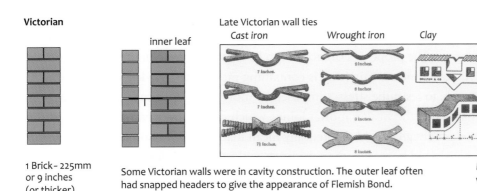

Late Victorian wall ties

Cast iron　　*Wrought iron*　　*Clay*

1930s - 1950s

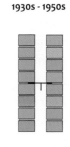

1 Brick– 225mm or 9 inches (or thicker)

Some Victorian walls were in cavity construction. The outer leaf often had snapped headers to give the appearance of Flemish Bond.

Modern walls are usually formed with a lightweight block inner leaf.

The early devices used to tie together the two leaves of the cavity were made from a variety of materials including, bricks or, variously, cranked, glazed, perforated and hollow stoneware. However, the popular approach in the mid-Victorian era, and the norm towards the end of the century, was to use metal ties. Metal ties were either cast or wrought iron. Bitumen was sometimes added as a protective coating against rust. The ties were quite bulky, perhaps because of an assumption that rusting would occur. Because of their size and the type of metal used, the ties were relatively stiff and did not easily accommodate movement between the two skins. By the time that cavity walls were becoming the preferred method for external wall construction mild steel ties were the norm. Again, protective treatments were rather hit and miss, although galvanising with zinc was used as an alternative to bitumen.

Modern ties are made from galvanised steel, stainless steel or plastic. Copper ties have also been used.

Brick categories

Brickwork is generally a durable material with modern methods of manufacture ensuring good quality control. Problems experienced in the past, such as underfiring (leading to softness) or overfiring (leading to brittleness), deterioration through reactive materials, lamination, distortions, and differing sizes, are now rare.

Modern bricks are categorised by strength, salt content and frost resistance. The category of frost resistance considers whether or not a brick is suitable for use where there is repetitive subjection to both water saturation and freezing temperatures. At present Clay bricks are categorised as:

F0 – which means the bricks are not frost resistant and should not be used externally

F1 – which means that the bricks are moderately frost resistant

F2 – which means that the bricks are frost resistant in all normal situations and degrees of exposure.

Soluble salt content is categorised as:

S0 – which means there is no restriction on the amount of soluble salts that they contain

S1 – which means that there are limits set on the amount of soluble salts that they contain

S2 – which means that there are lower limits set than for category S1.

By no means do all the soluble salts that may cause problems come from the bricks. They can be present in the sand used in mortar, and they can enter through rainfall, or from the ground.

Mortar

Mortar helps to distribute the load through a wall and it seals against water ingress. It is made from a fine aggregate (usually sand) and a binding agent (nowadays, in new buildings at least, this is usually cement). When mixed with water a chemical reaction called hydration occurs and the mortar sets. Before cement came into common usage mortars were based on lime and sand. Modern mortars use cement as a binding agent although lime should be included in order to make the mix more plastic and workable. Lime also gives better adhesion to the brick and it improves the mortar's ability to cope with thermal and moisture movement. The properties of mortar can be changed by varying the proportions of the constituent materials. A weaker mortar mix (that is containing proportionately less cement) is more flexible and therefore more able to resist the stresses imposed by movement. With such a mix if cracking does occur it will tend to happen at the weaker mortar joints, which can be more easily and cheaply repaired than the bricks. Modern design should theoretically mean that there is no unforeseen movement but the adage, that the mortar should be weaker than the brick, still generally holds good. This is particularly so in older buildings, where repointing in a strong mortar can lead to damage in brick or stone where the mortar has not performed as a sacrificial item or has failed to allow water to evaporate adequately. However, if a mortar is too weak it will not be sufficiently durable. Stronger mortars are durable because they will absorb less water and are more resistant to frost attack. Paradoxically, a stronger mortar may reduce durability because of its tendency to shrink and crack thus allowing water into the wall.

Liquid plasticisers are sometimes used in place of lime. They usually work by introducing air into the mix in order to break down internal friction and therefore increase workability. They do not, however, impart the other qualities to the mix that lime does. In fact they can reduce the strength of the bond between the brick and the mortar.

DEFECTS IN BRICKWORK

Defects can arise in new brickwork because of poor design or specification, the use of sub-standard materials and poor standards of workmanship. These factors, which depend on quality assurance at the design and construction stage, are not specifically addressed in this chapter.

Frost attack

This is a problem that usually occurs in older bricks, and those that were underburnt on firing. In newer construction, failure through frost attack tends to be confined to areas of severe exposure, or where the frost resistance of the brick was incorrectly specified.

The ability of bricks to resist frost attack is determined by their pore structure (in particular the percentage of fine pores). Frost attack occurs through a combination of excessively wet brickwork and freezing temperatures. When water turns to ice there is a 9% increase in its volume. This expansion can produce stresses within the brick, which cause spalling, with the brick face flaking off and/or crumbling. Mortar is also subject to frost attack. In the deteriorating state both elements readily absorb water which in turn increases the rate of frost attack. Although the risk of frost attack is increased where the saturated brick is subjected to particularly low temperatures, it is the rapidity of the freeze-thaw cycle that causes the damage. Because the process is progressive, frost attack can lead to total disintegration of the brick.

Four examples of frost attack to bricks in exposed situations.

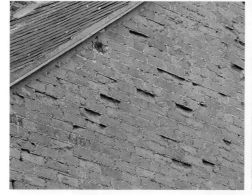

Frost attack is a potential problem where brick walls become saturated. Saturation can occur because of the failure of the design to protect brickwork or where an inappropriate brick type has been selected in an exposed position. It is also possible that individual bricks may be of poor quality due to having been inadequately fired or because they contain impurities.

Poorly applied pointing of a strong, cement-rich mortar has trapped water behind it. This has led to a localised build up of wetness and subsequent damage to the bricks by frost action.

Boundary walls are particularly prone to frost attack. The sides and the top of the wall are exposed to the elements and they are therefore easily saturated, particularly if the wall has inadequate copings. Their exposure means that they are also subjected to extremes of temperatures, including freezing conditions. The tops of the walls often suffer the most because of radiation heat losses to the night sky.

The clay bricks of moderate frost resistance perform satisfactorily on the house where overhanging details protect against wetting at roof verge and eaves and at sills to window openings. But frost damage has occured where the same brick has been used in areas where there is no such protection – such as the brick on edge coping to the boundary wall and the unprotected brickwork below it.

The external walls of a house should be less vulnerable than boundary and retaining walls to frost attack. Roof overhangs and other key design details will offer protection from saturation. This protection will also guard against very low temperatures and in many cases the heat losses from inside the house will keep the temperature of the brickwork above freezing. There have been cases where frost attack has occurred in older brick cavity walls following the insertion of thermal insulation into the wall cavity. Cavity insulation, by its very nature, lowers the temperature of the external leaf and restricts evaporation into the cavity.

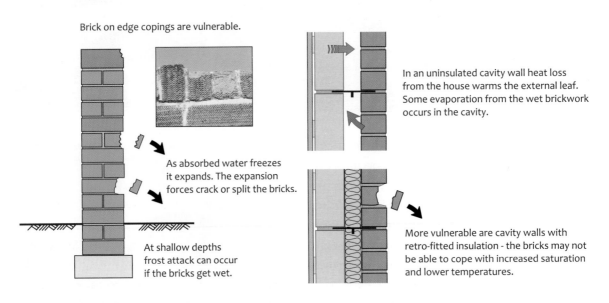

Brick on edge copings are vulnerable.

As absorbed water freezes it expands. The expansion forces crack or split the bricks.

At shallow depths frost attack can occur if the bricks get wet.

In an uninsulated cavity wall heat loss from the house warms the external leaf. Some evaporation from the wet brickwork occurs in the cavity.

More vulnerable are cavity walls with retro-fitted insulation - the bricks may not be able to cope with increased saturation and lower temperatures.

Wall treatments such as the use of silicone (as a remedy for rain penetration) can also cause problems because, when applied inappropriately, they may inhibit the drying out of the brickwork and therefore increase the likelihood of frost attack.

Efflorescence

Efflorescence is a common sight in new brickwork. It is caused by soluble salts in solution being brought to the surface as water in the wall dries out. It is usually a harmless, temporary problem, often occurring in spring following a wet winter. The main concern is the unsightly appearance caused by the white staining that it produces. Persistent efflorescence may indicate a design or construction fault which allows the brickwork to become, and to remain, saturated.

Efflorescence is caused by a number of soluble salts including the sulfate or carbonate compounds of calcium, sodium, potassium or magnesium. The salts may originate in the bricks or they may be introduced through the mixing water, the cement or sand used for the mortar mix, or even from the ground on which the bricks were stacked. Additional sources may include sea air and unfortunate site practices such as the use of washing-up liquid as a mortar plasticiser (it usually contains sodium chloride – common salt). The salts may therefore emerge from the bricks and mortar. As the salts are water soluble they are often removed by rainfall, although they can usually be brushed off if their appearance is causing concern.

White soluble salts deposited on the surface of the newly completed brickwork that was not given adequate protection against rainfall during construction and the same building about six weeks later. The efflorescence has been washed away by the natural weathering action of the rain.

Although it is usually a harmless problem there have been cases of damage caused by efflorescence. Crystalisation of salts just below the surface of the brick can cause spalling. This is known as cryptoflorescence. The problem is often associated with magnesium salts. Cryptoflorescence is associated with a large build-up of salts and usually occurs where old, relatively weak, bricks are re-used inappropriately, particularly in areas of excessive dampness. It can also occur through salts deposited by the run-off from limestone or from air pollution. Damage may also occur if the brickwork has been covered by a surface treatment because the salts may crystallise behind the treated surface and force it off. The effect on the bricks is similar to that caused by frost.

Lime run-off

Where excess water flows through cementitious material another type of staining can occur. Water can dissolve calcium hydroxide (free lime) which is then deposited on the brick face. The calcium hydroxide is a soluble form of lime which is created as Portland cement hydrates. The source of the lime may be the cement from mortar joints or it may come from concrete or cast stone elements; for example a coping above a brick wall or a floor slab built into the brickwork.

Insoluble lime (calcium carbonate) deposits on the surface of newly completed brickwork. The lime has originated from soluble lime material (calcium hydroxide) washed from the mortar due to lack of adequate protection against rainfall during construction.

Once the calcium hydroxide has carbonated removal can only be affected with acids.

The run-off is often seen 'dribbling' from weep holes. The calcium hydroxide reacts with carbon dioxide in the air producing a hard crystalline formation of calcium carbonate. The initial staining can be removed with water and brushing before it carbonates but once the reaction has occurred an acid solution will be necessary.

Other stains

Brickwork can be stained by a variety of materials including:

Vanadium salts These salts produce a yellow or green efflorescence on new brickwork. They are generally best left to weather away naturally. The salts occur naturally in certain clays (usually, but not exclusively those that are used to produce buff/lighter coloured bricks).

Iron staining When it occurs it usually appears as a stain to the mortar joint. The staining can come from metal imbedded in the structure, or it can be derived from the bricks or the mortar sand. Iron staining can be removed mechanically if the mortar is still relatively weak; otherwise chemicals may have to be employed. If it appears on the brick it should be allowed to weather away.

Manganese staining This manifests itself as a dark brown or black colour. The approach to treatment is similar to that used for iron staining.

Lime blow in bricks

This is caused when clay bricks contain small amounts of lime. When the bricks are fired the lime is converted to calcium oxide (quicklime). When the bricks become wet the calcium oxide begins to slake. The process of slaking is vigorous and in this situation can cause an eruption on the face of the brick.

Lime blow in plasters and renders is discussed in Chapter 9: External Rendering.

Sulfate attack

Sulfate attack is a serious problem as it can cause crumbling of the mortar joint and expansion and instability in the wall. It is caused by a reaction between sulfates in solution and a constituent of Ordinary Portland Cement known as tricalcium aluminate. Sulfate attack relies upon a number of conditions occurring simultaneously, that is, it requires water saturation over a relatively long period, a source of sulfates and reasonable amounts of tricalcium aluminate. Even where these factors exist simultaneously the attack will take a relatively long time to develop. The rate of deterioration is affected by the quantity and the type of sulfate – the sulfates of magnesium and potassium are the most aggressive.

The reaction between the sulfates and the tricalcium aluminate forms a compound known as calcium sulfo-aluminate. This compound expands as it forms, leading to cracking in the mortar joints, followed by general deterioration and the loss of the mortar's 'bonding' function, as the face of the joint spalls and the mortar cracks and crumbles. The cracks can be at the edge of the mortar joint or through the middle. The expansion, which leads to leaning and bulging, exacerbates the instability in the brickwork caused by the deterioration of the mortar. Sometimes the face of the bricks spall, most commonly around their edges.

The source of the sulfates can be the bricks them-selves but they can also be introduced from the ground, or from air pollution. An additional source is the exhaust gases from slow-burning fuel appliances.

Sulfate attack usually occurs in situations which are particularly exposed to relatively large amounts of water.

Parapet walls, being exposed on both sides, are always at risk from sulfate attack. Sulfate attack can be recognised here by the pronounced cracks in the bed joints. It should not be confused with wall tie failure (see later section).

The two photographs above show sulfate attack to a garden wall.

Sulfate attack in retaining walls

Saturation of the wall can be caused by high levels of ground water, defective drainage, ineffective damp proofing details and evaporation through the wall.

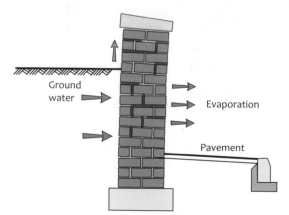

Ground water

Evaporation

Pavement

Sulfate attack has just started to crack the horizontal bed joints in this retaining wall.

A common consequence of sulfate attack in chimneys is a pronounced lean caused by the different wetting and drying cycles between elevations. In chimneys additional sulfates may be deposited by the combustion process and additional water may be introduced from condensation within the chimney itself.

Sulfate attack in chimneys

Rain

Additional sulfates are provided by exhaust gases from fires. Additional moisture is provided by exhaust gases condensing inside the cold upper parts of the chimney.

Sulfates in the bricks combine with cement by-products in the mortar to form calcium sulfo-aluminate. As this crystal forms it expands.

Tall chimneys can develop a noticeable lean because the wetting and drying cycle is uneven due to nature of prevailing winds. The wetter side of the chimney (through rainfall or condensation) suffers the most expansion.

The situation is often made worse by applying render to 'at risk' elements. Render which is too strong (ie has too much cement in it) may shrink and allow rainwater into the wall. Because of its density, a strong render restricts the rate of evaporation. As sulfate attack will also cause cracking in the render, more water can get in and this further exacerbates the situation.

Sulfate attack in a garden wall.

The horizontal cracking caused by sulfate attack can be distinguished from that caused by wall tie failure(discussed later) because it may occur in every joint. Also there is often a characteristic white colouring to the mortar as it deteriorates. Because sulfate attack involves water saturation it is often accompanied by frost attack.

Characteristic cracks such as these at the centre line of a bed joint are indicative of sulfate attack.

Wall tie failure

Failure of wall ties has become a significant problem in recent years. The main cause of failure is rusting of metal ties, although there can be other causes, such as a failure to properly bed the tie in the mortar joint, poor quality mortar reducing the bond between tie and mortar, or not installing the requisite number of ties.

The obvious danger with rusting wall ties is the possible collapse of the outer leaf of the cavity wall. Other consequences of rusting wall ties are set out below.

- The rust will have a significantly greater volume than the original metal. This expansion of the tie may cause cracking and distortion of the structure, particularly where strip ties have been used.

The rust-induced expansion in strip ties can lead to secondary damage, such as a redistribution of loads, buckling and bulging of walls, and damage to the roof as the external leaf increases in height.

A tie from a house built in the 1920s – the right hand side of the tie was in the outer leaf.

- The less bulky wire ties will not generally produce enough expansion to induce cracking unless the joint is abnormally thin or the mortar is very dense. Unfortunately, wire ties produced before 1981 *(see later)* had less rust protection than strip ties and therefore are likely to have a shorter life; a particular problem because failure can occur without the outwardly visible warning signs produced by the cracking.

- The cracking will also reduce the weather resistance of the wall, which in turn accelerates the rusting process

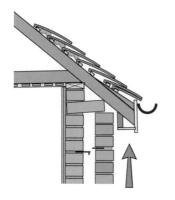

Expansion in the joints can lift the roof.

Ties like this (vertical twist ties) are likely to cause the widest cracks in the bed joints and the most vertical movement in the wall.

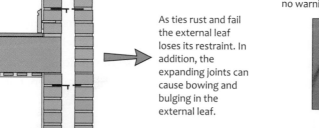

As ties rust and fail the external leaf loses its restraint. In addition, the expanding joints can cause bowing and bulging in the external leaf.

Wire ties do not cause as much expansion as twist ties but, because they can rust without causing cracks, there may be no warning signs of structural instability.

Cracking is not always easy to spot - here it's more obvious because of the mastic 'repair'.

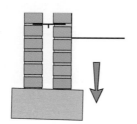

If there is extensive expansion in the mortar joints the loads from the roof can be transferred to the outer leaf. This can, in extreme cases, cause rotation in the foundation.

A Cox

The term corrosion refers to a partial or complete wearing away or a dissolving or softening of a substance by a chemical or electrochemical interaction between the metal and its environment. Rusting is a type of corrosion and it is caused by the interaction of water and iron or steel in the presence of oxygen. Rust is a hydrated iron oxide, which although solid in form, is weak and brittle. Galvanising is a means of protecting iron or steel by coating it with zinc. The zinc protects the steel electrochemically. Of the two metals zinc is the more chemically active and it will slowly corrode in preference to the steel. The thicker the original coating of zinc the longer it will last. The zinc protection reduces in thickness but it does not undergo a chemical change. Eventually the zinc coating will disappear.

The process of rusting can be speeded up by:

- aggressive chemicals: black ash is a product of coal mining which was sometimes added to mortar. The high sulfur content of black ash produces a weak sulfuric acid when wet for long periods. This sulfuric acid can attack a galvanised coating
- chloride salts, which may come from marine sands or may have been added to the mortar as accelerators, can, even in small amounts, speed up corrosion
- carbonation: where ties are well bedded in mortar, protection against rust is provided by calcium hydroxide, which is formed as the cement hydrates. Unfortunately, as the mortar slowly carbonates this protective alkaline layer is destroyed. Carbonation will occur more quickly in permeable mortars
- the building, or particular sections/elevations, being exposed to high levels of moisture ingress because of their exposure.

The portion of the tie in the outer leaf will usually be the most severely affected because of exposure to rain penetration. However, condensation may also produce enough moisture to allow rust failure to occur.

In 1945 a British Standard was introduced which set out, amongst other things, a minimum thickness for the galvanising layer. The standard identified the two basic shapes of tie – vertical twist and butterfly. The standard specified that the galvanising should be twice as thick on the vertical twist as it was on the butterfly. Unfortunately this minimum thickness was reduced in 1968, because it was thought that the standard was excessive. In 1981 the standard was increased and specified the same thickness for both types. The reason for this was the increasing evidence regarding performance in use and the realisation of the scale of the problem. The problem of rusting was thought at one time to be localised as it was believed that it occurred mainly in areas where a catalyst, for example black ash mortar, increased the likelihood of breakdown. It was later appreciated that the problem was more widespread and the causes more complex. The Building Research Establishment has suggested that all steel or iron wall ties inserted prior to 1981 are at risk of (premature) failure. (Premature relates to the notional 60 years life of a building.) Because cavity walls tended to be first introduced in areas with high rain penetration, including marine environments with their associated salts, there can be a potent combination of age and aggressive environments.

It is worth noting that stainless steel wall ties are available. Stainless steel is a ferrous alloy which contains at least 10% chromium. Since chromium is more reactive than iron, a 'self healing' impermeable layer of chromium oxide forms naturally on the surface of the steel. Chromium oxide is durable in a wide range of exposures and prevents the formation of rust. Where there is exposure to chlorides, for example in marine conditions, care must be taken in selecting the appropriate grade.

Unlike sulfate attack the cracks associated with wall tie failure correspond with the courses containing the ties. Ties normally occur at every sixth course of brickwork.

Wall tie failure can often be identified by the horizontal cracking that results from the expansion caused by the rusting process. Unlike sulfate attack the cracks will coincide with the position of the ties (and will obviously only occur in cavity walls). However, because butterfly ties will usually not cause cracking it may be necessary to carry out investigations where the problem is suspected or anticipated. Of course, ideally, strip ties should be identified before the cracking occurs. This might involve a visual inspection of the suspect tie, either by using an optical probe or by removing bricks. It should be borne in mind, however, that a tie can be in good condition in the cavity but poor in the outer leaf and therefore exposure of the end embedded in the outer leaf is a sensible step.

Ties can be assessed visually for rusting. The presence of red rust, the product of the oxidation of iron, indicates severe rusting, while the presence of white deposits or a blackening of the galvanising indicates that the zinc is corroding. Measurements of zinc thickness will enable the projection of the remaining life. (The Building Research Establishment gives some useful guidance on this in Digest 401.) Metal detectors can determine the exact pattern of tie placements and can help in establishing if a defect is caused by wall tie failure.

Disguising wall tie replacement is difficult, particularly where houses are rendered. Painting will help but it is very difficult to match the texture of the render. In plain brickwork extensive repointing will be necessary.

Weathering and disintegration of joints

Brickwork can be jointed as work proceeds or pointed after completion. The mortar can be finished with one of a number of different profiles.

The profile of a joint has an aesthetic effect – because of the way it casts shadows and reflects light – and it affects the durability of a wall. Some profiles are more effective than others in preventing water ingress. The nature of the joint can, in fact, be more significant than the porosity of the brick in terms of weather protection.

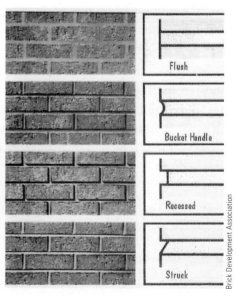

Flush

Bucket Handle

Recessed

Struck

Brick Development Association

A tooled joint using, say, a bucket handle is effective. The tooling compresses the mortar slightly, thus reducing its permeability. At the same time tooling ensures a good seal between mortar and brick. Struck joints are equally effective although they require slightly more skill.

Recessed joints are problematical because they do not easily shed rainwater but allow it to sit on the edge of the brick and, although tooling may compress and seal the joint, the brick is at risk. Flush joints often leave small cracks between the brick and the mortar. Also, they are often 'buttered over' the joint, increasing the likelihood of water penetration. Chapter 14: Dampness, explains some of these aspects in more detail.

Deterioration of mortar will reduce the wall's resistance to rain penetration and may, in an extreme situation, lead to structural instability. A deeply recessed bed joint could reduce both lateral and compressive strength. The deterioration may be due to inadequacies in the mortar mix or damage by frost attack or sulfate attack.

Repointing in the incorrect mortar can be an aesthetic problem in that it alters the way a wall will look. It can also be a technical problem because, for instance, a dense (ie too much cement) mortar will not allow the brickwork to 'breathe' adequately and/or it may shrink and crack and so allow water ingress.

SECTION 3

STONEWORK

The existence of many stone buildings, both as single examples, or coherent settlements such as in the city of Bath, is a testament to the durability of stone, as well as a reminder of the aesthetic and symbolic appeal that made it such a high status material.

There is evidence of stone buildings in this country dating back to the 1st century BC. The Romans used stone and brick but, as mentioned, there was a decline in the general use of both materials after they left. Nevertheless, the Saxons used stone for significant buildings, as evidenced by the simple churches of the period. The Normans brought their own craftsmen and trained the Saxons to build monumental buildings such as cathedrals and castles, establishing stonemasonry as a recognised craft. However, stone was not used for poorer dwellings until much later, mainly because wood was readily available.

Gradually, stone became available to a wider spectrum of society (or perhaps more accurately there was a broadening of strata in society and more people were able to afford stone), until by the 1500s it was in relatively common use.

Today stone remains a material of status – if is often still used for significant buildings, particularly those that are supposed to signify power, gravitas, or trustworthiness.

Although brick became a popular material in the Victorian period, stone was still used to convey status. During this period brick became the cheaper material and was often used for structural work. While it is true that brick was also used for high status dwellings, and went through phases of popularity because of changing architectural styles, stone was predominantly the material that was aspired to. In practice stone was often just used for the front elevation of a house with brick being used on side and rear walls and often rendered.

The nature of stone

Stone is a natural material which is derived from one of three basic rock formations: igneous, sedimentary or metamorphic.

Igneous These rocks were originally formed from the cooling and solidification of molten magma – the name is derived from ignis, the Latin for fire. Building stones derived from igneous rocks are generally durable. Granite is an example of an igneous building stone.

Metamorphic As the name suggests these rocks were produced by changes brought about in existing rocks. The changes mainly occurred through the influence of pressure or heat or both. Examples of building stones derived from metamorphic rocks include marble and slate.

Sedimentary These were created by the laying down of sediment produced from the wearing away of existing rocks, or from the shells and skeletons of living organisms, or by chemical deposition. These particles were deposited in layers on the floors of lakes, rivers, seas and deserts. The sediment particles were cemented together by other minerals and compacted by self-weight and sometimes by earth movements. Limestone and sandstone are examples of sedimentary rocks. As they are the most widely used stones for building this chapter will focus on these two materials.

Although limestone is still being formed today the stones that are most commonly used in buildings come from the Jurassic age and were created between 136 and 195 million years ago.

Limestone as a building material has been used extensively in the southern part of the UK for many centuries. It has helped to define the character of significant monuments such as the cathedrals of Wells, Ely, Exeter and Salisbury, the Palace of Westminster and the Tower of London, as well as cities such as Bath and geographical areas such as the Cotswolds. Limestone is found in the Jurassic belt, which covers an area from the coast of Yorkshire, through Lincolnshire and Northamptonshire and then through Oxfordshire, the Cotswolds and on down to Dorset. There are also older deposits of limestone in Wales and Cumbria.

Sandstones can be found in quite a wide geographical area from Sussex to the north of Scotland. The most important sandstones are those produced in the Carboniferous period; between 280 and 345 million years ago. These sandstones can be found in large areas of the north of England and the Midlands, as well as Gloucestershire, Devon and Cornwall. Sandstones were used in, for example, the creation of Edinburgh New Town, important houses such as Chatsworth and cathedrals such as Durham, Hereford, Liverpool and Coventry (new and old).

Sedimentary rocks were formed in parallel, mainly horizontal, layers or beds (diagonal bedding sometimes occurred due to changes in wind or water direction). Bedding planes are the defining points of an episode of sedimentation – they mark where sedimentation stopped before recommencing at a later time. Joints in rocks are formed by movements, tensions or shrinkage in the rocks. They will divide the horizontal bedding planes vertically, as well as providing further horizontal delineation. Sometimes the shrinkage forces will have been strong enough to pass through one bedding plane into the next. The blocks defined by the joints and bedding planes make quarrying easier but they may cause problems in later repair or extension work because they determine the size of stones that can be extracted.

Limestone

Limestones are made up mainly of calcium carbonate which accumulated at the bottom of seas and lakes. However very few limestones consist entirely of calcium carbonate. Many also contain clay, sand, or organic material.

Limestones can be categorised according to their mode of formation.

Chemical Produced by water passing through rock formations and picking up calcium carbonate. Water contains carbon dioxide and this reacted with the calcium carbonate to form calcium bicarbonate, which is soluble. The water containing the calcium bicarbonate then flowed into a larger body of water, such as the sea, where the precipitation of calcium carbonate took place.

Organic The main constituents of these stones are the fossilised shells of once living organisms.

Detrital (Clastic) These are derived from the decomposition and erosion of existing limestones.

As mentioned above limestones are all made of similar material – with the exception of magnesium (dolomitic) limestones which consist of calcium and magnesium carbonate – but their structure can vary considerably in ways that will affect both how easy they are to 'work' and their durability. Some are very coarse while others are very fine; some contain very significant and obvious amounts of original material which has not been ground down, such as shells. Unfortunately, some of the more durable limestones are not very easy to work or they come from narrow beds.

Limestone has been quarried at Doulting (Somerset) for nearly 2000 years. In the 13th century it provided the stone for Wells Cathedral. Sometimes fissures don't become obvious until the stone is cut.

Sandstone

Sandstones were formed from components derived from the breakdown of other rocks. The deposits formed were transported by water, and to a lesser extent wind and ice, and deposited at the bottom of larger bodies of water such as lakes and seas. Sandstones are made up of a variety of materials but the essential material is quartz. Sandstones were produced when grains of quartz were cemented by various compounds. They are, in fact, categorised by the cementing medium or matrix. Because quartz is largely inert, the cement largely determines the weathering resistance of sandstones.

The types of cement found include:

- Siliceous. These are silicas, they are inert and therefore produce durable stones. Effectively, these are quartz grains cemented by a quartz matrix
- Ferruginous. These are iron oxides and have poor durability
- Calcareous. These consist of calcium carbonate and are reasonably durable except in urban environments where they are exposed to air pollution
- Dolomitic. These are calcium and magnesium carbonates and, as with calcareous matrices, they are vulnerable in polluted environments
- Argillaceous. These are clay and have poor durability. They are particularly susceptible to frost action.

Construction of stone walls

Stone walls are defined by the manner employed in dressing the stone.
The two general categories are ashlar and rubble.

Ashlar (right)

This is the description given to squared stone which is presented by being dressed to a smooth face. Ashlar consists of mathematically precise blocks separated by very thin joints to produce a 'classical' appearance.

Rubble

This was a popular approach in the Victorian era. It was used because it was cheaper than ashlar but, in many instances, it was deliberately chosen because of its appearance. In part, the preference for the natural rustic look of rubble was a generalised reaction against the plainness of classically derived Georgian architecture. There are a number of variations, two are shown below.

Often, rubble and ashlar were used together, with rubble being framed by ashlar work at the corners of elevations, or ashlar bay windows being inserted into rubble walls. Brick was also commonly used as the dressing around the windows and doors of rubble faced houses.

DEFECTS IN STONEWORK

Stone is a natural but complex material. Within the general headings of sandstone and limestone there are many stone types, with different durability and weathering characteristics. The discussion of defects is therefore inevitably generalised.

Pore structure

There is still much debate about the key factors affecting durability in stone and empirical evidence is somewhat contradictory. However, it would appear that while natural durability can be affected by a number of factors, probably the most significant is pore structure. Pore structure is significant mainly because it affects the amount of water entering and moving through the stone. Also, salts that may damage the stone can be transported by water and accumulate in the pores. The critical factor is not the total amount of space created by the pores but how they are structured. Stones with low porosity will not allow much water penetration and are therefore less likely to suffer from salt and/or frost attack. Stones with low porosity are therefore generally more durable. Stones with a high porosity will allow more water in but, if the pores are large, the water will tend to be able to evaporate reasonably quickly. However, if there is a large network of fine pores capillary action will be high – but evaporation will be relatively low. In addition, a stone with large pores is less likely to suffer salt damage than one with small pores. This is because the larger space is more likely to be able to accommodate the expansion pressures of salt crystallisation.

Incorrect bedding

As mentioned previously sedimentary rocks were laid down in beds. Defects can arise in use if a stone is incorrectly placed in a building in relation to its bedding plane.

When it is placed in a wall the stone should generally lie in its natural bedding position. That is, the layers should run horizontally, in the manner in which the stone was originally formed. The stone is stronger in this position and is also less vulnerable to defects. If the stone is face bedded (the layers are vertical), it is more vulnerable to damage through crystallisation of salts and/or frost action. This is because the mechanical actions involved find it relatively easy to push off the bedding layers as there is no restraint from the adjoining stones.

Stone delaminating because of face bedding.

Bedding sedimentary stone

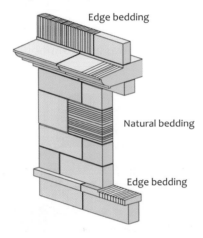

Edge bedding

Natural bedding

Edge bedding

Sedimentary stone should normally be laid in its natural bed, ie, with the layers horizontally, the way it was formed over thousands of years. However, copings and cills are usually edge bedded because they are exposed on their top side.

The damaged stone above has been incorrectly laid. It has been face bedded and has started to delaminate. Failure occurs over time through the wetting and drying cycle, possibly exacerbated by frost attack.

The laying of stone in a face bedded manner is not uncommon. It may have happened through ignorance of the consequences, or through difficulty in determining the layering pattern – it is easy to distinguish with some stones but is less so with others. For some features of a building, such as cornice or coping, laying the stone in its natural bed will be incorrect because it will be exposed on its top side. In these situations the stone should be edge bedded. With a string course, the corners of a coping or carved stone, no solution will be entirely satisfactory. In these situations attention to protective detailing will be particularly important.

Coping stone delaminating.

In 1753 Francis Price, examining Salisbury Cathedral, built some 500 years earlier, wrote...

THE pillars and ſhafts, both for uſe and ornament, are of Purbeck marble, yet with this difference; thoſe for uſe, I mean the pillars that bear the weight, lay in their natural bed, as found in the quarry, while thoſe ſhafts for ornament have their bed inverted, or turned perpendicularly; and by that means they are ſubject to ſplit or cleave aſunder, where they ſupport a weight.

Naturally bedded

The main columns are made up of several parts, all of them naturally bedded for strength. The size (height) of each stone is often determined by the depth of the stone seam. The ornamental shafts, often made from Purbeck marble, a soft limestone which can be polished, are edge bedded. This is because the seams are not very deep and edge bedding allows a single shaft to be made without any joints. As Francis Price states, this affects the strength of the shaft; if it was loadbearing it would split.

Salt crystallisation

This involves salts in solution passing into the stonework and entering the pores of the limestone or sandstone. There are a number of salts which can cause problems and they may come from a variety of sources. Sources may include building materials such as concrete, bricks or mortar, as well as soil and the air. As the stone dries out, salt is deposited at the surface (efflorescence) or within the stone.

The curved pattern staining is caused by different rates of evaporation

In this column salts from the ground have migrated upwards through capillary action. Crystallisation and damage are most visible where the rising damp ends.

Crystallisation within the pores will exert pressure, often resulting in damage. The possibility, and extent, of the damage will be related to the type of stone, the type of salt, and the characteristics of the pores – mainly their size and their arrangement.

The potential for damage is increased because of the repetitive nature of the mechanical action caused by the cycle of (re)dissolving and (re)crystallising. The damage usually shows as a powdering of the surface. In some cases, it can look more dramatic by causing splits in the stone as the mechanical expansion works against an existing weakness, such as incorrect bedding or a fissure.

Because some of the salts will absorb water from the air (ie they are hygroscopic – *see Chapter 14 on Damp*) they will cause damage even if there is no direct wetting from the more obvious sources such as rain, rising damp or plumbing and drainage faults. The relative humidity at which they will absorb water will vary with the salt, but for some it will be below the normal level for an occupied building. In a coastal area there will be a potent combination of salts from the sea and a high relative humidity.

Road salt has crypto-fluoresced within the pores of the stone shattering it.

Acid rain

A significant cause of damage is the interaction between limestone and the products of pollution in the form of acid rain. Acid rain is produced by industrial and domestic activities and, increasingly, by road traffic. Acid rain is a generic term which includes other weather conditions, eg fogs, mist etc.

There is a popular impression that damage to limestone through acid rain is rather like the process involved in dissolving an antacid tablet in water. In reality, although acid rain can directly dissolve stonework, this action is relatively insignificant compared to the reaction between the acidic rainwater and the calcium carbonate in limestone.

The major problem is the production of soluble gypsum. The main way in which this occurs is when sulfur gases go into solution with rainwater and form sulfurous acid (H_2SO_3). Sulfurous acid then combines with oxygen from the air. The resultant sulfuric acid reacts with the calcium carbonate ($CaCO_3$), which is the main constituent of limestone, and forms calcium sulfate ($CaSO_4$). The calcium sulfate then takes up water and crystallises as gypsum ($CaSO_42H_2O$). The resulting pattern of decay of the limestone is determined, essentially, by its exposure.

- On exposed areas of the building rainwater will wash away the gypsum. Gypsum is slightly soluble in water but more soluble in acidic water or rain that contains common salt – as might be the case in a marine environment. What occurs is a gradual weathering away of the stone, although the visual effect can, from a distance, look like cleaning or renewal.

- In sheltered areas of the building a relatively impermeable skin will start to form. The skin can be up to 12mm thick and in polluted urban environments it will contain dirt from the atmosphere. The skin may stay in place for a long period (this might be affected by the characteristics of the particular stone), or it may form blisters that burst to reveal a powdery decayed interior. Crystallisation and recrystallisation may also be going on underneath this skin creating large areas of damage. In sheltered areas damage to magnesium limestones can be particularly severe because both calcium and magnesium sulfate are produced. The calcium sulfate again creates a skin but below it the more soluble magnesium sulfate will often be found forming crystals. This crystallisation often

results in relatively deep caverns of decay in the stone, which are revealed when the skin breaks down. This is sometimes referred to as cavitation.

Although the durable sandstones that are cemented by a quartz matrix are inert, and therefore resistant to sulfur based gases, those sandstones which are cemented together by a calcium carbonate matrix (calcareous sandstones) can be severely damaged by the chemical reaction with acid rain. The reaction between the calcium carbonate and the acidic gases is the same as described above, that is

sulfurous gases in solution react with calcium carbonate to produce relatively soluble calcium sulfate. However, as in this case the cementing matrix is being attacked, the result is the release of large amounts of otherwise stable quartz crystals. The damage is often worse than in limestones placed in a similar environment.

Dirt on stone buildings may cause problems because it can contain soluble salts. The dirt comes from particulate air pollutants which are 'cemented' onto a limestone wall by the gypsum produced from chemical attack. The particulates may be derived from various animal, mineral or vegetable sources but the most common are tar, diesel residue, oil, and carbon deposits. As stated previously the gypsum, and therefore the dirt, will be removed by rain on exposed surfaces of limestone, but it will remain on sheltered surfaces. However, with sandstone the particles become 'bonded' to the stone's surface and will not be easily washed away by rain.

Limestone run-off

The soluble salts produced by the reaction between acid rain and the calcium carbonate in limestone can cause problems in adjacent materials. When the soluble salts are formed they may run off say, a limestone coping, onto a brick or sandstone element below. As these salts crystalise they can cause decay in materials that would otherwise not be particularly vulnerable. The salts involved may also cause lime staining similar to that described in the section on brickwork.

Damage caused to brickwork due to the expansion of calcium sulfate washed into the bricks following its production cause by the interaction between the limestone coping and acid rain.

Frost attack

Frost will tend to be a problem in those areas which get both excessively wet and are subject to freezing. The vulnerable areas have been identified at the start of this chapter, but it is worth reiterating that frost damage is not very common in sheltered plain walls, except, perhaps at below damp proof course level. As with brickwork, pore structure is a significant factor in determining susceptibility to frost attack and the process of attack is the same.

Where it does occur frost damage can dislodge quite big pieces of stone either in vulnerable areas, such as parapets or copings, or where the stone has been damaged by other agents.

Frost attack has damaged the stonework. The effect of the lack of a damp proof course has been aggravated by paintwork which has slowed down the rate of water evaporation. It would appear that the most badly damaged stone is face bedded.

Contour scaling

This is another problem associated with salts. Contour scaling occurs in sandstones and is thought to be the result of the pores of the stone being blocked with calcium sulfate. This seems to happen even when it is not a calcereous sandstone. The effect is a breaking away of a rather thick crust from the face of the stone. The effect produced by contour scaling resembles that caused by incorrect bedding, but it occurs irrespective of the way the bed lies. It is thought that the damage is caused by the different reaction of that part of the surface material with the blocked pores compared to the underlying body of stone when subjected to thermal and moisture movement stresses.

Expansion of metals

Iron and steel cramps have traditionally been used as fixing devices in stonework for many centuries. When these metal fixings rust, they expand and can fracture the stone. Additionally, stone cavity walls may suffer from the cavity wall tie problems discussed earlier in the chapter.

Many external boundary walls have stone copings with metal railings bedded in them. Damage may also occur where external stone features surround metal service pipes.

Extraction and dressing

Stone can be damaged in the quarry if it is extracted by the use of explosives which may cause internal fractures. Damage can also be caused by excessive tooling of the surface of the stone.

Organic growths

Plants and lesser organic growths on stones are common. Sometimes they are seen as adding to the charm of stone buildings, sometimes they are regarded as unsightly. The different perspectives depend to some extent on context – random growths may be thought of as detracting from a classically designed building for example, or, in other situations, valued as part of the patina of age.

Lichens Lichens are a symbiotic association of fungi and algae. They are hardy in the sense that they are able to derive sustenance from a number of sources: the fungi element seeks out water and salts and the algae element manufactures food by photosynthesis. Lichens can produce acids which may etch into the surface of soft stones such as limestones or calcereous sandstones. Finely carved stonework may be particularly vulnerable. The fine rootlets of some lichens may cause minor mechanical damage. The potential for damage is reduced by the fact that lichens are extremely sensitive to air pollution.

Algae Algae are a common form of vegetation on buildings. They prefer conditions where damp, warmth and light are available as they manufacture food by photosynthesis. They are usually green in colour, but can be red, brown and blue. Like lichens their acidic secretions can damage stone. They also collect both water and dirt and can affect the transpiration rate of walls.

Mosses Mosses can act as 'sponges' and retain moisture; perhaps enough in a concentrated area to cause frost attack. Like algae their presence will tend to indicate excessive dampness. They can secrete acids which may cause etching in some stones. They have also been known to develop enough to cause root damage to very soft stones.

Fungi Fungi can grow on stone if they can find a food source (they cannot manufacture their own, unlike plants). Fungi can produce acids and therefore possibly contribute to the decay of limestones.

Higher plants

The biological growths mentioned above could create the conditions for other plants, which may be more damaging to buildings, to grow. Biological growths do this by retaining moisture and by providing humus when they die.

Clearly any plant or tree could potentially seed and grow in stonework. Some may cause damage, some may not. For example:

Ivy The roots of ivy can cause mechanical damage. Usually the damage is to mortar joints that they have lodged in, but the fine rootlets can split stone if they penetrate through crevices.

An example of Virginia Creeper.

Virginia creeper This is usually harmless as the plant merely attaches itself to the wall by suckers rather than through a root system.

Bacteria are thought to be able to affect decay but from a practical viewpoint the damage caused is generally insignificant. Bacteria cause damage by either starting or adding to the production of chemicals that then attack stone. An example is by oxidising sulfurous acids in polluted atmospheres thus forming sulfuric acid.

Birds and bees

Birds can damage stone (or brick) in two ways: they can cause mechanical damage by pecking for salt or grit. They can also cause decay in limestone and calcareous sandstone because their droppings release acids. Starlings and pigeons are the main culprits.

Masonry bees can cause a small amount of damage as they bore into soft stone and brick to make a nest (photo right).

Damage caused by masonry bees.

POOR REPAIR/CONSERVATION TECHNIQUES

Although repairs are not discussed in this book there are instances where poor repairs can cause further defects.

Incorrect pointing As with brickwork an inappropriate mortar mix can cause problems in both the mortar and the stone itself. The main problem is usually related to having too strong a mortar. Because a strong (dense) mortar allows less evaporation any moisture in the wall will have to evaporate through the stone. This may increase the likelihood and severity of salt related defects. Additionally, as mentioned before, stronger mortars are relatively brittle and may be more susceptible to shrinkage cracking. This may increase the likelihood of rain penetration, which in turn may increase the potential for salt related defects and frost attack.

Making the jointing stand proud of the face of rubble stonework is another common mistake. This is sometimes known as ribbon pointing. Ribbon pointing is usually formed with a very strong mortar. Aesthetically, it is wrong because it is so dominant – the impression is often of a mortar wall with stone inserts. Technically, the ribbon pointing performs poorly because it shrinks and allows water ingress, it acts as a ledge for retaining water, and it is brittle and so easily cracks and falls away.

An example of ribbon pointing. Note how parts of it have failed and fallen away.

Elevation

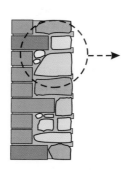

Section through traditional rubble wall with ribbon pointing

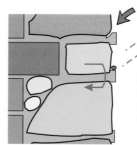

Shrinkage of mortar allows water penetration

Rainwater cannot be shed efficiently and dense pointing impedes evaporation

In freezing conditions ice forming in joints can force pointing away from wall

Failure of protective coatings

Coatings are often applied to stone in an attempt to protect it. They are usually applied in order to:

- prevent water penetration
- prevent stone decay.

Additionally, given the number of bay windows where the stonework is painted, some coatings are presumably added because they are judged to improve the look of the stone. While there may theoretically be good reasons for applying a silicone based water treatment to stone (ie where penetrating damp is occurring), they should not be used to try to arrest the effects of stone decay. Where they have been used there is evidence that the rate of decay is actually accelerated. This is probably because the water takes longer to evaporate when a coating is applied. Also, because salt crystals are often trapped behind the treatment, they can cause greater damage to the stone. Eventually, the forces exerted by the salts will probably break down the coating.

Extensive damage can occur where bitumen based solutions are applied. These are a particular problem as they are very effective in preventing the stone from 'breathing'.

Above: Defects in this limestone bay have been exacerbated by the application of paint.

Right: Damage to the stonework has been accelerated by the use of a bitumen based treatment.

CONCLUSION

The common element in brick and stone defects is water. When excessive water combines with freezing temperatures or soluble salts then defects may result. In practice these situations often occur where insufficient consideration is given to the relationship between design, detailing and materials selection.

Some of the defects discussed may occur irrespective of whether the wall is stone or brick. Others may only occur in one material or the other. There is potential for confusion because the naming of defects uses similar phrases to describe different problems. For instance the term sulfate attack, used to describe the reaction between sulfates and ordinary portland cement, should not be confused with the problem caused (in limestone) by the conversion of calcium carbonate to calcium sulfate.

CHAPTER 5

Ground Floors

TYPES OF FLOOR

Most Victorian and Edwardian houses of the middle and upper classes had timber ground floors. These same houses often featured small areas (usually circulation areas) of decorative tiling laid on early forms of concrete. However, in housing built for the working classes, many dwellings had ground floors made in whole, or in part, from stone flags, or large heavy clay tiles, laid directly on compacted earth or a thin layer of broken stone.

During the first half of the 20th century most ground floors were built in timber and are usually referred to as raised timber, or suspended, ground floors. Nowadays, most ground floors are formed from in situ (ie cast in place) or pre-cast concrete.

Clay tiles such as these were popular from the 1850s until the 1930s. They were bedded in mortar and laid on a screed or direct on a concrete slab. Their use was often reserved for those areas visitors would see (and admire) ie halls and lobbies.

This example is from the Metal Agency Company Catalogue 1932.

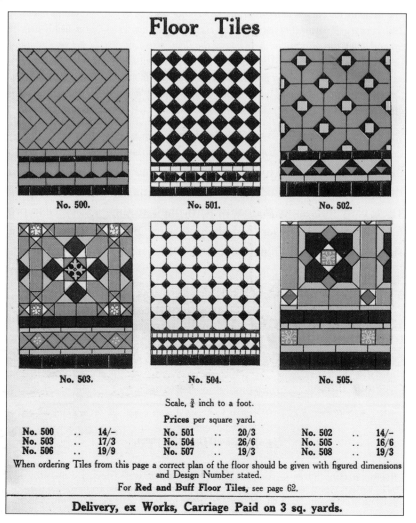

Floor Tiles

No. 500. No. 501. No. 502.

No. 503. No. 504. No. 505.

Scale, ¾ inch to a foot.

Prices per square yard.

No. 500	..	14/-	No. 501	..	20/3	No. 502	..	14/-
No. 503	..	17/3	No. 504	..	26/6	No. 505	..	16/6
No. 506	..	19/9	No. 507	..	19/3	No. 508	..	19/3

When ordering Tiles from this page a correct plan of the floor should be given with figured dimensions and Design Number stated.

For **Red and Buff Floor Tiles**, see page 62.

Delivery, ex Works, Carriage Paid on 3 sq. yards.

RAISED TIMBER FLOORS – CONSTRUCTION SUMMARY

Timber floors, correctly built, will last virtually indefinitely as long as insect attack and rot can be kept at bay. Unfortunately, many early timber floor structures were poorly designed and poorly constructed. If they become, and remain, wet, decay can be swift.

At its simplest a timber ground floor comprises a series of beams or joists covered with timber boarding or, nowadays, a manufactured particle board such as chipboard.

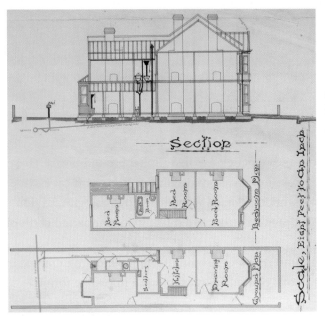

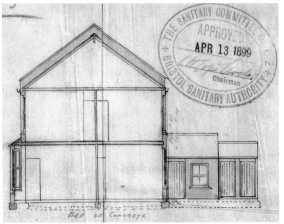

The drawing on the left dates from the very end of the 19th century. It's typical of many artisan class terraced houses up and down the country. In these houses the floor joists can be found running front to back or side to side (party wall to party wall). It was common practice to lay a concrete floor or flags in the scullery because this area, by its very nature, was often fairly wet.

Several towns and cities adopted bye-laws during the 1880s or 1890s which required a concrete slab or layer of asphalt below a ground floor. On this drawing (1899) the concrete slab ('oversite') has been added: whether it was added by the builder or by the Sanitary Authority we shall never know. Either way, many houses were built with nothing but bare earth below the timber floor joists.

In a typical late Victorian floor the floor joists are supported by the external walls, the internal load bearing walls between rooms, and intermediate sleeper walls. Timber floors require ventilation below the floor to keep the underfloor space dry. This was achieved by providing vents in the external walls. These early floors vary in quality, particularly in their resistance to problems of dampness. Late Victorian houses, for example, were usually built with DPCs in the external walls, internal walls and intermediate sleeper walls. However, many of these DPCs are no longer effective and the timbers are, therefore,

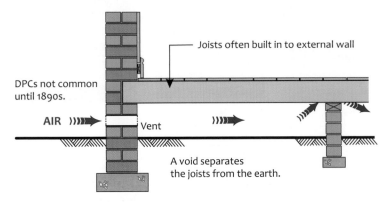

The foundations (if they existed) could be of various types; see Chapter 1: Building Movement - Foundations

Intermediate walls (sleeper walls) provided support for the joists. This avoided the need for expensive, deep section, timbers.

Air flow limited by internal wall although the odd missing brick sometimes helped to improve air flow.

susceptible to rising damp. In addition, penetrating damp through the solid external walls, together with inadequate sub-floor ventilation, increases the risk of floor failure. The example shown on the previous page has bare earth below the floor.

In the 1920s and 1930s timber floors were constructed to higher standards. There were a number of improvements which included provision of better quality DPCs in all walls, keeping the joists clear of the external walls, and the provision of honeycombed sleeper walls to aid ventilation. In honeycombed sleeper walls a number of bricks are omitted to encourage through-flow of air. In addition, provision of the concrete 'oversite', sometimes found in earlier floors, became standard practice. The oversite provides a base for the sleeper walls and reduced the incidence of rising damp; it was normally laid just above external ground level. These early 20th century floors usually contained additional vents (compared to their Victorian counterparts) to ensure adequate ventilation of the underfloor void.

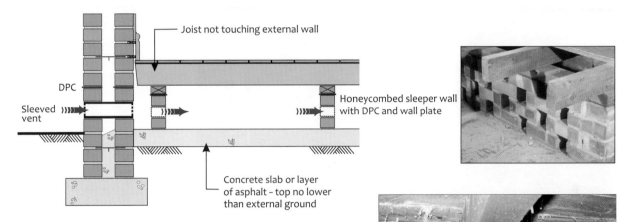

This photograph shows the underfloor void from a house built in the early 1930s. The sleeper wall, built on a concrete oversite (covered with later builder's debris) does not have a wall plate on top but there is a DPC. A few missing bricks at the base of the wall help ventilation.

Modern floors can still be formed in timber and a typical modern example is shown below. Nowadays, sleeper walls are still found although floor joists are usually supported by hangers built into the external loadbearing walls. In many small houses the use of deeper timber joists precludes the need for intermediate sleeper walls. Periscopic vents keep the outside vents well clear of ground level.

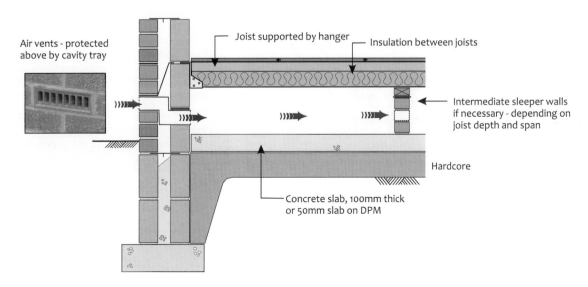

CONCRETE GROUND BEARING FLOORS – CONSTRUCTION SUMMARY

Although there are examples of concrete ground bearing floors from the1930s and even before, it was not until after the Second World War that they became the most common method of forming ground floors. This was, in the main, due to a shortage of timber caused by government restrictions on imports (eventually relaxed in the mid 1950s). Nowadays, although timber floors can be found, concrete ground bearing slabs, and more recently, suspended concrete slabs, are much more common.

At its simplest the floor comprises a slab, or bed, of in situ concrete supported by a hardcore sub-base. The floor is independent of the external walls and loadbearing internal walls to allow for slight differential movement.

If the hardcore is over 600mm deep a ground bearing slab should be avoided. Long term settlement in the hardcore is likely to cause cracking or movement in the slab. Suspended concrete floors, usually pre-cast, can be used in this situation.

Hardcore

The hardcore below the slab fulfils a number of functions. It provides a cheap bulk filler to make up levels and prevents the concrete from being contaminated by the sub-soil below. It helps spread the load evenly across the ground and should reduce the incidence of rising damp (a good quality hardcore has an open, porous structure which limits capillary action). A good hardcore should be chemically inert, easy to compact and should not be affected by water. It should be compacted in layers to prevent subsequent settlement. The most effective hardcores are usually by-products from aggregate quarrying. Incorrect choice of hardcore, or poor site practice, are common causes of floor failure.

Concrete slab

The concrete slab is usually about 100–150mm thick. Nowadays the required strength of the concrete is set out in the Building Regulations. Prior to this it depended on local bye-laws and generally accepted good practice. In some cases the slab is reinforced with a steel mesh, for example where soft spots exist in the ground or in garage floors. The mesh prevents the slab from cracking due to uneven loading or minor settlement. The surface of the slab can be power floated to produce a very smooth surface, usually smooth enough for vinyl tiles or sheet materials. However, power floating has only recently become common and most ground bearing slabs have a separate surface finish added towards completion of the building (see later section).

The example on the right was laid in 2007. The slab is 100mm thick and laid on a 200mm layer of limestone hardcore (75mm 'down' – all the hardcore must pass through a 75mm sieve – the 'fines' aid compaction). The garage slab (reinforced) is in the front of the picture.

Damp proof membranes

Although some early ground bearing floors incorporated waterproofing agents in the slab or screed (the most common finish) most ground bearing slabs contain a separate damp proof membrane. They can be formed from a variety of materials, some of which are no longer readily available. They include:

Hot applied liquid membranes
- bitumen (distilled from oil)
- asphalt (bitumen and aggregate)
- pitch (distilled from coal tar)

Cold applied liquid membranes
- bitumen solutions
- bitumen/rubber emulsions

Sheet materials
- polythene sheet (post 1960s)
- bitumen sheet

To be effective these materials need to be continuous and need to form a good joint with the DPC in the external walls. The sheet materials can be found above or below the slab; liquid materials can only be used above the slab. DPMs are designed to prevent capillary action; they are not effective against direct water pressure. The drawing below shows a number of common DPM options.

Membranes below the slab are relatively cheap and will protect the concrete from aggressive chemicals emanating from below. They should be laid on a sand blinding layer to protect them from any sharp stones in the hardcore. When the membrane is below the slab thinner floor screeds can be used because

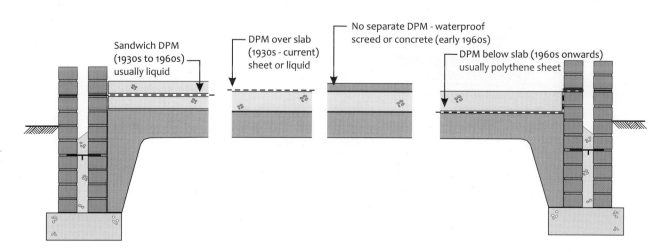

Sandwich DPM (1930s to 1960s) usually liquid

DPM over slab (1930s - current) sheet or liquid

No separate DPM - waterproof screed or concrete (early 1960s)

DPM below slab (1960s onwards) usually polythene sheet

a screed laid directly onto a slab will partially bond to it. However, in this situation the floor slab will take a long time to dry, delaying the fixing of moisture sensitive floor coverings. Some sheet membranes are not completely vapour proof, either because of inadequate joints or because the material is not completely impervious, and are therefore not suitable under moisture sensitive floor coverings, eg chipboard sheeting and cork tiles.

Membranes above the floor do not protect the slab from potentially harmful chemicals in the hardcore or ground but can offer the very best in terms of damp proofing, particularly if hot applied liquid DPMs are used. If sheet materials such as polythene are used the screed cannot bond with the slab and will therefore have to be thicker to prevent it from cracking. With both materials, achieving continuity with the DPC can be difficult.

Early floor finishes

In the 1940s and 1950s timber floor finishes were popular. These often took the form of floorboards fixed to dove-tailed battens secured by a screed or rectangular battens fixed by clips or nails. This gave the floor the appearance of a traditional raised timber floor and was suitable for a period when few people had fitted carpets. The membrane below the screed was most likely to be formed from hot liquid materials such as pitch or bitumen although cold proprietary materials, applied in two or three coats, were also used.

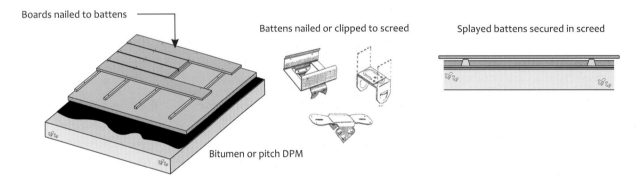

Block floors were also popular. These could be laid in a basket weave, stretcher bond or herringbone pattern. The design of the blocks could incorporate a dovetail to provide a good bond with the DPM. Other blocks were tongued and grooved. The blocks were usually laid in hot mastic (bitumen with added chemicals to improve flexibility) which also acted as the DPM.

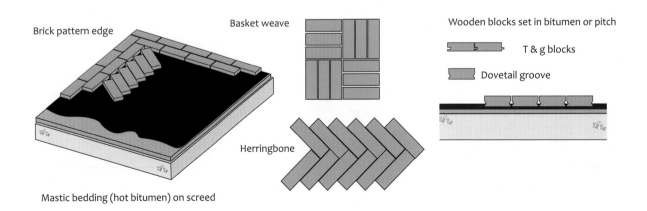

More recent finishes

From the late 1950s to the early 1990s screeds were the most common floor finish. Screeds are formed from a mix of coarse sand and cement (usually in the ratio of about 4:1). Screeds could be carpeted direct although, at first, fitted carpets were rare and the screed was normally covered with asbestos tiles, or later, vinyl tiles. During the early 1960s polythene membranes became more readily available. Although, as indicated earlier, these do not offer the best protection from capillary action they are the most common method, nowadays, of forming the membrane. Unlike liquid materials, polythene membranes can be used above or below the slab. However, liquid materials are more likely to be found where moisture sensitive floor coverings were specified. The thickness of the screed depends on the position of the membrane. Where the membrane is below the slab thinner screeds are acceptable because there is some bond with the slab.

Modern floor slabs are insulated; the insulation, depending on its type, can be laid above or below the slab. A number of DPM, screed and insulation combinations are shown below.

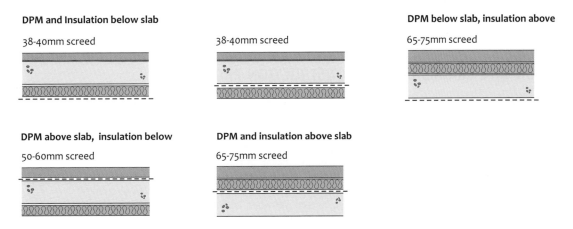

DPM and Insulation below slab
38-40mm screed

38-40mm screed

DPM below slab, insulation above
65-75mm screed

DPM above slab, insulation below
50-60mm screed

DPM and insulation above slab
65-75mm screed

A modern alternative to the screed is particle board; chipboard and strandboard are the two most common products. They are laid on a resilient layer of insulation boards laid directly onto a clean concrete slab. They offer fast construction and good insulation. Where a membrane is positioned below the slab an additional membrane is required above it to prevent trapped construction water from affecting the particle board. Problems of durability are discussed in a later section.

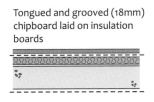

Tongued and grooved (18mm) chipboard laid on insulation boards

Ideally a polythene membrane should be sandwiched between the insulation and the chipboard. This prevents construction water trapped in the slab from affecting the chipboard. It also acts as a vapour control in the reverse direction, ie reducing the risk of condensation on the cold slab. If a DPM is laid over the slab several manufacturers suggest that the additional membrane is not required – although the risk of condensation will be increased.

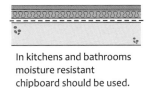

In kitchens and bathrooms moisture resistant chipboard should be used.

SUSPENDED PRE-CAST FLOORS

In many cases the use of a concrete ground bearing slab is not suitable. These include:
- where there are aggressive chemicals in the ground
- where water tables are high
- on sloping ground where uneven depth of hardcore may lead to differential settlement on ground likely to heave
- where the hardcore is likely to be more than 600mm thick.

In these situations a pre-cast floor is more appropriate. They usually comprise a series of pre-stressed beams with a block infill. The floors are supported by the external and internal load bearing walls and are not in contact with the ground. DPMs are not required as long as minimum recommended gaps between floor soffits and sub-soil are maintained.

Screed or particle board floor finishes can be specified as outlined earlier in the chapter.

The suspended floor shown near right can be finished with a board insulation material covered with a cement:sand screed or particle boarding. In early examples, ie when insulation requirements were less onerous than today, the screed was laid directly on the floor structure. The floor, far right, is a recent development; the polystyrene infill blocks preclude the need for additional insulation. The floor is covered with a concrete topping and power floated. Suspended concrete floors are very common these days and are often used in situations where ground bearing slabs would be quite acceptable. They are not necessarily cheaper but do provide a quick and seemingly reliable ground floor.

COMMON SITE ERRORS DURING CONSTRUCTION PROCESS

PROCESS/STAGE	TYPICAL SITE ERRORS	FUTURE IMPLICATIONS
Slab substructure	Some industrial wastes expand if wet. Hardcore not compacted in layers, or hardcore too deep. Trees felled or growing in clay soils	Settlement or lifting of slab
Concrete strength etc	Cracking caused by shrinkage. Concrete not laid in bays. Concrete too wet when laid. No surface protection on hot days. Frost attack	Loss of strength
Uneven finish to slab	Concrete too dry when laid. Concrete finished with lightweight tamp (arches in middle)	Problems with partition walls. Floating chipboard floors on insulation cannot provide level finish
Screed	Screed too dry or too wet. Screeds can crack if they dry too quickly or are laid too thin. Screed on wrong grade of insulation, joints not taped	Cracking screeds, surface dusting. Damage to floor finishes
Chipboard	No perimeter gaps. Incorrect grade of chipboard. No vapour check below chipboard	Lifting. Breakdown of chipboard
Integrity of DPC and DPM	Poor bond between DPM and DPC. Inadequate laps in DPM. No blinding on slab	Rising dampness
Choice of membrane not appropriate to floor finish	Polythene DPMs not always suitable for moisture sensitive floor finishes	Rising dampness and damage to floor finish
Concrete oversite (suspended timber floors)	Risk of ponding in wet spells and potential problems with high water table if oversite too low	Dampness and mould growth, rot
Timber joists (suspended floors)	Joists built into cavity. Build up of mortar in cavity. Joists undersized or at wrong centres	Dampness Deflection
Sub floor ventilation	Blocked or insufficient number of vents	Dampness, build-up of methane gas or leaked piped gas
Insulation	No edge insulation around slab. Gaps	Cold bridging and condensation
Radon gas protection	Substructure not sealed	Health risks
Pre-cast floors	Pre-cast floor not grouted. Beams too long No sub floor void	Cracking to screed Damp penetration

GENERAL DEFECTS IN TIMBER FLOORS

Dampness

There are a number of defects which occur in raised timber floors. It is the early floors which are particularly at risk. Most of these defects are caused by dampness. Rising damp and penetrating damp can occur in a number of circumstances and these are shown in the diagram below.

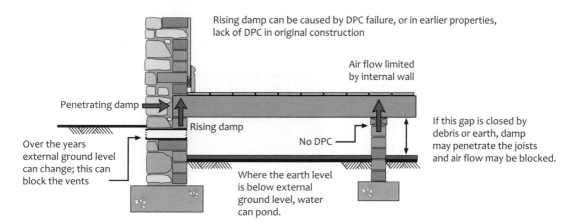

Rising damp can be caused by DPC failure, or in earlier properties, lack of DPC in original construction

Air flow limited by internal wall

Penetrating damp

Rising damp

No DPC

If this gap is closed by debris or earth, damp may penetrate the joists and air flow may be blocked.

Over the years external ground level can change; this can block the vents

Where the earth level is below external ground level, water can pond.

If the cause of the dampness is not addressed promptly dry or wet rot may occur. Adequate ventilation of the sub-floor void is vital if problems of rot are to be avoided. If moist air lingers in the sub-floor void, the moisture content of the floor timbers will increase until it is in equilibrium with the moist air. This problem will be exacerbated if there is no concrete oversite and if the ground is naturally damp, for example where the water table is high. If the moisture content of the timber exceeds 20% for long periods it is at risk. It is, unfortunately, quite common to find builders' rubble and other rubbish tipped below the floor. These can block effective ventilation and, in some cases, provide a direct path for rising damp to reach the floor joists.

In extreme cases condensation may occur on the floor timbers. Warmth from the house will normally ensure that the timbers remain above dew point, in other words the temperature at which condensation occurs. However, the use of an impermeable floor covering such as sheet vinyl will restrict evaporation into the house and condensation may occur on its underside. Any trapped water will slowly soak into the timber floor boarding.

Many terraced houses built during the Victorian and Edwardian periods do not contain adequate provision for ventilation. The diagram to the right shows a typical early example and shows how the addition of extra vents can improve air circulation.

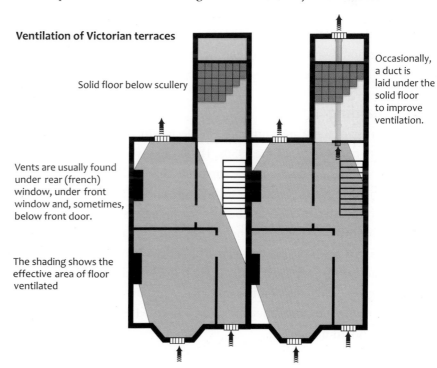

Ventilation of Victorian terraces

Solid floor below scullery

Occasionally, a duct is laid under the solid floor to improve ventilation.

Vents are usually found under rear (french) window, under front window and, sometimes, below front door.

The shading shows the effective area of floor ventilated

Adding extra vents is a fairly straightforward task although ducting may be necessary where some of the ground floor is concrete. Levels of ventilation should be at least equal to 1500mm^2 per metre run of wall (equivalent to 225mm by 150mm clay airbricks every 1.5 metres or so).

When inspecting a property some of the problems above mentioned may be obvious. Where the house has no effective DPC, where vents are blocked, or non existent, floor problems can be expected.

A distinctive 'mushroomy' smell inside the building will warrant further investigation. A damp meter *(see page 263)* will confirm the moisture content of the floor. If the levels are higher than 20% there is increased risk of rot, and lifting some of the floor boards to enable a more detailed inspection would be a wise move. If the floor feels soft and spongy when jumped on, or if there is a gap between the bottom of the skirting and the floor boarding, the joists or wall plates may already be affected by rot.

Over the years vents can become blocked with soil and debris: they can also be blocked due to changing levels of gardens and footpaths. This new tarmac is just starting to encroach on the vent. If the tarmac does not 'fall' away from the vent, water can run through into the holes during heavy rain.

Movement

Although rot may often account for movement in a floor there are other causes which are related to movement of the structure. If the house has joists built into the external walls differential movement of the external and sleeper walls will result in an uneven floor. In most houses built in or before the 1950s there is an internal load bearing wall, usually the wall between the dining room and lounge, carrying the load from purlins (via one or two pairs of struts) in the roof. These internal walls often have inadequate foundations and may settle. This can be a particular problem where roofs are recovered with heavier materials. If this is suspected it is worth closer examination of the partitions at right angles to the suspected load bearing wall. If the loadbearing wall has dropped the doors will be out of square with the frames; a tapered gap above the door is a typical manifestation of this type of failure.

Joist problems

If the floor feels quite springy when jumped on there could be a problem with undersized joists. Furniture or appliances which do not sit level, or crockery rattling in cabinets whenever someone walks across the floor, both suggest a floor structure which is undersized. Another possibility is that remedial or improvement works to wiring and heating installations have, over the years, resulted in excessively deep notches which reduce the effective depth of the joists.

Floor boarding

Dampness apart, there are a number of defects which occur in floor boarding. Square edged timber boards were common until the 1950s. These can give sterling service. Although square edged boards do give rise to draughts because of the gaps between the boards (in centrally heated houses shrinkage can be quite high) they are easy to lift for maintenance purposes and generally of substantial section. However, if incorrectly fixed, the boards can damage surface coverings; a rippled appearance to a carpet or sheet covering is usually evidence of this. Floor boards can be converted from timber logs in two main ways; flat sawn or quarter sawn. The former is more efficient at using all the timber but these boards readily twist and curl, particularly if the boards have been laid as shown opposite.
Square edged boards were traditionally fixed with brads (two nails to each joist) but, when lifted for rewiring or other purposes, they are often refixed with lost-head nails. These are not so effective at restraining board movement.

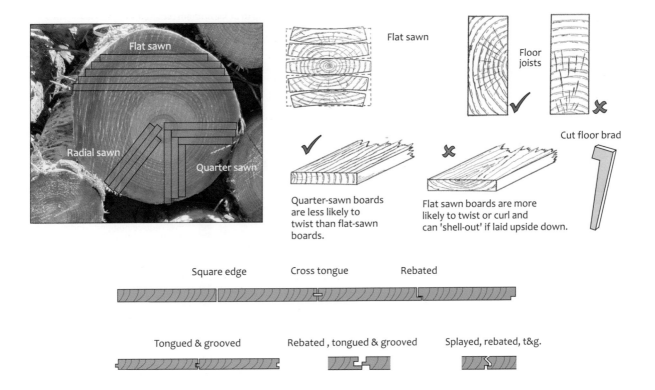

Quarter-sawn boards are less likely to twist than flat-sawn boards.

Flat sawn boards are more likely to twist or curl and can 'shell-out' if laid upside down.

Cut floor brad

Square edge   Cross tongue   Rebated

Tongued & grooved   Rebated , tongued & grooved   Splayed, rebated, t&g.

Tongued and grooved boards are less likely to twist and curl, and if correctly cramped before fixing, should not shrink to a point where gaps occur. However, if tongued and grooved boards are lifted the tongues will be broken off; subsequent renailing of the boards may lead to deflection or twisting and damage to the surface covering. Not all boards are square edged or tongued and grooved – there are a number of other types, none of them very common.

Square edged boards, (left) tend to be wider than tongued and grooved boards (right). If correctly fixed and problems of rot can be avoided, boards should last indefinitely.

Chipboard

Chipboard is common in modern construction. It is made from timber waste products, at the point of timber conversion, bound together in a resin and formed into large sheets. It is available in a range of grades. Standard grade chipboard is very sensitive to moisture. It can therefore be very vulnerable in a ground floor, not just because of the risk of rising and penetrating damp, but also because of spillage and leakage. A leaking washing machine can ruin chipboard in a few months. If the floor remains wet for any length of time the chipboard swells and breaks apart. Because of this risk some organisations insist that where chipboard is used in houses it should always be a moisture resistant grade. Whatever the grade, chipboard is likely to expand as its moisture content increases. Moisture movement can be as much as 6mm on a 2400mm board. This is more than enough to crack the floor finish above if it is glued in position, eg vinyl tiles.

107

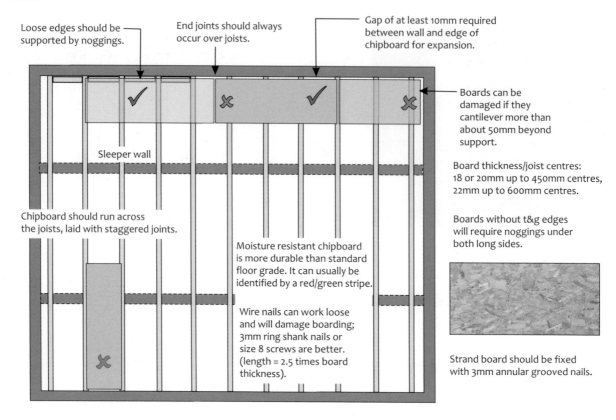

Loose edges should be supported by noggings.

End joints should always occur over joists.

Gap of at least 10mm required between wall and edge of chipboard for expansion.

Boards can be damaged if they cantilever more than about 50mm beyond support.

Sleeper wall

Board thickness/joist centres: 18 or 20mm up to 450mm centres, 22mm up to 600mm centres.

Chipboard should run across the joists, laid with staggered joints.

Boards without t&g edges will require noggings under both long sides.

Moisture resistant chipboard is more durable than standard floor grade. It can usually be identified by a red/green stripe.

Wire nails can work loose and will damage boarding; 3mm ring shank nails or size 8 screws are better. (length = 2.5 times board thickness).

Strand board should be fixed with 3mm annular grooved nails.

Nowadays it's recommended that all particle boards are glued to joists to prevent squeaking.

There are other problems related to chipboard, most of them due to poor workmanship. Most of these problems are related to inadequate support. These are summarised in the diagram above. Strandboard is similar to chipboard but made with shavings rather than fine particles.

Insulation

If the floor feels uneven but slightly resilient (usually with accompanying creaks) it may be because an insulation quilt has been laid over the joists. This is most likely to occur where older timber floors have been upgraded. However, quilts draped over the joists prevent the floor from lying properly and this uneven surface can soon damage surface coverings (very soon if nails work themselves loose). More importantly, there is a risk of cold bridging. This occurs where heat can escape through some areas more easily than others. It encourages condensation and is often confused with rising or penetrating damp. Where timber floors are insulated a better method is to support the insulation on synthetic netting as shown on the right. Even with this method there is a risk of cold bridging at the perimeter or edge of the floor.

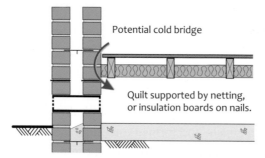

Potential cold bridge

Quilt supported by netting, or insulation boards on nails.

The risk of cold bridging (at the wall adjacent to the floor) depends on the nature of any wall insulation and the insulating properties of the internal leaf.

Resilient layer (quilt) laid over joists. This can lead to cold bridging; it is also difficult to get an even, creak free, finish.

Potential defects

Whether or not these photos show defects is, perhaps, open to debate. However no-one can deny they show less than satisfactory practice. The first photo shows the sub floor void of a semi detached house built in the 1930s; the wall running left to right is the party wall. At some point in the past the cavity wall has been injected with mineral wool. Unfortunately, the contractor did not check whether the vents were sealed where they cross the cavity – hence the pile of insulation in the bottom right corner. The other photos show the ground floor of a detached house built in the late 1950s. Although the floor has been trouble free, you can

see that the floor void is, at times, flooded (note the 'tide mark' further up the wall). In addition, the joist support at the internal loadbearing takes the unusual form of brick corbels and timber wedges.

DEFECTS IN GROUND BEARING CONCRETE FLOORS

Concrete floors can be very durable. However, in certain circumstances their failure can be quite dramatic and very expensive to resolve. Fortunately, failures of this type are comparatively rare.

The BRE has established that many of the problems are caused by poor site practice. Where appropriate these practices have been summarised.

SETTLEMENT OF HARDCORE AND SUB SOIL

Settlement or consolidation of hardcore can occur where the hardcore has been inadequately compacted. Where the hardcore is over 600mm thick compaction can be difficult even when carried out in thin layers. Particularly vulnerable is the edge of the floor where the hardcore runs down to the foundation and can therefore be of considerable depth. On sloping sites the settlement may be uneven thus

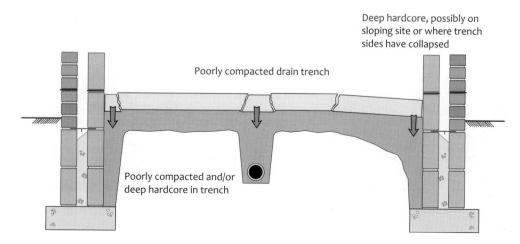

Deep hardcore, possibly on sloping site or where trench sides have collapsed

Poorly compacted drain trench

Poorly compacted and/or deep hardcore in trench

accounting for differing size gaps under skirtings either side of rooms. Partitions built on the floor can also crack. Where poor quality, soft ground is not removed before commencing the floor structure the risk of settlement is increased. However, this would normally manifest itself fairly quickly, possibly even before the building is complete.

SWELLING GROUND AND HARDCORE

Some hardcores will swell in moist conditions. These hardcores, such as steel slags which include unhydrated lime, and colliery waste which can contain large proportions of clay are, thankfully, comparatively rare. In many cases suspected problems of swelling turn out to be sulfate attack of the concrete slab. Swelling can also be confused with ground heave. This occurs in clay soils and results in expansion of the clay as it absorbs water.

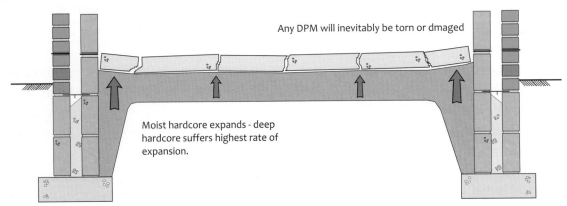

Any DPM will inevitably be torn or dmaged

Moist hardcore expands - deep hardcore suffers highest rate of expansion.

Ground heave commonly occurs where large trees have been removed from clay soils. The zone of swelling is concentrated around the tree roots and, at its worst, can lift the surface of the ground by as much as 150mm. Ground heave can also be caused by changes in water courses or broken drains. Heave caused by frost may also affect floors in unheated outbuildings but should not be a problem in heated dwellings.

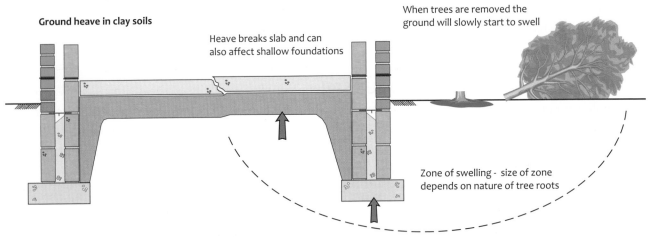

Ground heave in clay soils

Heave breaks slab and can also affect shallow foundations

When trees are removed the ground will slowly start to swell

Zone of swelling - size of zone depends on nature of tree roots

The house shown here was built in the 1950s. In the 1990s a large tree, situated just to the left of the outhouse was felled. Within 12 months the outhouse floor cracked and lifted. The main floor of the house is suspended timber and there were no signs of movement when this photo was taken.

Slab failure

In many cases defects in the slab are caused by errors during the construction stage. These include excess water in the concrete, exposure during curing to frost, inadequate thickness or incorrect mix. These problems may not manifest themselves in any tangible form because of the relatively low loads on the slab. A much more serious defect is sulfate attack.

Sulfate attack

The phenomenon of sulfate (an acknowledged change from sulphate) attack has been explained in an earlier chapter. Sulfate attack occurs when sulfates in the hardcore, or possibly the ground below, migrate in solution to the underside of the slab, react with a by product of the cement in the slab and form tri calcium sulfo aluminate. The resulting crystal expands as it forms. The slab lifts slightly and domes, leaving a small gap between the hardcore and slab. In extreme cases the resultant expansion can break the back of the slab and even cause lateral movement of the surrounding walls.

The process of attack and expansion is quite slow and can take several years to visibly manifest itself. It is unlikely to occur where DPMs are effective as considerable amounts of moisture are required to complete the chemical reaction. A number of hardcores are known to be responsible for sulfate attack and these include colliery shale, old bricks containing gypsum plaster (calcium sulfate) and a variety of industrial wastes. Sub-soils which are liable to cause sulfate attack include London clay, Oxford clay and Keuper Marl.

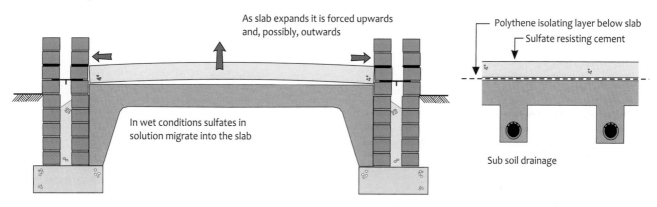

As slab expands it is forced upwards and, possibly, outwards

In wet conditions sulfates in solution migrate into the slab

Polythene isolating layer below slab

Sulfate resisting cement

Sub soil drainage

It can be avoided in the first instance by careful choice of hardcore, the use of sulfate resisting cement in the slab, isolating the slab from the hardcore with a sheet of polythene, and good subsoil drainage. Once it has occurred the only effective solution is to break up the slab and renew the hardcore.

MEMBRANE FAILURE

Membrane failure is most likely caused by poor site practice during construction or damage caused by floor or hardcore movement later in a building's life. Poor site practice includes:

Liquid membranes
- inadequate thickness (or number of coats)
- poor joint with DPC
- dusty slab surface prevents good bond
- areas omitted

Sheet membranes
- torn during concrete or screeding
- poor laps and joints
- using DPMs of inadequate thickness

Fortunately, minor defects will not necessarily result in damp penetration. This is just as well as, in practice, few polythene DPMs can be regarded as impervious membranes. They are designed to resist capillary action, not direct water pressure. This is why they are usually positioned just above ground level. Permeable surface finishes such as carpets will allow small amounts of moisture to evaporate safely through the slab and disperse into the atmosphere of the room. Sheet vinyl, on the other hand,

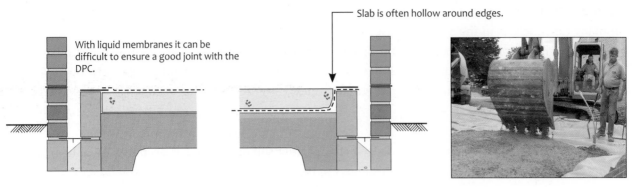

Slab is often hollow around edges.

With liquid membranes it can be difficult to ensure a good joint with the DPC.

Polythene should be laid on sand or stone-dust blinding (25mm) to reduce risk of hardcore perforating membrane.

If the slab settles, the polythene can tear at junction with wall.

A polythene membrane is difficult to join effectively and damage is always likely during concreting.

will resist the passage of water vapour resulting, possibly, in moisture collecting below the sheet covering. Testing with a damp meter will give some idea of the level of water present.

Any of the structural problems related to heave and settlement are likely to cause failure of the damp proof membrane. These are relatively easy to identify, if not diagnose. Perhaps more difficult is the damp patch which appears on a floor screed mysteriously and with no obvious associated defects. The drawing below shows some typical problems.

Where vinyl tiles are starting to curl and lift the problem may be a failed membrane but, in new buildings, it could be due to tiles laid before the screed is dry.

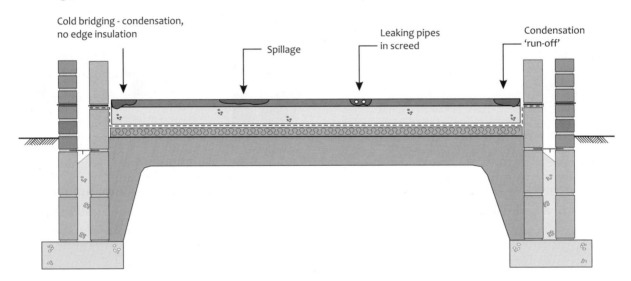

Cold bridging - condensation, no edge insulation

Spillage

Leaking pipes in screed

Condensation 'run-off'

SCREED FAILURE

Although, as explained in an earlier section, some timber finishes are found with ground bearing slabs, most slabs are finished with a floor screed. The screed is a fairly delicate element and will normally be laid towards completion of a building. It is easily damaged and the slightest of imperfections can affect the appearance and adhesion of any thin sheet covering. A number of defects can occur at the construction stage. These include insufficient thickness, inadequate mix, excess water, poor compaction, and incorrect aggregate (sand too fine – encourages shrinkage and therefore cracking). Screeds should be allowed to dry slowly (possibly under a polythene sheet in hot weather) if they are to achieve their maximum strength and successfully resist any shrinkage.

Problems of ground heave, swelling hardcore or sulfate attack will obviously have a dramatic impact on the screed. Vertical movement in all or part of the floor screed, or wide cracks in the screed, are almost certainly caused by problems from below.

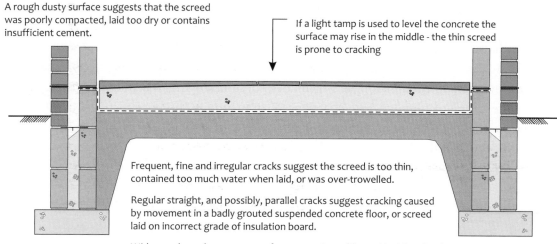

A rough dusty surface suggests that the screed was poorly compacted, laid too dry or contains insufficient cement.

If a light tamp is used to level the concrete the surface may rise in the middle - the thin screed is prone to cracking

Frequent, fine and irregular cracks suggest the screed is too thin, contained too much water when laid, or was over-trowelled.

Regular straight, and possibly, parallel cracks suggest cracking caused by movement in a badly grouted suspended concrete floor, or screed laid on incorrect grade of insulation board.

Wider cracks and an uneven surface suggests problem with slab or hardcore.

Parallel long cracks in a screed, over a block and beam floor, suggest that the floor has not been grouted properly (*see Pre-cast floors below for more detail*). Similar cracking can occur if a screed is laid over insulation boards without first taping the joints or where pipes have been bedded in the screed.

In some parts of the country asphalt screeds are common. When laid in two coats (25mm) on a well prepared slab an asphalt screed provides a durable and smooth surface, quite suitable for carpets etc. However, if they are laid too thin or on a wet or dusty base they are likely to lift in places: they will sound hollow when tapped or walked over.

CHIPBOARD FAILURE – GROUND BEARING SLABS

Chipboard flooring is becoming an increasingly popular alternative to traditional floor screeds. Chipboard is very sensitive to moisture and this has obvious implications for the type and position of a damp proof membrane. If the chipboard is covered with an impervious material such as sheet vinyl minor defects in the membrane can lead to wetting of the chipboard. Trapped construction water in the slab can cause similar problems which is why the chipboard manufacturers recommend an additional membrane above the slab (unfortunately this is often omitted). The position of this membrane requires careful thought. This is explained in the diagram below. With laminated boards, ie where insulation is bonded to the chipboard, the additional membrane has to be positioned directly on top of the slab. If insulation is separate, a better membrane position is above the insulation, this will also prevent condensation from above.

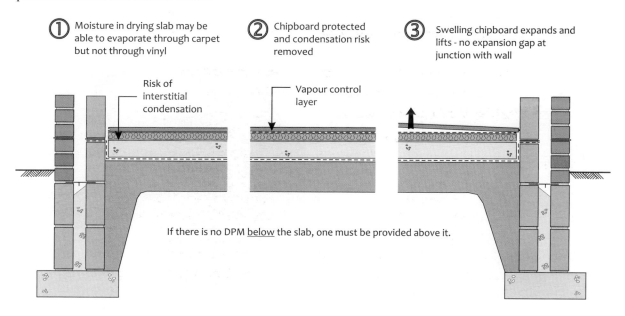

① Moisture in drying slab may be able to evaporate through carpet but not through vinyl

② Chipboard protected and condensation risk removed

③ Swelling chipboard expands and lifts - no expansion gap at junction with wall

Risk of interstitial condensation

Vapour control layer

If there is no DPM <u>below</u> the slab, one must be provided above it.

Water from leaking washing machines can quickly destroy chipboard. Even moisture resistant grades will deteriorate if they remain saturated for long periods.

Chipboard, if it remains dry, is a durable material. Its durability depends, to some extent, on the way it is fixed. It is important that the tongued and grooved joints are tightly fitting. On site there is always the temptation to cut off a tongue which does not fit neatly into the groove. This reduces the stiffness of the floor. In addition, where boards need cutting to fit round obstructions they will need to be butt jointed and the tongued and grooved joint is lost. A stiletto heel can open up a hole where two or more boards meet and this can be difficult to repair.

Chipboard flooring will also expand slightly once laid. To allow for this expansion a gap of about 10mm should be allowed at its perimeter. If a chipboard floor has lifted and bowed in the centre of a room this should be the first item to check.

OLDER FLOOR FINISHES

Wood block floors

Wood block floors were common until the 1950s, not just in houses but also in schools, offices, village halls, etc. They are usually about 20–25mm thick and bedded in hot pitch (coal based) or bitumen (oil based) by dipping the block into a tray of adhesive and then laying it in place. The adhesive also formed the damp proof membrane.

Wood blocks are very durable, and when the surface has worn over many years, can be sanded or planed down to provide a new, blemish free surface. There are however, a number of potential problems, some more serious than others.

Blocks which, when laid, contained too much moisture would shrink and this may account for wider than normal joints.

Wetting, caused by washing machine leakage, for example, can ruin a block floor. Even if the blocks settle back into place the bond with the adhesive will have been lost and the blocks will rock slightly. Persistent and high humidity levels can cause similar problems.

Some modern polishes and sealers are too strong for block floors. They tend to build up in the joints, and because they don't have the flexibility of wax polish, they limit the ability of the floor to accommodate expansion and contraction. This sometimes means that movement is concentrated across one part of the floor (see diagram below).

Most adhesives will break down after 50 years or so. This may mean re-bedding the blocks although careful use of a blow torch on the underlying pitch or bitumen may provide some respite.

If the loose blocks are around the perimeter of a room (ie against external walls) the possibility of moisture transfer from the walls should be investigated.

Blocks will lift if they stay wet for long periods

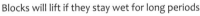
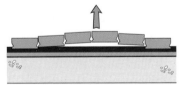

'Split'

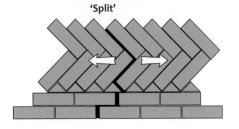

Loose blocks need to be re-fixed as soon as possible to prevent adjacent blocks from lifting.

Blocks should be finished with flexible coatings - oils and waxes are more durable and less prone to cracking than varnishes.

Thermoplastic tiles

These were introduced just after the Second World War. They became very popular, particularly on solid floors (timber was rationed after the war and the amount of timber used in a house was limited – hence the concrete ground floor). The tiles were made from thermoplastic resin binders, mineral fillers, pigments and sometimes asbestos. In the mid 1950s vinyl tiles became available and these slowly overtook thermoplastic tiles in popularity. Their advantage lay in increased flexibility and in an improved range of colours. Both types of tile were normally laid in a solvent bitumen adhesive on a screed or trowelled slab.

At this time it was not standard practice to provide damp proof membranes in ground floors and, over several years, evaporation through the joints of the tiles could leave behind a white crystalline material (sodium and potassium carbonates) alongside the tile joints. This can often be cleaned off although in severe cases the evaporating moisture is absorbed into the edges of the tiles. As the moisture evaporates and the carbonates crystalise the resultant expansion can break down the tile edges.

Quarry tile floors

These are usually laid on a screed and are sometimes found in kitchens, utility rooms etc. There are three main causes of failure:

- irreversible expansion as the tiles absorb moisture (when new) from the atmosphere or from cleaning and leaks
- thermal movement – enough to break the bond with the screed
- shrinkage of new screeds.

Failure usually manifests itself as arching along two or three rows. Before the tiles lose their bond they may sound hollow when walked over.

DEFECTS IN PRE-CAST CONCRETE FLOORS

These floors are comparatively recent and, as such, their long term durability is unproven. However, there are a number of problems which have manifested themselves, most of which are due to inadequate site preparation or specification rather than failure of the floor components themselves.

Beam and block floors are not usually designed to carry internal load bearing walls. Deflection in a floor below a wall or horizontal cracks in the wall itself may suggest that over loading has occurred. Similarly, but less significant, cracking of garage floors is commonly caused by heavy point loads, ie car jacks, damaging the thin concrete topping surface of the floor (even though it may be reinforced).

If a precast floor is damp and the obvious problems of spillage or leakage are discounted condensation or rising dampness may be affecting the floor. Rising damp can occur where the subsoil is in contact with the soffit of the floor. There should be a minimum gap of at least 75mm between the underside of the floor and the ground surface although this is obviously impossible to check without lifting part of the floor. In clays likely to heave this gap should be increased to 150mm

Damp patches round the edge of the floor could be attributable to condensation around the edge of the floor due to cold bridging (especially near air vents), failure of the DPC or problems with beams of incorrect length which project into the cavity. The risk of general surface condensation will depend on the type of infill blocks, the moisture permeability of the floor finish and the temperature and relative humidity above the floor. Note: these floors do not require damp proof membranes where the minimum gaps specified between floor and ground are provided and where the floor void is ventilated.

Ventilation is provided not only to reduce the humidity of the subfloor void but also to prevent the build up of gases. This is particularly important where landfill gas or methane might be present.

Where beams run parallel to walls there should be some restraint strapping to tie the wall back to the floor. In extreme cases a wall may bow slightly (*see the next chapter for more details*). Finally, narrow parallel lines in the screed suggest that the floor, ie the gaps between the beams and blocks, has not been grouted properly. This stage is sometimes omitted because its importance is not clear to the operatives. However, it can be very expensive to resolve and may require complete removal of the screed.

RADON GAS

Radon is a colourless gas which is radioactive. Although it is present throughout the UK, in some locations its concentrations are a health hazard. To protect occupants a house should be sealed to prevent the gas rising up through the subsoil and entering the building. The Health Protection Agency (which now includes the former National Radiological Protection Board) publishes guidance showing which areas are most at risk, how the gas can be detected and measured, and appropriate steps for safeguarding a house against it. A detailed examination of radon is beyond the scope of this text. However, as far as ground floors are concerned there are two levels of protection, passive and active. Both may be required where the risks of radon gas are high. The passive system, which provides basic radon protection, can usually be achieved by increasing the 'airtightness' of the damp proof membrane and extending it across the external leaf of brickwork. Additional, active protection, where risks are higher, usually requires some form of ground depressurisation under a ground bearing slab, or air vents (sometimes fan assisted) below a suspended floor slab. The diagram below shows two methods of providing basic protection below a ground bearing concrete slab. If this membrane is damaged, either during or after construction, the barrier will not be effective and additional measures (ie active protection) may be required.

Damp proof membrane providing radon barrier

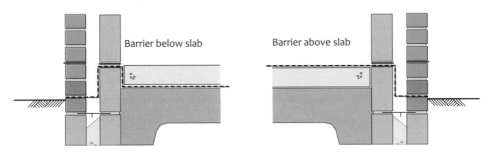

Barrier below slab   Barrier above slab

Working details will vary but the principle is to make the construction gas-tight

Damage to radon barrier, often through service connections or later additions etc, will breach barrier.

A heavy duty radon-proof membrane laid over the hardcore of a new development of flats in the south west.

CHAPTER 6

Upper Floors

TIMBER FLOORS

CONSTRUCTION SUMMARY

Most houses have upper floors made from timber. The construction principles have changed very little in the last two or three hundred years although, in modern construction, prefabricated joists have become more common. Nevertheless, most floors still comprise a series of timber joists covered with softwood floorboards or sheet materials such as chipboard. The floor structure is usually hidden from view below by a ceiling, originally made from timber lath and plaster but, nowadays, almost always made from plasterboard. Besides concealing the floor structure the ceiling provides a surface for decoration, improves the floor's fire resistance and reduces, to some extent, the transfer of sound.

In modern construction timber joists are normally at 400mm centres (16 inches) and are usually made from imported softwood. The depth of a joist resists its tendency to bend or deform under load and the width of a joist resists its tendency to twist. Ideally the joists should span the shortest distance across a room to keep the joist depth to a minimum.

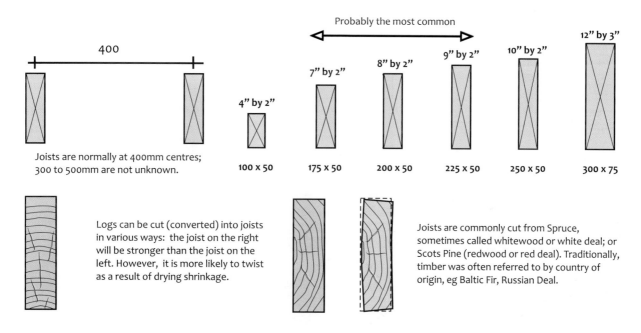

Joists are normally at 400mm centres; 300 to 500mm are not unknown.

Probably the most common

4" by 2" — 100 x 50
7" by 2" — 175 x 50
8" by 2" — 200 x 50
9" by 2" — 225 x 50
10" by 2" — 250 x 50
12" by 3" — 300 x 75

Logs can be cut (converted) into joists in various ways: the joist on the right will be stronger than the joist on the left. However, it is more likely to twist as a result of drying shrinkage.

Joists are commonly cut from Spruce, sometimes called whitewood or white deal; or Scots Pine (redwood or red deal). Traditionally, timber was often referred to by country of origin, eg Baltic Fir, Russian Deal.

Although joist centres of 400mm have been common for the last 150 years they can vary from 300mm to 600mm. In earlier floors the construction is even more haphazard with joist widths and centres varying considerably.

A typical timber floor from a small Victorian house is shown below.

Late 19th century small terraced house

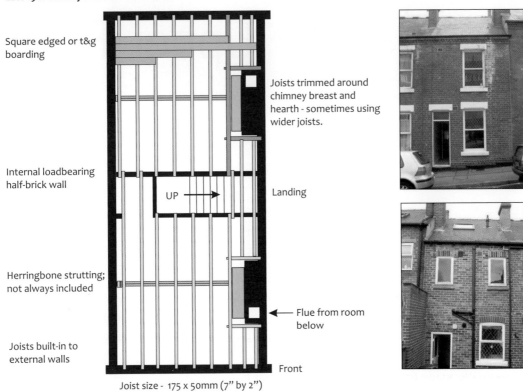

Square edged or t&g boarding

Joists trimmed around chimney breast and hearth - sometimes using wider joists.

Internal loadbearing half-brick wall

UP → Landing

Herringbone strutting; not always included

← Flue from room below

Joists built-in to external walls

Front

Joist size - 175 x 50mm (7" by 2")

Until the 1920s solid walls were the norm. In small and modest houses joists were usually built into the one-brick thick solid walls as shown below although cantilevered iron brackets and brick corbels can sometimes be found in better quality houses. In larger buildings (18th and 19th centuries) walls for the lower storeys were often more than one-brick thick, diminishing in thickness towards the top of the building.

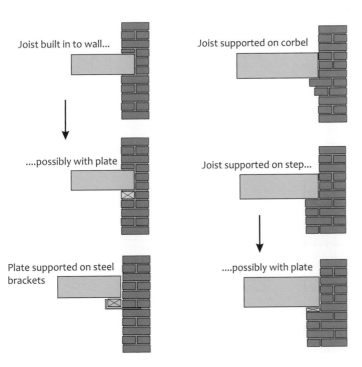

Joist built in to wall...

....possibly with plate

Plate supported on steel brackets

Joist supported on corbel

Joist supported on step...

....possibly with plate

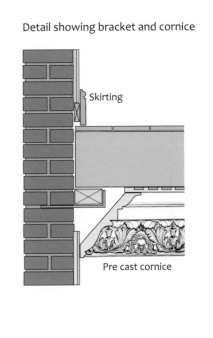

Detail showing bracket and cornice

Skirting

Pre cast cornice

Some of these 'grander' houses had double floors. In a double floor one or more beams provide intermediate support for the floor joists. These main beams are often supported on pad stones to help spread the load safely into the wall.

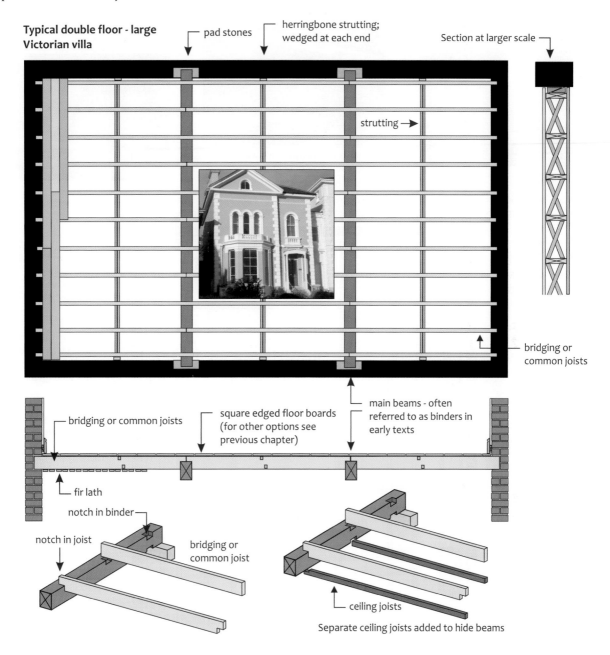

Typical double floor - large Victorian villa

pad stones

herringbone strutting; wedged at each end

Section at larger scale

strutting →

bridging or common joists

bridging or common joists

square edged floor boards (for other options see previous chapter)

main beams - often referred to as binders in early texts

fir lath

notch in binder

notch in joist

bridging or common joist

ceiling joists

Separate ceiling joists added to hide beams

In modern construction prefabricated joists with metal webs are becoming increasingly popular; they are usually fixed at 600mm centres. They can span large distances and can easily accommodate heating, plumbing and electrical services.

Traditional joists are normally supported by the internal leaf of the cavity wall, either built into the inner leaf or supported on joist hangers. The drawing below shows typical construction of a modern timber joist floor.

A modern timber (joist) floor

Joists are normally softwood and at 400mm or 600mm centres.

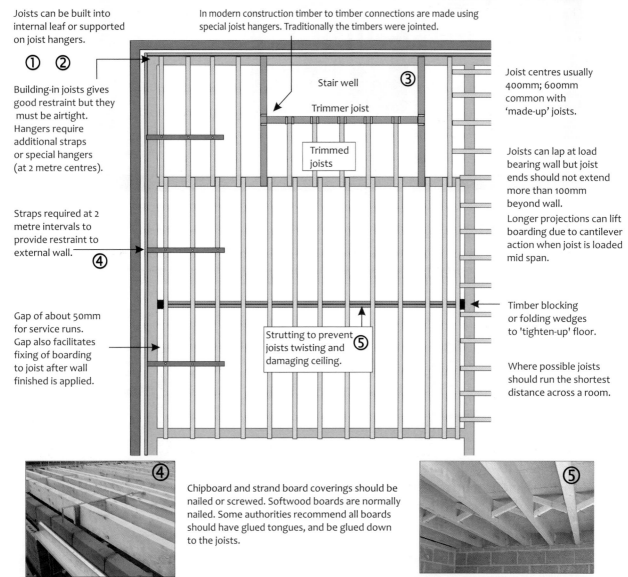

Joists can be built into internal leaf or supported on joist hangers.

① ②

Building-in joists gives good restraint but they must be airtight. Hangers require additional straps or special hangers (at 2 metre centres).

Straps required at 2 metre intervals to provide restraint to external wall.
④

Gap of about 50mm for service runs. Gap also facilitates fixing of boarding to joist after wall finished is applied.

In modern construction timber to timber connections are made using special joist hangers. Traditionally the timbers were jointed.

Stair well

③

Trimmer joist

Trimmed joists

Strutting to prevent joists twisting and damaging ceiling. ⑤

Joist centres usually 400mm; 600mm common with 'made-up' joists.

Joists can lap at load bearing wall but joist ends should not extend more than 100mm beyond wall.

Longer projections can lift boarding due to cantilever action when joist is loaded mid span.

Timber blocking or folding wedges to 'tighten-up' floor.

Where possible joists should run the shortest distance across a room.

④

Chipboard and strand board coverings should be nailed or screwed. Softwood boards are normally nailed. Some authorities recommend all boards should have glued tongues, and be glued down to the joists.

⑤

Restraint and strutting

Since 1985 the Building Regulations have required joists to be tied or secured to the walls at intervals of not more than two metres to provide restraint for the comparatively thin external walls. Where joists run parallel to the external walls galvanised metal straps can be fixed as shown above and opposite.

If the joists are built into the inner leaf and have at least 90mm bearing no additional restraint is necessary. However, where standard joist hangers are used additional restraint is required. This can be provided by straps or by the use of special joist hangers. These restraint hangers are, again, required at two metre centres (usually every sixth joist). Intermediate joists can be supported by standard hangers. Timber blocking below the strap ensures the strap also works in compression.

In all but the shortest of spans strutting is required to prevent the floor joists from twisting and cracking the ceiling below. This twisting can be caused by loading or by shrinkage as the new joists dry out. A single set of struts, at right angles to the joists, is normally provided mid span. The strutting can be herringbone pattern or straight, using joist off-cuts (in practice the straight strutting is normally staggered to facilitate nailing). Herringbone strutting is probably the more traditional method although straight strutting is commonly found in Georgian buildings.

Timber blocking or timber wedges, fixed at both ends of the strutting between the end joists and the wall, tighten-up the floor and help minimise movement.

Openings in floors

Upper floors usually contain openings for stair wells. To form the opening the floor needs to be trimmed as shown below. Similar techniques were used to trim floors around chimney breasts to ensure that the joists were not directly at risk from fire.

The trimming and trimmer joists are usually of thicker section timber (or two standard joists bolted together) to cope with the increased load. Until the 1960s special carpentry joints were used to join these timbers. A tusk tenon jointed the trimmer to the trimming joist and a housed joint was used to fix the trimmed joists to the trimmer. Nowadays timber to timber galvanised hangers are more likely to be used.

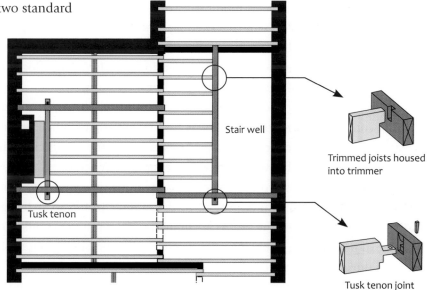

Stair well

Tusk tenon

Trimmed joists housed into trimmer

Tusk tenon joint

Floor coverings

In the Georgian and Victorian periods most upper floors were covered with square edged boards. These were usually made from softwood although hardwoods such as oak were sometimes found in the reception rooms of grander houses. At the end of the Victorian period the introduction of better wood-

121

working machinery meant that tongued and grooved boards became more readily available. An alternative form of construction, sometimes found in better quality housing, used grooved boards (grooved along both edges) secured with continuous metal tongues. In the 20th century square edged boarding and tongued and grooved boarding were both common although, since the 1970s, tongued and grooved chipboard has become the most popular material.

Ceiling finishes

Until the 1930s ceilings were usually made from lath and plaster. Thin strips of softwood were nailed across the underside of the joists and finished with three coats of lime plaster. Nowadays ceilings are usually formed from plasterboard. The plasterboard can be self-finished, plastered with gypsum plaster or finished with flexible materials such as Artex.

The photograph on the left shows a typical mid-Victorian working class house. The joists are quite thin, probably about 38mm. There is no ceiling so fire protection and sound insulation are minimal. Most floors in working class houses were covered with square edged, softwood boards.

CONCRETE FLOORS

CONSTRUCTION SUMMARY

Low-rise purpose-built flats from the 1950s onwards usually have upper floors made from precast concrete. The floors in flats separate dwellings and, as such, require higher levels of fire protection and sound insulation than their timber counterparts in houses. Concrete floors are also capable of greater spans than timber floors and are ideally suited to flats where large floor areas may be required without the complication of unnecessary internal load bearing walls.

For the last 20 years or so plank floors and beam & fill floors have been the two most common forms of construction. In both cases the main loadbearing members are usually made from pre-cast, pre-stressed, reinforced concrete. In a pre-stressed beam the reinforcing wires are stretched to a specific tension before placing the wet concrete in the moulds. When the concrete has cured the tension in the wires is released and this produces a very strong but lightweight beam. The beams are usually pre-cambered; in other words the tensioned wires cause the beams to arch slightly (usually no more than a 350th of the span). When the beams are under load the camber should not be noticeable.

Once in position the floor is usually covered with a brushed grout to tighten up the floor and fill any gaps thus improving fire protection and sound insulation. Concrete floors usually have sufficient mass to provide adequate resistance to airborne sound (noise from radios, etc) but require additional surface coverings to reduce impact sound (the noise from people walking across the floor). This is usually achieved by creating a structural break in the floor. One option is to use a sand cement floor screed laid on a resilient layer of paper backed mineral wool. Another option is to use tongued and grooved chipboard laid on insulation boards. Note that in 2004 sound insulation requirements became more onerous for new buildings – some beam and block floors are no longer suitable.

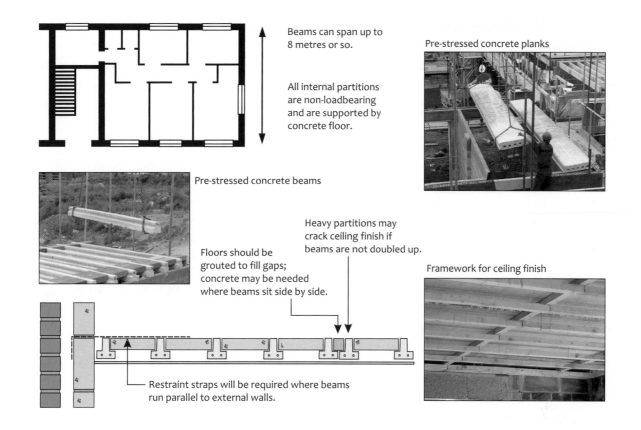

Beams can span up to 8 metres or so.

All internal partitions are non-loadbearing and are supported by concrete floor.

Pre-stressed concrete planks

Pre-stressed concrete beams

Heavy partitions may crack ceiling finish if beams are not doubled up.

Floors should be grouted to fill gaps; concrete may be needed where beams sit side by side.

Framework for ceiling finish

Restraint straps will be required where beams run parallel to external walls.

Depending on the nature of the floor structure the soffit can either be plastered directly or can be made from a timber or metal framework with plasterboard fixed to its underside.

From the 1920s to 1950s traditionally built (ie with load bearing external walls) low rise and medium rise flats often had intermediate floors constructed from in situ concrete. The floors could be constructed in a variety of ways; three are illustrated below.

Most of these floor types also act as diaphragms, in other words they help to distribute wind loads from walls running in one direction to walls running in the opposite direction. To do this successfully the joints or connections between walls and floors are of vital importance. If correctly built, in situ concrete floors provide good loadbearing capacity, good fire resistance and high resistance to airborne sound. With appropriate detailing and surface finishes they will also provide good resistance to impact sound. Many early concrete floors, in fact, had no provision for resistance to impact sound and the floor finish often comprised a sand cement screed or quarry tiling laid directly onto the floor structure.

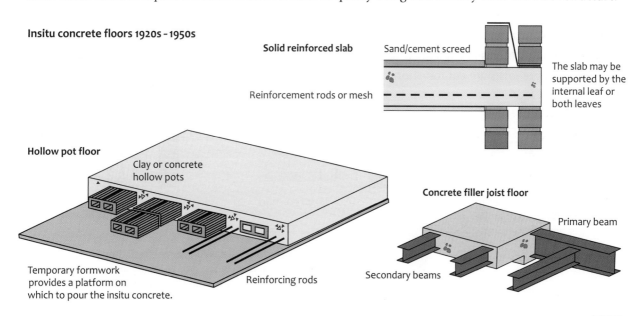

Insitu concrete floors 1920s - 1950s

Solid reinforced slab

Sand/cement screed

Reinforcement rods or mesh

The slab may be supported by the internal leaf or both leaves

Hollow pot floor

Clay or concrete hollow pots

Concrete filler joist floor

Primary beam

Secondary beams

Temporary formwork provides a platform on which to pour the insitu concrete.

Reinforcing rods

123

COMMON SITE ERRORS DURING CONSTRUCTION PROCESS

Upper floors in houses and flats should be trouble free. However, there are a number of common construction errors, most of them caused by incorrect site practice or poor site supervision. At best, these defects have cosmetic implications, at worst, they could affect the health and safety of the occupants.

PROBLEM AREAS/STAGES	TYPICAL SITE ERRORS	IMPLICATIONS
Joist support	Inadequate bearing, usually caused by joist hanger not tight against wall or joists too short. Joists should be nailed to hangers	Possible floor collapse
Junction of walls and joists	Incorrect joist hangers, straps omitted. Most concrete beam and block floors require straps along parallel walls	Lack of restraint may permit lateral wall movement
Joists generally	Undersized joists, incorrect centres, strutting and/or blocking omitted, holes and notches incorrectly cut. Incorrectly sized trimmers etc	Deflection causing cracking in partitions, gaps under skirtings, cracking in ceiling, 'bounce' in floors
Joist protection & strutting	Wet joists, lack of strutting and/or blocking at strutting ends	Warping joists, damaged ceilings, uneven floors.
Damp protection	Joists bridging cavity, or partial bridge and mortar droppings	Rot and damp penetration, damage to decorations
Floor coverings (boarding etc)	Flooring undersized ie too thin for span, boards incorrectly cramped (t&g). Badly warped joists. Chipboard not t&g or inadequate noggings	Uneven floors
Fire (concrete floors in flats)	Pre-cast floors not grouted, service ducts not protected, inadequate ceiling or ceiling fixings	Smoke or fire with risk to life and building
Sound (concrete floors in flats)	No floating floors, pre-cast floors not grouted, sound insulation too thin, floor of insufficient mass, screeds too thin. Beam and block floors won't always meet new Regs	Airborne and impact sound transfer, neighbour disputes
Uneven floor finish (flats)	Screed problems, lack of reinforcement over resilient insulation. Poor grouting of beams/blocks. Sloping (crowning) chipboard finishes caused by camber in pre-cast beams	Damage to coverings, possible effects on sound insulation

DEFECTS – TIMBER UPPER FLOORS

INTRODUCTION

There are a number of potential problems with upper floors. Disregarding, for the moment, defects in the floor coverings or ceiling finish, these problems include damp penetration, excess deflection and lack of restraint for external walls. Problems of pest attack are dealt with in a later chapter. Perhaps one of the most common problems is inadequate support. This can occur where joist hangers are not tight against the wall or where joists are cut too short.

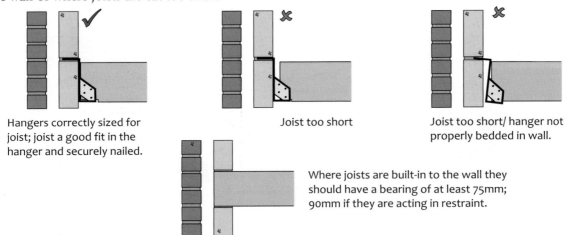

Hangers correctly sized for joist; joist a good fit in the hanger and securely nailed.

Joist too short

Joist too short/ hanger not properly bedded in wall.

Where joists are built-in to the wall they should have a bearing of at least 75mm; 90mm if they are acting in restraint.

Damp penetration

In cavity walls problems of damp penetration are usually due to faulty workmanship. This can occur in a number of ways. One common problem is caused by bridging of the cavity which can allow moisture to penetrate the joist ends. This can occur where careless bricklayers allow mortar to accumulate in the cavity. The problem is exacerbated if the joists project into the cavity and form a ledge on which mortar droppings can collect. The cavity can also be bridged where insulation has been fitted incorrectly, where cavity trays have been omitted or where mortar has collected on ties.

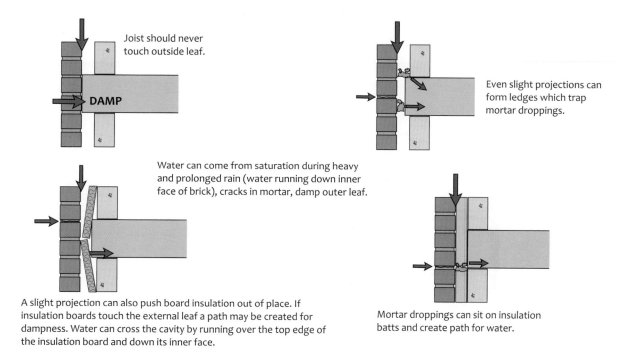

Joist should never touch outside leaf.

DAMP

Even slight projections can form ledges which trap mortar droppings.

Water can come from saturation during heavy and prolonged rain (water running down inner face of brick), cracks in mortar, damp outer leaf.

A slight projection can also push board insulation out of place. If insulation boards touch the external leaf a path may be created for dampness. Water can cross the cavity by running over the top edge of the insulation board and down its inner face.

Mortar droppings can sit on insulation batts and create path for water.

In solid walls the joist ends must always be considered at risk unless the external walls are quite thick. In the majority of speculative housing built during the Victorian and Edwardian periods external walls are usually one brick thick. This effectively means that the protection to the joist end never exceeds 100mm or so (and is often as little as 50mm). Deterioration in the mortar or poor quality brickwork can result in the joist ends becoming damp. In some cases joists are supported on timber wall plates built into the wall. If the wall plates rot the stability of the wall itself may be affected.

Damp penetration will manifest itself in a number of ways. There may be damp patches or staining around the edge of the floor; a moisture meter may show that the moisture content of the timber is above 20%. There may be obvious signs (or smells) which suggest rot is occurring, the floor may feel soft when jumped on and there may be wide gaps under the skirtings.

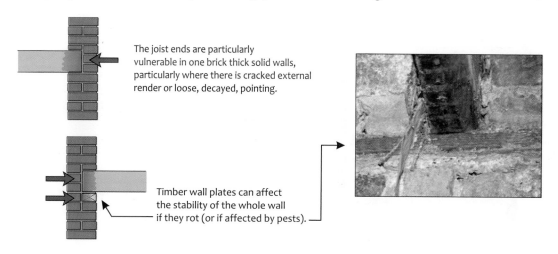

The joist ends are particularly vulnerable in one brick thick solid walls, particularly where there is cracked external render or loose, decayed, pointing.

Timber wall plates can affect the stability of the whole wall if they rot (or if affected by pests).

Sloping floors

If a floor slopes towards a loadbearing external or internal wall, movement in the main structure is likely to be the cause. Torn wallpaper or cracking in the plaster or wall itself will help to confirm this. Possible causes have been covered in the Chapters 1 and 2 on Foundations and Wall Movement.

If a floor sags or deflects the problem is more likely to be related to the floor itself.

Deflection

Deflection or sagging can occur if the joist ends have started to rot. It can also occur for a number of other reasons and systematic inspection is necessary if the cause is to be correctly identified. Deflection can be caused by any or all of the following:

- undersized joists or joists at incorrect centres
- excessive loading from furniture or partitions bearing on the floor
- excessive notching for services
- rot or insect attack.

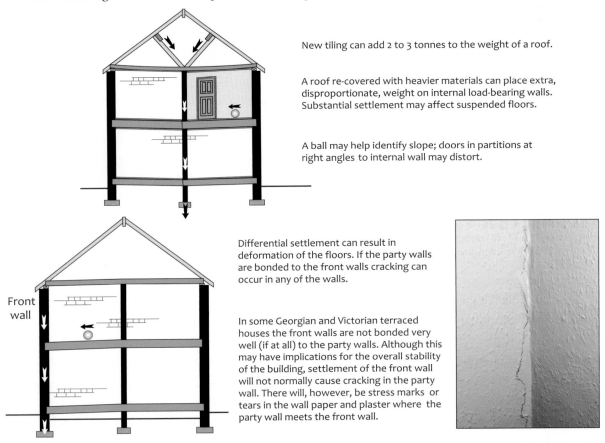

New tiling can add 2 to 3 tonnes to the weight of a roof.

A roof re-covered with heavier materials can place extra, disproportionate, weight on internal load-bearing walls. Substantial settlement may affect suspended floors.

A ball may help identify slope; doors in partitions at right angles to internal wall may distort.

Front wall

Differential settlement can result in deformation of the floors. If the party walls are bonded to the front walls cracking can occur in any of the walls.

In some Georgian and Victorian terraced houses the front walls are not bonded very well (if at all) to the party walls. Although this may have implications for the overall stability of the building, settlement of the front wall will not normally cause cracking in the party wall. There will, however, be stress marks or tears in the wall paper and plaster where the party wall meets the front wall.

In extreme cases deflection can affect the stability of the external walls. This occurs where the load from the sagging floor exerts lateral pressure on the wall.

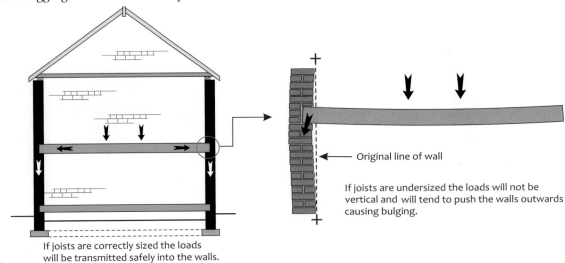

If joists are correctly sized the loads will be transmitted safely into the walls.

Original line of wall

If joists are undersized the loads will not be vertical and will tend to push the walls outwards causing bulging.

Visual inspection may indicate deflection; the floor may have an obvious dip or sag. This can sometimes be confirmed by the use of a straight edge or by looking at the ceiling from the room below. In other cases the 'heel test' (stand on the floor, lift the heels and bring them sharply down) will reveal excess spring in the floor. If glass in cabinets or window glass rattles the floor joists may be undersized (or overloaded).

Where deflection occurs, the first stage in the diagnosis should be to ascertain the depth and centres of the joists. The joist depth may be difficult to assess unless floorboards are lifted. Many joists in older houses will not conform to the current requirements of the Building Regulations although reference to them should give some idea as to their preferred size. An alternative approach, which should only be used as a general guide, is to apply a simple 'rule of thumb' method; take the span in feet, divide by two and add two; this will give the required depth in inches. So, a span of 14 ft requires a joist 9 inches deep. Alternatively take the span in millimetres, divide by 24 and add 50 to give the depth in millimetres; a span of 4000mm requires a joist depth of 216mm, say 225mm (joist sizes are available in increments of 25mm). These rule of thumb methods assume joist centres of approximately 400mm.

In many older houses central heating may have been installed or rewiring may have taken place. Both these activities may result in careless drilling or notching of the joists. The Building Regulations contain a number of simple rules to ensure that notching and drilling do not affect the stability of the joists. These are shown in the diagram below.

Notches in top of joist;

Notch zone - depth no more than 0.125 of joist

0.10 of span

0.20 of span

Holes along centre-line; and '3 x diameter' apart.

No hole within 100mm of notch and not greater than 0.25 joist depth.

0.25 of span

0.40 of span

Where notches lie outside these zones of safety the effective depth of joist should be regarded as the actual depth less the depth of the notch. In extreme cases the installation or renewal of wash hand basins or showers may result in notches as deep as 75mm to take the waste pipes. Thus a 200mm joist effectively acts as a 125mm joist.

If the depth of joist is satisfactory and notching is within the guidelines the joist support needs to be investigated. If one or two joist ends are rotten or not adequately supported by the loadbearing walls, it will have the same effect as increasing the joist centres.

Some of the notches in these joists are quite deep, particularly the one on the right. The notch on the left is original – it was cut for the lead gas pipe (lighting).

If the joist support is adequate another defect which may cause deflection is lack of effective strutting. This could be because strutting was omitted in the original construction or, more likely, strutting has become ineffective because individual struts have been removed for service runs. Another possibility is that the folding wedges or timber blocking pieces at either end of the struts have worked loose or were omitted from the original construction.

Partitions built directly onto the floor can also cause deflection. These partitions may have been part of the original construction or may have been added to change room layouts. If the floor has adequately sized joists a partition running at right angles to the joists will not necessarily cause problems. However, where the partition runs parallel to the joists, or in some cases, in between the joists supported by the floor boarding alone, deflection may occur. In all the above cases the problems will be exacerbated if the partition is carrying heavy loads. Several rows of book shelves for example, can increase the load dramatically. Where the partition is on the highest floor it may be carrying loads from the roof not intended in the original construction. This may be due to structural alterations to the roof itself or the load from stored materials.

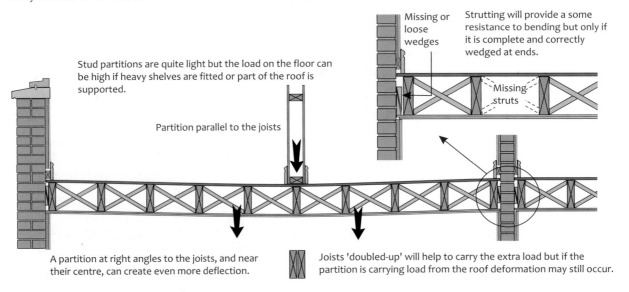

Stud partitions are quite light but the load on the floor can be high if heavy shelves are fitted or part of the roof is supported.

Partition parallel to the joists

Missing or loose wedges

Strutting will provide a some resistance to bending but only if it is complete and correctly wedged at ends.

Missing struts

A partition at right angles to the joists, and near their centre, can create even more deflection.

Joists 'doubled-up' will help to carry the extra load but if the partition is carrying load from the roof deformation may still occur.

Failure of the ceiling finish

Any deflection in the floor is likely to induce cracking in the ceiling finish. Cracking can also be caused by lack of floor strutting, inadequate nailing, plasterboard of insufficient thickness for the joist centres, lack of noggins between joists and plasterboard fixed without staggered joints or with joints incorrectly taped. Wallboard, which tends to be in large sheets (often 2400 x 1200mm), is more likely to crack than plasterboard lath (not to be confused with traditional lath and plaster). Lath, usually in sheets 1200 x 400mm, does not require taped joints and, because of its smaller size, stresses can be spread evenly across the ceiling. It's rare in modern construction but was common from the 1940s to 1980s. Failure of plasterboard, plaster skim or artex-type finishes are discussed further in a later chapter.

Horizontal gaps between floor and external walls

If there is a clear gap between the floor and the wall there may be insufficient restraint to the external walls. This is most likely to occur where the wall runs parallel to the joists. In extreme cases it's possible to stand in a room upstairs and see into the room below. The problem is most common in end of terrace properties, particularly those with more than two storeys, and where the floor joists on each floor run in the same direction.

These straps are in the right position and will offer restraint to the gable wall. However, straps should work in compression as well as tension. This requires timber blocking in between the joists under the straps.

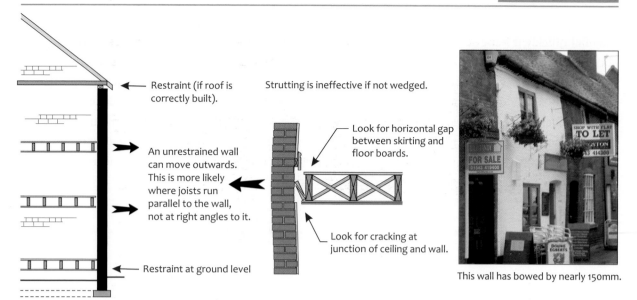

Restraint (if roof is correctly built).

An unrestrained wall can move outwards. This is more likely where joists run parallel to the wall, not at right angles to it.

Restraint at ground level

Strutting is ineffective if not wedged.

Look for horizontal gap between skirting and floor boards.

Look for cracking at junction of ceiling and wall.

This wall has bowed by nearly 150mm.

Building conversion

Many large Georgian, Victorian and Edwardian buildings have been converted into flats. In many cases these conversions will have been carried out in the past when Building Control requirements were less onerous than today. It is the higher floors that are normally most at risk. In many of these early properties joist size varied from floor to floor. The top floor, for example, was possibly built as servant accommodation and would contain joists of modest size to reflect the light loading. Change of use can result in these floors being overloaded; this is a particular problem where layout changes have required the installation of additional partitions. The situation is compounded by the need for adequate fire protection and sound insulation, both of which increase the load on the floor. Victorian and earlier buildings often had floors formed with joists of various widths and varying centres – sometimes the centres were wider than current practice requires. Such floors are easily overloaded unless extra joists have been added.

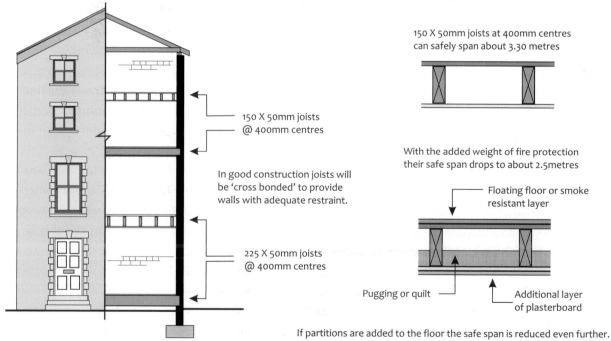

150 X 50mm joists @ 400mm centres

In good construction joists will be 'cross bonded' to provide walls with adequate restraint.

225 X 50mm joists @ 400mm centres

150 X 50mm joists at 400mm centres can safely span about 3.30 metres

With the added weight of fire protection their safe span drops to about 2.5metres

Floating floor or smoke resistant layer

Pugging or quilt

Additional layer of plasterboard

If partitions are added to the floor the safe span is reduced even further.

Fire protection

In a modern two storey house the use of plasterboard ceilings and tongued and grooved boarding both contribute to the floor's resistance to the spread of smoke and fire. In many older properties lath and plaster ceilings together with square edged boarding are more likely to be found. Square edged boards are not effective in limiting the spread of smoke (the main risk to life) and lath and plaster ceilings may fail quickly if a fire starts. Early construction does not necessarily require renewal although it should be recognised that the performance of a pre Second World War floor may be less than satisfactory if a fire breaks out. The risks are more significant where properties have been converted into flats. In these cases the performance requirements of the floor are much more onerous; the higher the building the greater the level of performance required. The purpose of the Building Regulations is not to make the floor non-combustible but to provide a reasonably 'safe time' to ensure that occupants can escape. In converted buildings there may be some 'trade-off' in standards of fire protection depending on the level of fire alarm equipment and emergency lighting.

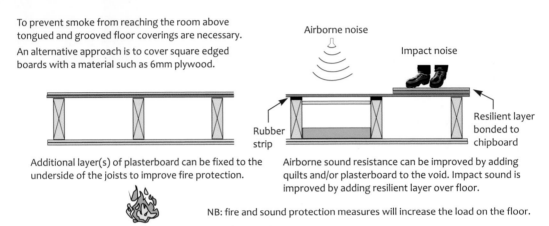

To prevent smoke from reaching the room above tongued and grooved floor coverings are necessary. An alternative approach is to cover square edged boards with a material such as 6mm plywood.

Airborne noise

Impact noise

Rubber strip

Resilient layer bonded to chipboard

Additional layer(s) of plasterboard can be fixed to the underside of the joists to improve fire protection.

Airborne sound resistance can be improved by adding quilts and/or plasterboard to the void. Impact sound is improved by adding resilient layer over floor.

NB: fire and sound protection measures will increase the load on the floor.

Sound insulation

It is only in recent years that converted properties have had to conform to the Building Regulations in terms of sound insulation. Buildings altered before this (and even after), however, may still fall within the boundaries of the Environmental Protection Act 1990. Many local authorities, housing associations and private landlords are struggling to cope with the financial implications of this Act which can require them to provide adequate sound insulation even if the property conformed to the relevant legal requirements at the time of conversion. As mentioned earlier there are two discrete problems; providing resistance to airborne sound and providing resistance to impact sound.

Meeting these requirements can dramatically increase the load on the floor. In many cases existing joists will not cope with the increased load and this may require additional joists, new joists of deeper section or intermediate support to convert the floor structure to a double floor. In the latter option fire protection of the supporting beam will need to be considered.

Nowadays sound insulation systems depend, to a large extent, on plasterboard. Prior to this it was common practice to use sand pugging spread between the joists. This was a heavy material even when dry – if it becomes wet, through say a plumbing leak, the ceiling may collapse. Unfortunately, however good the level of sound insulation in a floor, the transfer of sound may still be a problem and this is usually caused by flanking transmission.

Services penetrating separating floors

Where drainage or water services penetrate a separating floor there are strict rules to ensure adequate sound insulation and fire protection. These rules are quite complex and are therefore beyond the scope of this book. However there are some general defects, which, if found may require more specialist investigation. These include airborne sound transfer between flats where the pipes penetrate the floor, annoying sound from waste water in soil vent pipes and lack of adequate fire protection. In modern construction the pipework either side of the floor should be protected by some form of ducting or casing which will provide a minimum of half hour fire resistance. Access panels fitted in the ducting or casing should not be located in bedrooms or circulation space. The floor should be sealed or fire stopped where the pipe penetrates it. No services other than water pipework or drainage should share this duct. At the same time, to reduce the risk of sound transfer, the pipe should be wrapped in mineral wool (or similar) and the casing boarding should have a minimum density as set out in the Building Regulations.

DEFECTS – CONCRETE UPPER FLOORS

PRE-CAST FLOORS

These floors are comparatively new, they first became common at the end of the 1950s in low rise flats. As such, their longevity is unknown. However, some problems have come to light, these include:
- inadequate end bearing
- cracking of ceiling finishes
- poor sound insulation
- lack of fire protection where services penetrate the floor
- lack of restraint to external walls
- failure of the floor finish
- sloping chipboard floors.

End bearing

Most manufacturers recommend a minimum end bearing of 100mm. Where brickwork has been badly set out or built out of plumb there is the danger that the end bearing can be reduced. In practice this can be avoided by careful site supervision or insisting that the beam manufacturers provide beams to suit the as-built dimensions rather than the dimensions shown on the drawings. While this will inevitably cause some delay the associated costs outweigh the potential costs of remedying incorrectly sized beams.

Cracking of ceiling finishes

This is most likely to occur where plaster finishes have been applied directly to the soffit of the beams. It may be caused by incorrect or inadequate grouting of the floor surface, the load from heavy partitions built directly on a single beam, or the use of wet infill blocks which will shrink slightly as the floor dries out. It could also be caused by using undersized beams which suffer excessive deflection although this defect should not occur if the manufacturer's design guidance is followed.

Poor sound insulation

If the manufacturers' recommendations have been followed modern pre-cast floors will conform to requirements of the Building Regulations with regard to sound insulation. In practice, poor site workmanship or inadequate site supervision can result in number of problems which reduce the floor's resistance to the passage of impact and airborne sound. Airborne sound resistance can be reduced if the floors are not grouted properly. Impact sound resistance will be reduced if there are defects in the floating floor. This can be caused by inadequate thickness of the separating medium (usually a paper backed quilt under screeds or insulation boards under screeds or chipboard), or where the screed has penetrated the joints between insulation boards. Sound insulation will also be reduced if holes cut for service runs (obviously through the blocks, not the beams) are not adequately filled. This will also reduce the fire protection of the floor.

Failure of the surface finish

Screeds laid on a resilient layer will crack if they are not reinforced (the usual recommendation is galvanised chicken wire) or if they are of inadequate thickness; they should be at least 65mm thick. Screeds can also crack if they are laid too wet or are not sufficiently compacted thus reducing their bond with the wire reinforcement.

Chipboard can fail if it becomes wet and this can occur through spillage or where wet flooring components are used and a polythene vapour check is omitted above the insulation boards (the resilient layer). As with solid ground floors the chipboard should be tongued and grooved, with glued joints, and with a minimum gap of 10mm at the perimeter to allow for moisture movement.

If a chipboard floor has been correctly laid but still appears to rise slightly towards the centre of a room it could be due to the pre-camber of the beams. A floor spanning 8m may rise by as much as 20mm mid-span. Before laying the resilient layer the floor should have been checked for level and, if necessary, a thin levelling screed should have been applied.

Lack of restraint to external walls

Some beam and block floors require restraint straps where beams run parallel to external walls. As with timber floors strapping is required at 2m centres.

IN SITU FLOORS

Three in situ floors were described earlier in the chapter. These floor types are normally very durable and serious structural problems are rare; where cracking and deflection are evident the advice of a structural engineer should be sought. Some structural problems such as carbonation and the use of high alumina cement are described in a later chapter. More common are problems related to fire resistance where services penetrate the floor and poor impact sound insulation (see above). Other defects include cold bridging and poor detailing where floors cantilever to form balconies.

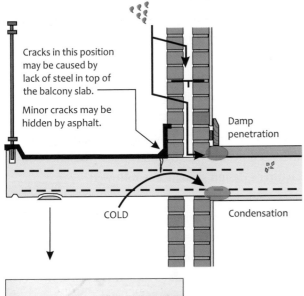

Cracks in this position may be caused by lack of steel in top of the balcony slab.

Minor cracks may be hidden by asphalt.

Damp penetration

COLD

Condensation

Dampness in the screen or on the ceiling can be difficult to diagnose with any certainty. Both can be caused by condensation or damp penetration. The latter is more likely where cavity trays have been omitted above the floor slab or where cracking in the cantilevered section of the slab has torn the asphalt.

Balconies 'guarded' by railings or brick walls.

Spalling of soffitt caused by carbonation; as steel rusts, it expands, forcing away the thin concrete cover.

CHAPTER **7**

Pitched Roofs

INTRODUCTION

The proper performance of a roof is often considered to be more important than that of any other building element, except perhaps the foundation. This is because whole or, more usually, partial failure of the roof inevitably has an adverse effect upon one or more elements elsewhere within the building.

A roof can fail for a number of reasons ranging from poor design and/or poor construction to lack of maintenance. Failure may also be due to the roof having outlived its designed life expectancy or to the effect of other failed building elements upon its performance, eg failure of a supporting wall.

CONSTRUCTION SUMMARY

FUNCTION OF THE ROOF

The primary function of the roof is to provide protection to the building beneath from the weather. It should be constructed in a form that will meet the following five key performance criteria.

1. Strength

It must be capable of supporting both its own weight and any loads that might be imposed upon it, eg snow, standing water, water tanks and other plant.

2. Stability

The various components should not be subject to excessive movement nor should they create instability in other elements.

3. Durability

The materials used, especially coverings, must be capable of performing for their designed life expectancy.

4. Weathertightness

The roof covering must prevent moisture penetration. This is achieved by it being either thin and totally impervious (eg sheet roof coverings, slate) or by the use of a thicker, less dense material that is sloping and allows evaporation plus water shedding to take place (stone tiles, clay tiles).

5. Thermal efficiency

In recent years it has become recognised that a roof must also provide good thermal insulation.

PITCHED ROOFS – BRIEF HISTORY

Protection from the elements has been essential to human survival for many thousands of years.

The first roofs, once man moved out of caves, were probably formed by covering a sloping framework of branches with a covering of animal skin or vegetation – direct descendants of these forms of construction can still occasionally be found in this country, eg thatched roofs of reed, wheat or heather. Sloping, or pitched roofs, were found to be the most efficient form of weatherproof construction in all but the driest climatic conditions and are the traditional form in the British Isles.

A traditional pitched roof is expensive to construct. It involves the use of skilled craftsmen and a considerable quantity of materials. Earlier roofs involved cutting the individual timbers to size on site (a country-cut roof), while a modern roof is commonly constructed with factory-formed trusses. Coverings were of thatch, slates, stone or clay tiles – naturally based materials – which required steeper pitches to ensure efficiency. It was only in the 20th century that man-made materials, such as concrete slates and tiles, with interlocking details that allow much lower pitches were introduced.

THE FORM OF PITCHED ROOFS

The simplest form of pitched roof is the lean-to or monopitch roof that is commonly found above many rear extensions of Victorian houses. It can also be found as the roof form on a great number of housing estates built since 1960. Originally such roofs were constructed on site from individual timbers but the modern version is normally formed with trusses and often has an extremely low pitch (17.5 degrees or less).

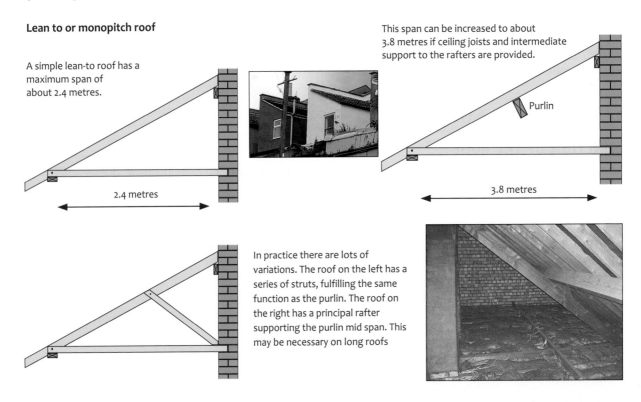

Lean to or monopitch roof

A simple lean-to roof has a maximum span of about 2.4 metres.

2.4 metres

This span can be increased to about 3.8 metres if ceiling joists and intermediate support to the rafters are provided.

Purlin

3.8 metres

In practice there are lots of variations. The roof on the left has a series of struts, fulfilling the same function as the purlin. The roof on the right has a principal rafter supporting the purlin mid span. This may be necessary on long roofs

A lean-to roof can only span a short distance but a simple double pitch or couple roof can span a wider length. Incorporating a collar to act as a tie at about one third of the height enables even larger areas to be successfully roofed over, as it reduces the tendency for the roof to spread.

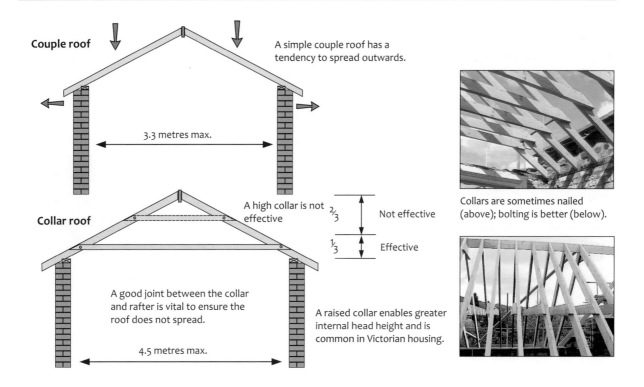

Couple roof

A simple couple roof has a tendency to spread outwards.

3.3 metres max.

Collar roof

A high collar is not effective

$\frac{2}{3}$ Not effective

$\frac{1}{3}$ Effective

A good joint between the collar and rafter is vital to ensure the roof does not spread.

4.5 metres max.

A raised collar enables greater internal head height and is common in Victorian housing.

Collars are sometimes nailed (above); bolting is better (below).

A collar roof can still only span a relatively short distance but is at its most efficient when the collar is positioned at the feet of the rafter. It is then known as a closed couple roof. Even greater spans are achieved by the inclusion of purlins, struts and braces. Such construction is typical of roofs formed during the last 150 years.

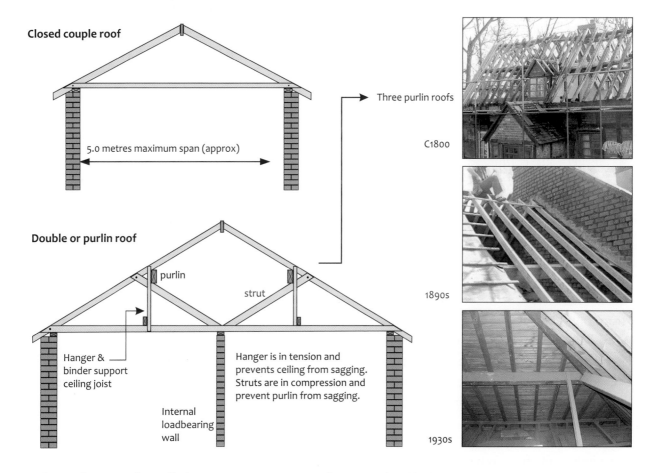

Closed couple roof

5.0 metres maximum span (approx)

Three purlin roofs

C1800

Double or purlin roof

purlin

strut

1890s

Hanger & binder support ceiling joist

Hanger is in tension and prevents ceiling from sagging. Struts are in compression and prevent purlin from sagging.

Internal loadbearing wall

1930s

The purlins span laterally between party or internal structural walls that support the loads involved. Where no such wall was available, one or more roof trusses, of the type known as king post trusses, would be constructed at intervals. These were formed of large timbers which acted as intermediate supports to the purlins and hence to the rest of the roof structure.

King Post Truss

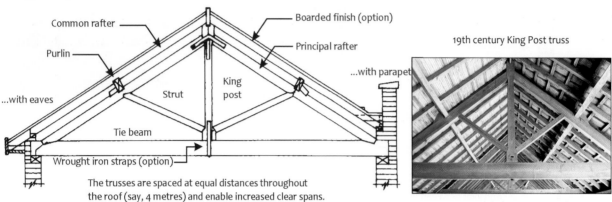

The trusses are spaced at equal distances throughout the roof (say, 4 metres) and enable increased clear spans.

19th century King Post truss

Often, the roof space would be required for accommodation or storage purposes. This could be achieved by the use of an open truss known as a queen post truss or, for smaller spans, a single mansard roof.

Mansard Roof with King\Queen Post Truss

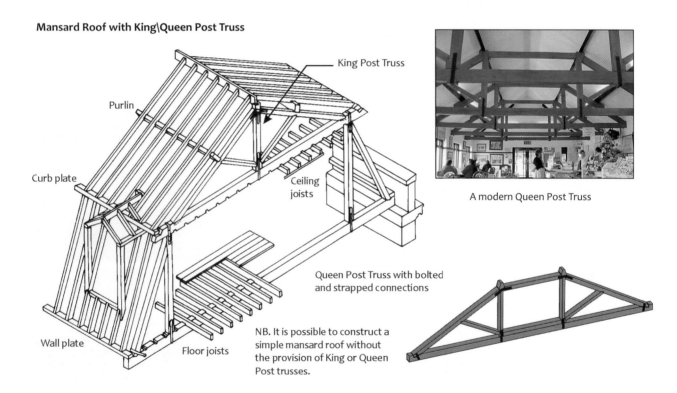

A modern Queen Post Truss

Queen Post Truss with bolted and strapped connections

NB. It is possible to construct a simple mansard roof without the provision of King or Queen Post trusses.

Historically, the quality of roof design, workmanship and materials was variable. The higher the social status of the occupant (and the greater the income) then, usually, the better the quality of the building. Speculative building was undertaken for all social levels and this often resulted in poorer quality construction, notwithstanding the eventual occupant's social status. However, it is generally recognised that traditional construction, in particular of the roof, achieved its highest quality during the years between 1900 and 1914.

Shortages of materials and skilled labour, combined with post-war economic problems, resulted in there being little domestic construction in the 1920s. This was followed by a period of greater activity until 1939 in both the public and private sectors. Pitched roofs were still of traditional 'cut' construction but the quality, although generally good, was usually lower than the peak reached just before the First World War.

The Second World War and the years immediately after was a period of little building activity except for some types of system building *(see Chapter 15)*. Between 1945 and 1954 timber was subject to rationing and this led to the extending or stretching of the centres of rafters and trusses in order to save materials.

Domestic construction commenced again in earnest in the 1950s. A great number of the roofs of the period between 1950 and 1970 were of 'cut' construction but this period also saw the introduction of modern trussed roof construction, so that by 1970 almost all roofs of low-rise domestic housing were formed with trussed rafters. The advantages of modern trusses are that they are relatively cheap to manufacture, easy to install and do not require skilled site labour. They are also lightweight and are capable of spanning wide distances. It is now uncommon to use anything other than trussed roof construction for new housing.

Modern trussed rafters

Connectors can be bolts and washers or 'Gang Nail' plates.

Double W truss

There are many different types/forms of truss but all work on similar principles

Bolted

'Gang Nail' plates

A truss of this design can span 15 metres or so without any intermediate supporting walls.

The Fink truss is capable of spanning up to about 10 metres.

In a simple fink truss roof five binders are required to tie the trusses together; a raking brace on each slope keeps the structure rigid. The lower end of the brace should be secured to the wall plate.

Where binders are not of sufficient length they can be joined by lapping them over 2 trusses.

Ends of binders should abut gable walls.

Wall plate strap

Gable strap

Gable ties are required every two metres along the pitch; on high roofs they are also required at ceiling level. The need for wall plate strapping depends on pitch, load etc.

The majority of low-rise domestic pitched roofs are of timber construction. However, since the Second World War some use has been made of pre-cast reinforced concrete members and steel members although these are often found to be installed as supports to what is essentially a timber roof (eg reinforced concrete trusses and purlins supporting timber rafters).

PITCHED ROOFS – THE DEFECTS

POOR DESIGN

Roof failures can occur as a result of designer error. Raised standards of education and professional training, increased knowledge of the performance of structure and materials, and regulatory control of building design, all mean that (in theory, at least) roofs have become increasingly less likely to fail. However, the introduction of new and not properly understood design approaches, the use of new materials, or of traditional materials used in new ways, even the loss of knowledge of how traditional materials perform, can each result in the roof structure and/or its covering failing to perform as required.

It also needs to be recognised that, historically, the standards of roof design before (and even after) the introduction of model bye-laws in the 19th century were variable. This is particularly noticeable in the buildings of cheaper, and probably speculative, domestic construction which were erected in the period of extensive urban expansion of the 19th century.

Generally, the older the building the less likely that there will have been any scientific design criteria

Common defects in pitched roofs at construction stage (new build)

Potential Problem Areas	Implications	Defects / Manifestation
Untreated timbers	Attack by wood-boring insects	Weakening of timbers due to infestation
Rafters, purlins and other roof components inadequately sized	Roof structure too weak to cope with roof covering and/or imposed loads and/or wind	Bowed and/or split timbers. Deflection of overstressed roof members. Dishing of roof
Rafters, purlins and/or other roof members omitted	Structural integrity of the roof compromised	Bowed and/or split timbers. Deflection of overstressed roof members. Dishing of roof
Roof trusses inappropriately stored or mishandled. (Physical damage, distortion, water saturation)	Physical damage or distortion leading to inadequate structural performance. Water saturation and risk of timber rot	Bowed, twisted and/or damaged trusses. Evidence of timber decay (wet or dry rot)
Inadequate bracing of roof trusses	Trussed roof likely to move or tilt sideways as a unit	Diagonal movement (racking) of roof. Gable walls leaning (leading to possible failure)
No allowance for weight of water tanks and other plant installed in roof space	Overloading of roof structure	Deflection of roof structure/components. Deflection and cracking of soffits and ceilings below
Too small an eaves or verge overhang	Lack of full weather protection to top of external wall	Penetrating damp affecting top of wall and wallplate. Damp staining on face of wall
Half tiles instead of tile-and-a-half tiles used at verges and valley gutters	Half tiles difficult to fix securely. Likelihood of loss due to wind especially on verges	Raised/missing tiles. Penetrating damp
Sarking felt improperly installed	Too limp leads to water ponding. Too tight leads to tearing. Poor attention to detailing leads to water penetration	Damp penetration of roof
Parapet and valley gutters with incorrect detailing (eg extended bay lengths, insufficient step height)	Gutter bays will not cope with effects of thermal expansion. Risk of capillary action at steps	Penetrating damp into roof void and ceilings below. Risk of wet or dry rot
Inadequate flashings/soakers at abutments with parallel walls, chimneys	Weather integrity of the roof compromised	Penetrating damp affecting roof structure and adjoining internal elements and components

Note: No reference is made in this table to ventilation and insulation problems. These are discussed in Chapter 13.

applied when considering structure and materials. The designer, or more likely, the master builder, would have based his construction on the accumulated knowledge of generations of previous similar builders using materials that had been 'proved by the test of time'. Furthermore, it was not uncommon for corners to be cut or design implications to be misunderstood, even in so-called quality work, resulting in the roof structure performing at the limit of, or possibly beyond, its proper structural capability. While not necessarily leading to total failure, it does mean that the roof will be over-stressed and, almost certainly, showing signs of distress.

The following are some examples of poor design.

Inadequately sized members The width and/or depth of the timber, concrete or steel member is too small to prevent it bending, bowing or buckling. This is a common problem of traditional timber roofs constructed before approximately 1890 – 19th century roofs often contain very narrow rafters while earlier roofs have extremely broad timbers which are often very shallow. In both instances, such roofs have a tendency to deform and this is frequently exacerbated by replacement modern tiles that are heavier than the original roof covering.

Insufficient members The individual members are too widely spaced apart and do not fully support any applied load.

Overloading of the structure The roof covering and/or contents are sometimes too heavy for the structure to support properly. This may be caused by the original design being inappropriate for the roof covering and other loads actually applied. Alternatively, the overloading may be due to subsequent maintenance or remedial works, eg slate or clay tiles being replaced by much heavier modern concrete slates or tiles, or the addition of water tanks and other plant. Post construction over-loading became such a problem that, since the 1991 Building Regulations, it has been mandatory to submit structural calculations to the local authority Building Control Department where a change of roof covering is being considered.

Inappropriate roof covering The covering can be unsuitable for a number of reasons, eg pitch is too shallow or steep for the product used; tiles are too heavy for the structure.

Incorrect detailing A great number of roof problems are caused by insufficient attention being paid to the design of 'minor' details, ie the mass of more intimate construction that needs to be carefully designed if the roof is to perform efficiently. Inadequate design is not a problem solely of the past. It still occurs in both new roofs and those that are undergoing refurbishment. It should be less likely in modern roofs because they would have been subject to greater design knowledge and regulatory control, eg the Building Regulations. However, roofs of older buildings undergoing refurbishment have been subject to much less strict controls until quite recently, eg the submission of a mathematically calculated structural assessment of such a roof has only been required since 1991.

The left-hand roof has been altered in an attempt to help remove the 'bow' in the rafters. The right-hand roof is late Georgian and additional rafters are being installed to strengthen the roof. The original rafters are being left in place, not for structural reasons, but in an attempt to conserve the original material.

POOR CONSTRUCTION

The problems of poor construction are often similar to those of poor design:

- incorrectly sized members
- insufficient members
- overloading of the structure
- inappropriate roof covering
- incorrect fixing
- incorrect detailing
- poor workmanship/supervision.

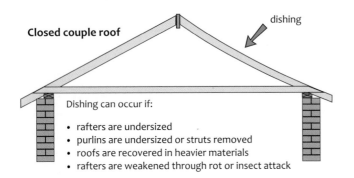

Closed couple roof

dishing

Dishing can occur if:

- rafters are undersized
- purlins are undersized or struts removed
- roofs are recovered in heavier materials
- rafters are weakened through rot or insect attack

Roof alterations

purlin

purlin, hanger and
binder removed

The owner of this roof (photo below left) removed the purlin strut, hanger and binder to provide an uncluttered roof space. Unfortunately, the ceiling started to sag - hence the hastily added hangers. Purlin deflection has also occurred and the roof has dished.

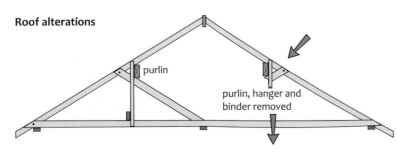

The photo on the right shows a trussed rafter roof. Two of the tension members have been removed to create space in the roof. A collar has been added to each truss in an attempt to compensate.

ROOF SPREAD

Each rafter of a pitched roof is subject to horizontal and vertical forces that result in it having a tendency to move outwards. This is overcome by the inclusion of horizontal members that tie the rafter to another element of the structure or to another rafter, eg collar or ceiling joist. The closer the horizontal member is to the foot of the rafter the more effective it is in preventing the outward movement or spread of the roof.

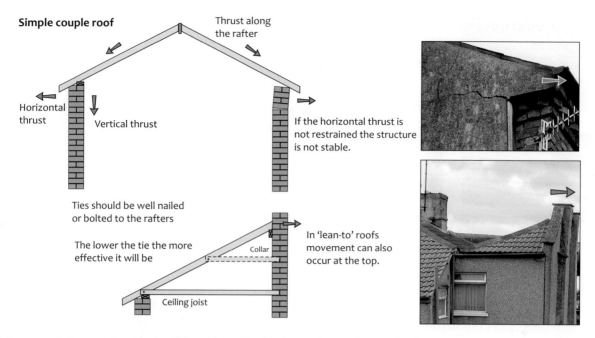

Simple couple roof

Thrust along the rafter

Horizontal thrust

Vertical thrust

If the horizontal thrust is not restrained the structure is not stable.

Ties should be well nailed or bolted to the rafters

The lower the tie the more effective it will be

Collar

In 'lean-to' roofs movement can also occur at the top.

Ceiling joist

However, it is not always possible to install a horizontal member at the base of the rafter, eg where the design of the building incorporates a habitable space at roof level. In this circumstance, roof spread can develop. It can also be a problem where ceiling joists are inadequately connected to the rafters and fail to provide restraint.

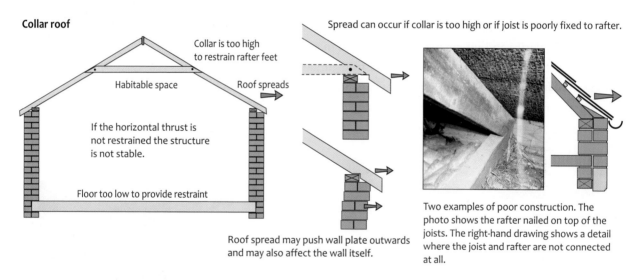

Collar roof

Collar is too high to restrain rafter feet

Habitable space

Roof spreads

If the horizontal thrust is not restrained the structure is not stable.

Floor too low to provide restraint

Spread can occur if collar is too high or if joist is poorly fixed to rafter.

Roof spread may push wall plate outwards and may also affect the wall itself.

Two examples of poor construction. The photo shows the rafter nailed on top of the joists. The right-hand drawing shows a detail where the joist and rafter are not connected at all.

THE EFFECT OF DAMP PENETRATION

The roof is particularly susceptible to the effect of weather and if there is a failure in preventing damp penetration both the roof itself and adjoining elements of construction may be affected. Initially, it can often be difficult to precisely identify the original defect causing the dampness, as moisture has a habit of travelling out from the original point of failure over a period of time. This may occur on the surface of, and within, adjoining materials. The precise cause can frequently be confirmed only after opening up of the roof construction has taken place.

Further problems may be caused where timber or steelwork is built into the adjoining structure, especially where the walls are of solid construction. If they are of any age, it is unlikely that the timber or steel will be protected by any form of damp proofing. Where the covering structure also lacks thickness, then the timber or steel truss, purlin, beam, etc may be adversely affected by damp penetration of the wall. This situation can be exacerbated in particularly highly exposed walls.

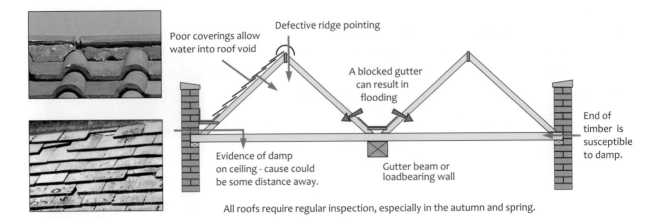

Poor coverings allow water into roof void

Defective ridge pointing

A blocked gutter can result in flooding

End of timber is susceptible to damp.

Evidence of damp on ceiling - cause could be some distance away.

Gutter beam or loadbearing wall

All roofs require regular inspection, especially in the autumn and spring.

Gutters should be appropriately sized so that they are capable of fully discharging all water received from the adjoining roof surfaces into rainwater outlets. However, an outlet can become partially or fully blocked by debris such as autumnal leaves or a tennis ball. Occasionally, a flash storm can result in more rainwater than the gutter's designed capacity. The result is that the water will rise above the top of the gutter and may flow into the roof itself. Unfortunately, as one of the effects of climatic change is an increase in the number, regularity and intensity of storms, there may, in the future, be increasing problems of existing gutters being unable to cope with roof rainwater run-off.

Snow can also be a problem. Wind-driven snow can find its way through the smallest opening and can be a serious problem with unfelted roofs.

Little protection from the weather is offered by this ancient stone tiled roof. Stone slates (as opposed to natural slate) are not always impervious to water. Heavy rain and cold winters can lead to frost damage.

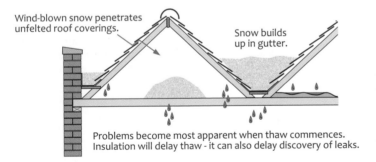

Wind-blown snow penetrates unfelted roof coverings.

Snow builds up in gutter.

Problems become most apparent when thaw commences. Insulation will delay thaw - it can also delay discovery of leaks.

INADEQUATE PITCH

The pitch of a roof is dictated by the type of roof covering. Recommended pitches, particularly for tiles, vary greatly and the manufacturer's recommendations should always be followed.

Modern inter-locking slates and tiles can be laid at much lower pitches than their traditional predecessors. However, if either type of covering is laid at too low a pitch, rain and snow may penetrate the roof covering, especially if the roof is in an exposed situation. Even with reclaimed slates and tiles that pre-date the issue of any manufacturer's advice, sensible adherence to traditional knowledge of appropriate minimum pitches is necessary. Laying the covering on steeper pitches (up to the vertical) is not normally a problem, although all tiles need to be nailed, as sole reliance on any nibs that are provided to give adequate fixings is insufficient.

Laying roof coverings on too shallow a pitch and/or extending the gauge, ie reducing the lap *(see next sub-section)* of tiles, slates and other coverings, are common problems. The reason for these approaches is either because of ignorance of what should be installed or, more likely, to save money as less tiles and slates are needed. Unfortunately, there is also reduced performance because such a roof is much more prone to water penetration and wind damage.

Examples of minimum recommended pitches for a range of roof coverings

Material	Type / Size	Minimum Recommended Pitch
Traditional slates	Large slate Small slate	21 degrees 33 degrees
Fibre-cement slates (man made)		20 degrees
Redland 'Cambrian' slates (man made)	Single standard	15 degrees
Traditional clay tiles	Pantiles Single lap tile Plain tile	25 degrees 35 degrees 40 degrees*
Redland 'Regent' tiles (man made concrete)	Through-coloured tile Granular finished tile	12.5 degrees 22.5 degrees
Cotswold stone tiles		55 degrees
Shingles	Western red cedar Oak	25 degrees 45 degrees
Thatch		45 degrees

* Note: Recent advances in manufacture and performance have led to the introduction of the Sandtoft New Generation 20/20 interlocking clay plain tile which can be installed at a recommended minimum pitch of 15 degrees.

The effect of inadequate pitch

If the pitch is too low for the covering, wind blown rain and snow will more easily penetrate the slate or tile covering.

Note; effective pitch is always less than that of the rafters.

The rafter pitch on this roof is about 12 degrees - the minimum pitch for these artificial slates should be about 20 degrees.

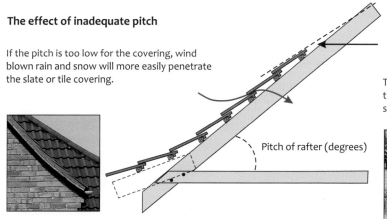

Pitch of rafter (degrees)

Some roofs have sprocket ends fixed to the rafters for aesthetic reasons. These can further reduce effective pitch and lead to water penetration.

GAUGE AND LAP PROBLEMS

The gauge or distance between the centreline of the battens determines the headlap (lap) or cover that a tile achieves when laid above another tile. The amount of headlap is important – too little and it will not prevent weather penetration. Unfortunately, many roofs suffer from this problem, either because of ignorance by a contractor or because an extended gauge/reduced lap can lead to savings in material, time and cost.

Occasionally, exactly the opposite may occur. The gauge may be reduced. This will increase the lap and probably give greater weather resistance but will also result in an excessive load on the underlying structure.

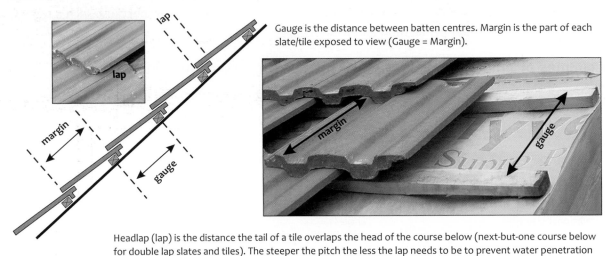

Gauge is the distance between batten centres. Margin is the part of each slate/tile exposed to view (Gauge = Margin).

Headlap (lap) is the distance the tail of a tile overlaps the head of the course below (next-but-one course below for double lap slates and tiles). The steeper the pitch the less the lap needs to be to prevent water penetration

The margin on these Double Romans (left) is too large. The subsequent reduction in lap may result in water penetration and increased risk of wind lift. In addition the lead gutter should be stepped (and laid in 3 or 4 sheets) to avoid problems of ridging and splitting due to thermal movement. The gutter on the right has been laid in separate sheets to accommodate thermal movement but the tiles have been 'stretched' to reduce the number of tiles required.

FAILURE OF ROOF COVERINGS

Roof coverings for pitched roofs fall into two distinct types:

    1: natural materials – eg clay tiles, stone tiles, natural slates, timber shingles
    2: man-made materials – eg concrete tiles, fibre slates, metal or fibre sheet.

Below are some of the problems associated with specific roof coverings.

Natural slate

Slate covered roofs can be found throughout the British Isles where they were extremely common in the 19th century. They can range in colour from red/black to green/blue depending upon the source which, historically, was within the UK – Cornwall, Cumbria, Devon, parts of Scotland and Wales. However, the popularity of slate roofing declined in the first three quarters of the twentieth century although there was a revival in the last quarter. This led to the re-opening of some formerly closed quarries to meet the increased demand. In recent years, foreign slates, in particular from Brazil, Canada, China and Spain, have also become commonly used because of their relative cheapness and greater availability. Some of these have been of poor quality and have caused problems but many have performed acceptably.

Natural slates come in a number of sizes, with the larger being used on shallower pitches or at the bottom of steeper roofs. They are produced in a range of thicknesses; the larger the slate, the thicker it is. Fixing is by means of head or centre nailing, the latter on larger slates or shallower pitches.

A number of slate-roofed houses are shown in the left-hand photograph. The right-hand roof shows the result when the fixing fails. One defective slate has already been supported by a tingle – an s-shaped length of lead or copper.

Life expectancy generally is up to 100 years, although it can be considerably longer. It depends upon source, thickness, cutting technique, etc.

Here are some of the particular problems of natural slate.

Reliability Natural slates are of varying degrees of durability and strength depending upon the following factors:

- source – country, quarry, bedrock
- method of manufacture – roofing slates are produced by hand splitting large blocks of slate (riven slates) and then each slate is dressed, either by hand or, increasingly, by machine
- installation – the grain should run parallel to the length.

Problems can be caused when slates contain impurities that accelerate decay, eg calcite, pyrite, or expanding clay materials.

Delamination Slate is a laminated material comprising thin layers of rock bonded together naturally. Poorer quality slate may contain carbonates or clay mineral impurities. With the former, a chemical reaction with the acids in rainwater creates calcium sulfate, the crystals of which force the thin laminations apart. In the case of the latter, water gradually penetrates between the layers and subsequent frost action can force these apart. Deterioration of this type more readily affects the underneath of the slate, where moisture has been drawn by capillary action. Each shedded layer weakens the slate and reduces durability. It can be identified by cracking between the layers appearing around the edge of the slate, and is more likely to occur in slates of inferior quality.

Problems with torching and bedding Torching is a traditional process that consisted of filling the gaps between slates and battens with a mortar pointing. This was commonly a mixture of lime mortar and animal hair. It was used to prevent rain or snow penetration into the roof space of inferior quality work where there was neither boarding nor felt. It lost favour between the two World Wars when it was realised that it was better to allow ventilation to the underside of the roof slates (or tiles) and to the battens.

Torching was usually a mix of lime mortar which helped reduce the incidence of draughts and the ingress of snow and dust. However, it also reduced ventilation resulting in premature failure of the battens.

Bedding (or shouldering) involved sealing between the tail of one slate and the slate above (often where loose) with mortar to provide both a weather-tight seal and a bond between the adjoining slates. Both torching and bedding are still found to be in occasional use as cheap 'repairs'.

The mortar in both cases tends to shrink and fall out, leading to water penetration and loose slates. Both techniques can lead to inadequate ventilation of the underside of the slate and encourage speedier deterioration of the roof covering because of moisture problems.

The batten adjoining any torching is also more likely to be adversely affected by moisture.

Split slates Slate is a naturally brittle material and this means that it can have a tendency to split down its length (along the 'grain'). This is often along the line of a nail hole. Each slate should be checked by tapping before installation but the defect may only appear after fixing.

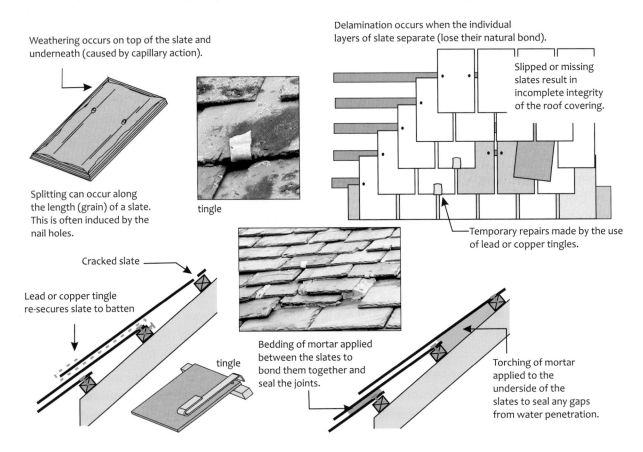

Weathering occurs on top of the slate and underneath (caused by capillary action).

Splitting can occur along the length (grain) of a slate. This is often induced by the nail holes.

tingle

Delamination occurs when the individual layers of slate separate (lose their natural bond).

Slipped or missing slates result in incomplete integrity of the roof covering.

Temporary repairs made by the use of lead or copper tingles.

Cracked slate

Lead or copper tingle re-secures slate to batten

tingle

Bedding of mortar applied between the slates to bond them together and seal the joints.

Torching of mortar applied to the underside of the slates to seal any gaps from water penetration.

Surface erosion or weathering Slates often suffer erosion of the surfaces that are covered by the overlap of each slate. This reduces the longevity of the slate. It is caused by surface water being drawn up between the slates by capillary action. Sulfur dioxide in the atmosphere mixing with rainwater creates sulfuric acid that affects the surface of natural slate and causes disintegration. Initially, this may appear as powdering, flaking or blistering of the surface. It is more likely to occur with poorer quality slates that contain carbonates in varying quantities.

Weathering may also affect the exposed upper surface of the slate for the same reason. Erosion may also be due to impurities in the slate, eg iron pyrites, in which case staining may be apparent.

Slipped slates Slates may partially slip or become completely detached from their original position. The cause is normally 'nail sickness' *(see later roof sub-section)*. Sometimes an attempt is made to re-secure such slates by the provision of lead or copper clips (often known as tingles) but this approach is unsightly.

Broken slates Broken or cracked slates may be due to a number of reasons, eg foot traffic, inherent defect. However, head-nailed slates are particularly prone to cracking as they are more susceptible to the effect of wind-lift.

Water penetration (or leakage) Well-laid slates lie close to each other. This encourages water seepage due to capillary action.

On the other hand, wide joints between the slates will often allow water penetration because the side lap is too small for the length. This problem is reduced as the pitch of the roof increases and water is shed more quickly. On flatter slopes, water tends to spread at a wider angle between the slates due to capillary action and may reach the nail holes of the course beneath (causing nail sickness) or even penetrating to the battens, felt, etc.

Verge failure This is often associated with incorrect selection of slates. Slate-and-a-half slates should be used rather than half slates, as the latter are extremely difficult to secure properly against wind uplift, which is often greatest at the verge of a roof.

Further problems can be posed where there is no oversail provided, so that water cannot drip away from the face of the underlying wall, or the undercloak is poorly formed, allowing moisture into the structure.

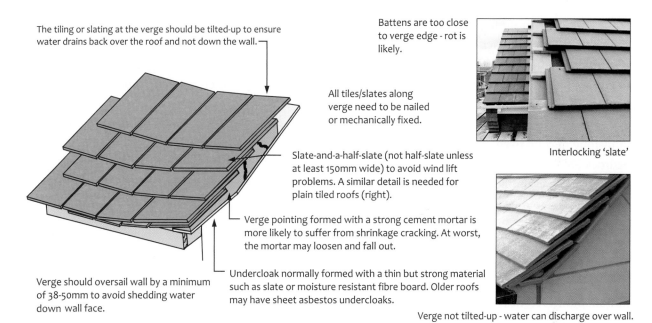

The tiling or slating at the verge should be tilted-up to ensure water drains back over the roof and not down the wall.

Battens are too close to verge edge - rot is likely.

All tiles/slates along verge need to be nailed or mechanically fixed.

Interlocking 'slate'

Slate-and-a-half-slate (not half-slate unless at least 150mm wide) to avoid wind lift problems. A similar detail is needed for plain tiled roofs (right).

Verge pointing formed with a strong cement mortar is more likely to suffer from shrinkage cracking. At worst, the mortar may loosen and fall out.

Undercloak normally formed with a thin but strong material such as slate or moisture resistant fibre board. Older roofs may have sheet asbestos undercloaks.

Verge should oversail wall by a minimum of 38-50mm to avoid shedding water down wall face.

Verge not tilted-up - water can discharge over wall.

Ridges and hips Forming a ridge or a hip in slate is extremely difficult. Ridges are rarely so formed but hips are occasionally constructed with mitred (cut) slate-and-a-half slates fixed over lead soakers. Securely fixing such a detail is difficult and failure is common, especially if the pitch is too low (less than 40 degrees). Ridges and hips are better formed with clay components or leadwork.

Left: Poor detailing at parapets will inevitably lead to rain penetration. Right: This slate roof is nearing the end of its life. There are also half slates used at the verge and damage to the verge pointing. Delamination can be seen on many of the slates.

Bitumastic oversealing A slate roof that is approaching the end of its life is sometimes given an extended life by the application of a proprietary over-sealing coat. The process normally involves re-fixing/replacing any loose/missing slates, applying a coat of hot bitumen to the whole of the roof followed by a layer of reinforcing fabric and finally a further top coat of more hot bitumen. This process will give some further years of life to an old slate roof. It is, however, limited in durability/longevity and will result in the eventual replacement of the complete roof covering including any good slates, as it will be impossible to distinguish these from those of poor quality.

This slate roof has been oversealed. This type of repair is not uncommon but it has a limited life and impedes ventilation

Stone slates

Roofs covered with stone slates are confined to those areas of the country where there is an easily accessible supply of either limestone (as in the Cotswolds) or sandstone (as in parts of Yorkshire).

They are sometimes referred to as tilestones. It is not unusual for such roofs, when properly installed, to last for 100 years without major overhaul. Even then, the slates are often perfectly capable of being re-used for another 100 years.

Limestone slates are normally riven and laid at a pitch of around 50–55 degrees, while those of sandstone may be riven or sawn, and are laid at pitches of between 24 and 50 degrees. Both types of bedrock are porous, and this means that the slates need to be of considerable thickness (12–25mm) in order to allow evaporation of absorbed moisture and avoid water penetration into the roof void. This results in an extremely heavy roof covering.

Below are some particular problems relating to stone slates.

Defective mortar fillets Traditionally, the junction between stone slates and the adjoining walls and chimneys was covered with an angled mortar fillet rather than a flashing. This would have been of lime mortar, but in recent years cement mortar is often used instead. Lime mortar is softer and will slowly weather, leading, eventually, to disintegration. Cement mortar has a tendency to shrink and crack, often within a very short space of time, and this can lead to penetrating damp problems.

Slipped slates Stone slates were originally hung on oak or deal pegs inserted through holes near the head of each slate. Sometimes brass screws were used and more recently galvanised or stainless steel nails. Once the fixings to a slate fail, slippage will occur.

Delamination The thin layers of stone making up the slate are often prone to delamination in much the same way as natural slates.

Lichen (top left) on this lean-to roof will impede ventilation. The other photos (centre and right) show mortar fillets at the junction with a vertical abutment. Both have been protected – the one above by a splayed weathering and the one on the right by the addition of slate drips. These traditional techniques are very effective.

Man-made slates

These are a cheaper alternative to natural slates and have been manufactured since the Edwardian period, when asbestos cement slates were first introduced. These consist of asbestos fibre woven into a fabric encased in Portland cement mortar. The slates are very light and extremely brittle. There are two types – 'tiles' which are square and laid diagonally; 'slates' which are rectangular and laid as ordinary slates. For much of the 20th century these were the only artificial slates manufactured but since the late 1970s asbestos-free mineral-fibre slates have replaced them. This was, in part, as a response to the health-related problems associated with asbestos-based products and the advice to discontinue their use.

Artificial slates are often nail fixed or riveted, although many manufacturers of mineral-fibre reinforced slates recommend the use of stainless steel hooks as these slates are thick and lightweight in form and tend to be more prone to wind-lift problems. Such slates come in a range of colours and shades and many of the earlier slates do have an artificial appearance. Many of the more recent types have a more slate-like appearance. In terms of anticipated life, manufacturer's guarantees are generally for 15–30 years but 50–60 years can normally be expected.

The man-made 'tiles' in the photograph above may still occasionally be found on the roofs of houses, more especially bungalows, of the early/mid 20th century. Notice the moss and lichen. Below are artificial 'slates' from a more modern house. They are showing signs of bowing and curling.

Below are some particular problems of man-made slates.

Dimensional stability Man-made slates often have a tendency to bow or curl after installation. A slight defect of this nature may merely cause aesthetic concern over the appearance of the roof, but bowing or curling of greater magnitude can lead to problems of water leakage or loosening of the fixing nails/hooks.

Discolouration The slates are artificially coloured and in some cases they may become discoloured. This may affect the edges of the slate and/or the remaining exposed surface. The cause can be the effect of sunlight, which leads to fading, or it may be for more complex reasons, such as water penetration of the material over long periods. The problem may eventually lead to performance difficulties and necessitate replacement.

Cracking or splitting The thinness of man-made slates, together with their relative lack of strength, means that they are prone to cracking. This can sometimes occur along the line of the fixing holes and can be induced by clumsy installation. This type of slate is also prone to impact damage as it ages because of the material hardening.

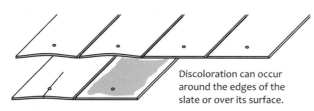

Bowing or curling of the slate sometimes occurs after installation

Discoloration can occur around the edges of the slate or over its surface.

Splits can occur along lines of weakness - often induced by fixing holes.

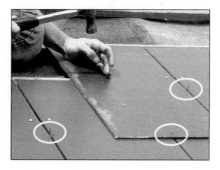

Most new slates are centre nailed; riveted tails prevent wind up-lift.

Verge problems The verge detail on a roof formed of man-made slates can be constructed in a similar fashion to that described earlier for natural slates. The problems experienced are generally similar.

The 1980s saw the introduction of dry verges. These are dry fixed components formed in man-made slate which wrap over the edge of the verge to form a water-tight detail. They avoid the problems associated with traditionally formed verges, eg shrinkage of the mortar bedding, inadequate over-sail, but can experience difficulties if the mechanical fixing, normally a nail or rivet, fails.

Future performance unknown More recently, slates formed with a mixture of crushed slate, resin and reinforcing material, which more exactly replicate the appearance of a real slate, have been produced. One major manufacturer indicates a probable life expectancy of 60 years for its slate of this type. The manufacturer's advice for a new product is given in all good faith but is normally based on laboratory and other controlled tests plus experience accumulated from the use of other similar products. The conditions of actual use are outside of its control (and liability). This means that the future performance of any new product is, to some degree, unpredictable.

The potential health hazards of asbestos slates Asbestos-cement products present no health problems as long as they are stable and not 'shedding' fibres. The microscopic fibres can become a hazard to health if ingested as they can lead to a form of lung cancer. Broken or cracked asbestos slates/tiles can present a potential health hazard but a more serious health problem is likely when the surface of the tile becomes roughened and allows fibres to be released into the air.

A roof containing asbestos slates should be monitored very carefully to assess whether it does present a health hazard. There is probably no danger to health where the components are stable and the roof may be retained, even if some of the slates are already cracked or broken. However, total removal and replacement of the complete roof area, rather than partial repair/replacement, is recommended if there is evidence of active deterioration occurring.

Note: The removal of any asbestos product is now required to be carried out under a very strictly regulated process (under the Control of Asbestos Regulations 2006) by a licensed contractor.

Clay tiles

Clay tiles have been used extensively in England and Wales for several hundred years. Originally introduced by the Romans, they fell out of use until the 13th century. Since then they have never been out of production. In recent years, increasing numbers of foreign-made clay tiles are being used.

They have been manufactured in a great number of shapes and sizes ranging from the small and simple (eg plain tiles) to the large and more complex (eg double roman). Tiles such as the former merely butt up to the sides of each other but are heavily overlapped vertically, in a similar manner to slates, to form a weather resistant covering. Other types of tile, like the pantile and double roman, are more intricately shaped to overlap each other at each side as well as the head. The side lap increases weather resistance and this was further enhanced by the introduction of interlocking side details in the second quarter of the 20th century. Shaped tiles for ridges and hips are produced for use on any roof, while shaped valley tiles are also manufactured for use with specific tile profiles (eg plain tiles).

The tiles are either hand-moulded or machine pressed from better quality brick clays. A good quality clay tile should be capable of lasting a hundred years or more, although a modern manufacturer's guarantee may be as short as 30 years. In the second half of the 19th century and the first half of the 20th a great number of low and medium cost houses were roofed over with double roman or other large profile clay tiles, while many better quality houses were finished with more expensive plain tiles. In the 1940s and 1950s clay tiles were, to a great extent, superseded by concrete tiles, and, although the former have undergone a renaissance since the early 1980s, they probably only account for about 20% of all current new roof installations.

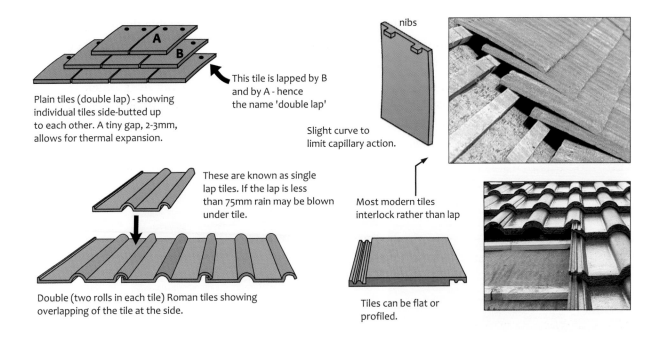

Plain tiles (double lap) - showing individual tiles side-butted up to each other. A tiny gap, 2-3mm, allows for thermal expansion.

This tile is lapped by B and by A - hence the name 'double lap'

nibs

Slight curve to limit capillary action.

These are known as single lap tiles. If the lap is less than 75mm rain may be blown under tile.

Most modern tiles interlock rather than lap

Double (two rolls in each tile) Roman tiles showing overlapping of the tile at the side.

Tiles can be flat or profiled.

Some of the particular problems with clay tiles are set out here

Delamination Clay tiles are not watertight. They absorb moisture but work on the principle that it evaporates back out through the upper surface before the moisture can penetrate completely through the tile. This means that the tiles are susceptible to the effect of frost, especially if they are of poorer quality. This increases the possibility of a laminated structure. Frozen moisture trapped in air pockets will expand and fracture the tile or its surfaces, particularly the uppermost. Delamination of the top surface is a common problem with older tiles and careful attention should be paid to these if they are being considered for re-use.

Pitted, cracked or broken tiles Defects such as pitting, cracks or breakage may occur for a number of reasons. Pitting of the top surface may be due to delamination *(as described in the previous section)* or to weak spots in the clay resulting in variable weathering over a prolonged period. Cracked or broken tiles may be as a result of weakness in the tile, often due to its age, or external activity, eg hit by a heavy object or walked over.

In the left-hand photograph of a double roman tiled roof, one of the tiles is showing signs of premature deterioration and the ridge pointing is disintegrating. The right hand photo shows damaged single roman tiles. Most of the defects are caused by frost action but the broken tile could also be as the result of clumsy maintenance work.

Cambered/non-cambered tiles Machine-pressed tiles are generally 'flat', ie non-cambered, across their width. This means that they can lie close to each other and this increases the likelihood of capillary action. Other tiles, including certain plain tiles, have a slight cross-camber, ie they have a concave profile across their width, with the result that driving rain or snow is more liable to penetrate a roof laid with such tiles.

Slipped tiles Most tiles are formed with nibs along their upper edge which enable them to be hung onto the battens. The tiles also need to be nailed to the battens when the roof is in an exposed position or has a steep pitch. Failure to carry out the additional nailing may result in tiles being lifted and moved by the effect of the wind.

A slipped tile may also be caused by the failure of the nibs or, with some types of plain tile which have no integral nibs, by failure of the fixing nail *(see later section on 'Other influential factors on the roof covering' which discusses 'nail sickness')*.

Typical problems of plain tiled roofs – slipped, cracked or broken tiles plus evidence of torching and bedding. Note the cross-camber of the industrial tiles (right) and the verge poorly formed with half tiles (left).

Torching Torching of tiles, in a similar fashion to that described for natural slating *(see earlier section)*, is sometimes discovered. The reasons for, and problems of, torching are also similar in both cases.

A plain tiled roof in a 1930s house. Torching has blocked the ventilation of the roof space and can also lead to rot in the battens. Note the insulation at ceiling level. Interstitial condensation due to lack of ventilation is a real danger in these conditions.

Efflorescence *See the next section on Concrete Tiles.*

Verge failure In traditional verges formed with lime or cement mortar infilling, the construction detailing is similar in principle to that of a slate roof except that tile-and-a-half tiles will be used and a tiled undercloak is common. The causes of failure may be due to problems similar to those referred to earlier in the section on natural slates.

In recent years, dry verge systems have been designed and manufactured for use on clay tile roofs and the comments made in the earlier section dealing with man-made slates apply equally here.

Ridges and hips Ridges and hips are commonly formed with clay components, although they may be constructed with other materials. Clay ridges are generally of a half round or inverted 'v' shape and many older ridge tiles also had fancy detailing along the actual ridgeline. Hips are often covered with ridge tiles formed in both the previous shapes as well as a variety of purpose-made shapes, eg bonnet hip tiles, which are more like an inverted dish.

Traditionally, ridge tiles are installed by bedding each tile in black or lime mortar, although in recent years cement mortar has often been used. The former types are inherently weaker and more flexible than the latter, resulting in greater tolerance of differential movement. However, they will lose their ability to bond to the ridge tile and the underlying structure as they age, leading to loose or missing sections of ridge. On the other hand, cement-based mortar is more prone to shrinkage cracking and this may allow moisture penetration of the ridge line. It also bonds to the ridges and adjoining tiles. As a result they cannot have any attached mortar cleaned off.

Clay hip tiles bedded traditionally in mortar suffer from problems similar to those affecting ridge tiles. It is also normal practice to install a galvanised metal hip iron or hook, screwed or nailed to the bottom of the hip board. This provides extra security in case the mortar bedding fails. It may work loose or be missing or might never have been originally installed.

Dry ridge and hip systems are also now manufactured in clay, for use with any type of roof where it is possible to install them. Correct installation in accordance with the manufacturer's recommendations is essential. This may be difficult on some hips, where there can be difficulties in marrying up the hip components and the tile profile, due to the angle or pitch of the actual hip not meeting the manufacturer's installation criteria.

Dry systems usually rely upon the integrity of their neoprene components to be watertight. Penetration may occur where these are damaged during installation or are missing, especially the sealing washers to nails/screws. There have also been questions raised about the life expectancy of the neoprene.

Valleys   Many tile-covered roofs have valleys formed by installing a metal lining of lead or zinc, which is laid over a timber or plywood board. This involves cutting the adjoining tiles and bedding them in mortar above each side of the valley lining. Often, these details are poorly formed, with clumsy cutting and crude pointing. The metal should be laid over a small tilting fillet at its junction with the tiles, and also have a welted edge. These are often missing and, if the mortar to the tiling fails, there is an increased likelihood of water penetration of the joint. (See Chapter 8, Flat Roofs for defects in leadwork and zinc).

Purpose-made clay valley tiles are available for certain tile profiles (eg plain tiles). These are specially moulded for a range of pitches and plan shapes. Such a valley presents few problems, if correctly installed. Each tile should be nailed and the roof felt properly dressed around the valley. Problems may arise if the battens are cut short and do not fully support the tiling.

Concrete tiles and slates

These were first installed on roofs before the Second World War and have subsequently superseded clay tiles to such an extent that, for the majority of the period since 1945, they have captured approximately 80% of the market for tiles. They are machine pressed, manufactured in a wide range of traditional and modern profiles and colours, and can be sand finished, smooth finished, or through coloured. The majority are formed with interlocking edges, although some, such as plain tiles, are not.
Concrete tiles are normally much heavier than comparable clay tiles, although there are specially manufactured lightweight versions of some profiles. Life expectancy for earlier concrete tiles is probably 50–75 years but for more recent tiles it may be longer.

Below are some of the problems associated with concrete tiles.

Overloading of the roof   The extra weight of a concrete tile means that due consideration must be given to the dead load imposed on the roof structure. This is a common problem on many older houses in which the original roof covering had reached the end of its life and was then replaced by concrete tiles for reasons of cost or ignorance.

Overloading of a roof inevitably results in bowing and buckling of the roof structure giving it a distorted appearance. In time, individual timbers can crack, split or break under the stresses created. Total roof failure is unlikely, although it may occur in extreme circumstances. The tiles on an overloaded roof will be adversely affected when the movement induced in the structure results in cocking and lifting of individual tiles and failure of the interlocking detail, all because of the unevenness caused in the roof. This can lead to leakage of the roof covering.

Loss of surface/colour Originally, the finish of concrete tiles, whether smooth or of sand/granules, was only surface deep. Once it is weathered away or it becomes chipped/damaged, the concrete beneath is revealed. This is unsightly but does not impair the performance of the tile. However, replacement tiles will not match the faded appearance of the original roof covering. For the last 25 years or so, completely through-coloured concrete tiles have been manufactured as well and these avoid this problem.

Efflorescence Earlier concrete tiles sometimes contained soluble salts, due to quality control problems during manufacture. These led to efflorescence on the underside of the tile, especially around the nibs that are gradually eroded. The presence of a white powdery deposit on the surface of any ceiling and felt beneath the tiles is an indicator of this problem. Note: this problem may also occur with under-burned clay tiles.

'Delaminated' tiles Incorrect manufacture can sometimes result in a tile that is not resistant to the effects of moisture and frost. This results in the gradual disintegration of the tile as the concrete softens and crumbles due to the action of the weather. The problem will normally affect the manufacture of a whole batch of tiles so, once detected, the complete roof should be examined.

Cracked and broken tiles These may be due to an inherent weakness in the tile due to poor manufacture or to a defect caused by poor installation. They may also result from external factors such as impact damage from foot traffic or balls.

Slipped tile, verges, ridges and hips, valleys *See earlier section on Clay tiles.*

Shingles

The use of wood shingles in the British Isles dates back some two thousand years to the Romans. Originally of oak, they fell into disuse as this particular home grown timber became scarce.

Re-introduction in the 1920s owed much to their continuing use in North America and the Scandinavian countries, although now they are more often formed from imported Western Red Cedar rather than oak.

Shingles are rectangular in shape (about 400mm long x 100–325mm wide) and vary in approximate thickness from 10mm for cedar to 20mm for oak. Both types are best laid on pitches exceeding 45 degrees, although cedar shingles can perform successfully on pitches down to 30 degrees or less. The appearance changes after a few months on exposure to the weather, both oak and cedar shingles developing a silver-grey colour.

A modern shingle roof with vertically-hung shingles on the elevation.

Durability of shingles is much greater than commonly supposed. Split (or cleft) shingles are more durable than those that have been sawn, as splitting ensures continuous fibres throughout the length of the shingle. Durability is also increased by installing the shingles with a 5–12mm side gap rather than the normal 3mm. This encourages efficient water run-off, quicker drying and allows for expansion. It also reduces the likelihood of buckling.

Perfectly sound examples of both oak and cedar roofs of 40–50 years of age can be found, although other materials, such as larch, may only last a maximum of 20 years. However, in this country, cedar shingles need to be treated with preservative if they are to be ensured a 50 year life, otherwise they have been known to fail within 10 years of installation. This practice has generally been followed since the early 1960s.

Problems experienced on shingle roofs include:

Splitting Older shingles were not pre-holed for nailing, the nails being driven directly into the wood about 150–200mm up from the butt (or bottom). This meant that the shingle could sometimes split when being installed and, if this was a hairline crack, it may not become apparent until later. Modern shingles are often pre-drilled to avoid this problem.

Loose or missing shingles These problems may be caused by the effect of wind lift. Shingles are extremely light in weight and require very careful nailing – it is normal for them to be twice nailed. A more secure and effective fixing is provided by ring-shanked nails, which have larger heads. *(See the next sub-section on Nail problems and to the later section on Nail sickness).*

Nail problems Corrosive acids in both oak and cedar will attack any zinc galvanised steel nail, resulting in 'nail sickness' and loss of the shingle. This may be avoided by the use of non-ferrous nails, although there may be some staining from those containing copper or brass. (Zinc gutters, flashings and ridge cappings should also not be used for similar reasons.)

Excessive dampness Earlier reference has been made to the need to treat shingles with preservative in order to prolong durability. Great Britain, particularly the western part, is renowned for its wet climate and any timber roofing material is particularly susceptible. The shingles are secured to battens that traditionally were fixed above an unfelted roof. The modern approach is to install a felt underlay that can restrict air circulation under the shingles and prolong drying out. This may lead to more rapid deterioration of the shingles. In some cases, air circulation and drying out are improved by the installation of counter-battens.

Surface deterioration The exposed surfaces of the shingle will weather and gradually erode over a period of time.

This shingle roof is showing signs of its age. Split, twisted and up-lifted shingles are evidence that replacement may be required.

Lichen Lichen may be found on the exposed upper surfaces of the shingles. It grows on the moist timber and although it does not directly attack the timber, it is unsightly and can adversely affect the efficiency of water run-off, leading to accelerated deterioration of the shingles.

Effect upon metals The water run-off from shingles is acidic in content and this may have a corrosive effect upon any metal gutters and flashings affected. Other materials should be selected for these components, where they come into contact with such run-off.

Thatch

In the British Isles, there is evidence of the use of thatched roofing for some 2500 years. For a period, during the Roman occupation, clay tiles were also used but thatch remained the roofing material until after the Norman Conquest. Large-scale fires in London and other cities in the 12th and 13th centuries, which resulted in thatched roofs being banned in many cities and towns, were followed by a gradual loss of popularity of thatch generally in urban areas over the next 300 years. Many thatched roofs in rural areas were replaced with other materials in the 19th and early 20th centuries but remained popular in some parts of the country, eg Devon and East Anglia. Thatched roofing has recently undergone a revival in rural areas.

The main thatching materials are:
* water reed – which is the most durable (giving a useful life of 50–80 years) or
* wheat straw – which is prepared as either:
    – long straw (15–25 years life)
    – combed wheat reed (30–40 years life).

Useful life varies because of regional/localised climatic and construction differences as well as any variable quality of both the materials and the installation. Natural deterioration is inevitable, however well constructed the thatch roof, due to the presence of moisture retained within the thatch. This can be delayed by increasing the pitch (it should never be less than 45 degrees) and encouraging more efficient water run-off. Roofs in drier or more exposed areas of the country will also tend to last longer. Life expectancy figures drop noticeably the further west the thatched property is situated and/or the lower the pitch.

The thickness and quality of workmanship in laying the thatch also plays an important part in the performance of such a roof. A properly constructed roof will eventually need replacement/repair because of natural deterioration, but it will frequently be found that the straw base coat of the thatch is still in perfect condition.

The steep pitch can be clearly seen in the left-hand roof; the right-hand roof unusually incorporates a lead gutter.

Some of the problems experienced with thatched roofs are listed below.

Atmospheric conditions The life expectancy of a thatched roof is determined by the rate at which the material breaks down. This is due to micro-biological activity which tends to be more rapid where the climate is wetter and more humid. The roof of a building in a sheltered position, or under trees, or a roof with a shallow pitch will also suffer from accelerated deterioration. This is because the lack of air movement (wind), shade and/or inability to shed moisture before it penetrates the thatch will increase the rate of decay.

Fungal attack Degradation of a thatched roof commences as soon as it has been installed due to the effect of fungi that are produced within the materials used. Increased fungal growth can be expected as the roof ages and also the further west the building is situated, as warmer, wetter and more humid climatic conditions encourage activity.

Gullies These are valleys running down the slope of the roof and are caused by water run-off wearing the surface of the thatch unevenly. The extent of the wear can often be determined by examining the depth of the valley at the eaves. If dealt with early enough, they can be patch repaired to enable the roof to reach its anticipated life-span.

The left-hand photograph shows gullies in the thatched roof formed by water run-off. In the right-hand roof, vegetable growth can be seen at the junction with the chimney. Note the lack of any rainwater gutter on both properties. This is quite normal. The thatch should oversail the external elevation by a considerable distance to avoid damp penetration.

Moss, lichen and algae Moss, lichen and algae can grow on the surface of the roof, especially if it is situated in a damp and shaded position, overhung by trees. Re-growth of previously treated outbreaks can often occur in such a situation. It is debatable whether the actual growths have an adverse effect upon the thatch. However, moss and algae indicate the presence of relatively high levels of moisture (which will accelerate decay of the thatch) and all three will prevent the free run-off of water as well as affect the appearance of the roof.

Weathering of the ridge Whatever the thatching material used, the ridge tends to deteriorate at a fairly rapid rate. Renewal within 10–15 years of installation is to be expected.

Rodent and bird damage Rats, mice and squirrels can all have an adverse effect upon thatched roofs. Birds, particularly in spring when they are seeking nesting materials, can cause considerable damage. Generally, the result does not immediately encourage water penetration but can lead to a weakening of the roof covering and increase the rate of deterioration. The presence of protective wire netting can initially help deter pest invasion, but, if they get beneath it, the result may be even greater damage as they attempt to escape.

The cause of the damage can normally be identified by examining the straw or reed on the ground below the affected area. Short, chewed lengths are left by rodents, while birds tend to pull out whole lengths.

Fire This is an obvious risk with a thatched roof but there is no evidence of it occurring any more frequently than with other roof coverings. Great care needs to be paid to the condition of internal electrical wiring in the roof void and any accumulation of stored materials or roof debris in the void should be avoided. Traditionally, lime-wash or lime and clay slurry was applied as a fire retardant but modern thatch is chemically treated. This can affect durability and may be washed out in a relatively short time (10–15 years).

Chimney stacks and central heating flues need careful construction and/or installation to ensure that heat within the flue does not affect the thatch. Regular maintenance of the flue is also essential, eg sweeping. Dirty flues causing sparks have led to thatched roof fires (as have garden bonfires). The increased use of wood burning stoves in the last few years has increased the risk because there are greater deposits of tar left on the inside surfaces of the flue. Spark arrestors are sometimes fitted to the top of a flue or chimney but they need regular cleaning to avoid clogging, which also affects performance.

Other influential factors on the roof covering

The overall performance of any roof covering will also depend upon factors other than the material itself. These include the following.

Nail sickness Man-made and natural slates, tiles and shingles are normally secured by nails (although man-made slates can be secured with stainless steel hooks and tiles can be hung on their nibs alone). Ferrous nails inevitably rust in the presence of the moisture always present between the roof covering and the sarking felt. Slipped or missing slates/tiles are evidence of such a problem.

Batten failure Modern roofing battens should be of preservative treated softwood, although untreated softwood may still be erroneously used. Untreated softwood and, occasional, hardwood battens may be found in older roofs.

Moisture leakage through the roof covering may adversely affect the battens and their fixings in even the best-constructed and maintained roofs. Over a prolonged period of time, this may result in the battens being infected with wet rot. Any ferrous nails securing the battens to the rafters may suffer nail sickness.

The size of the batten is also important. Battens that are not wide enough will split when they are nailed. On the other hand, too thin a batten will tend to 'spring' as the nail is driven in. Since 2003, BS 5534 has increased the minimum batten size for single lap tiling on 600 rafter centres from 25 x 38 to 25 x 50. Battens may be of insufficient length to fully span the full width of a roof. Butt jointing all of the battens above one particular rafter must also be avoided. It will create a structural weakness because they will not provide effective lateral bracing.

Sarking felt (underfelt) It became the practice in the 20th century to provide a pitched roof with secondary weather protection by means of an underfelt. This prevents both rain and snow penetration (especially when wind-blown) of the roof void. The felt is secured to the top of the rafters by the roofing battens. The provision of an underfelt was a gradual process and did not become a standard practice until the 1940s/1950s.

Care needs to be taken during the installation of the underfelt to ensure that it is laid with a slight sag between the rafters. This allows any rainwater/snow penetration of the tiles or slates to drain away and not collect against the battens. On the other hand, the felt needs to be properly supported at the eaves to avoid sagging, as this would allow any run-off moisture to pond. The trapped moisture can lead to wet rot in the battens, reduce the life of the felt and, if the felt fails, penetrate the underlying structure.

The felt should be laid with a slight sag between the rafters. This allows any leakage of the slates or tiles to run away to the eaves.

This felt is tightly drawn; any moisture trapped above the batten encourages wet rot in batten, and maybe in the rafter.

Most felts before the 1990s were bitumen based; they can become very brittle with age and are easily torn.

Absence of a tilting fillet at the eaves leads to a sag in the felt. This may result in ponding of any moisture and water leaking into the building as the fabric deteriorates. The eaves edge piece (right) acts as a tilting fillet; it also protects the outer edge of the underlay from ultra-voilet light and birds.

Prior to the 20th century only better quality buildings had any secondary protection and this was by means of close boarding laid over the rafters. The boarding also increased thermal insulation but was prone to moisture attack and did not fully prevent moisture penetration. To assist moisture run-off/ evaporation, counter-battens were often fixed between the boarding and the roofing battens. Alternatively, on some roofs, slates (but not tiles) were fixed directly to the boards. In cheaper housing and in the less important and ancillary roof areas of better housing, it was normal practice to make no provision of any form of secondary protection.

A number of different materials have been in use at various times. They include the following.

Hairfelt This was manufactured out of animal hair and was in use in the early part of the 20th century. It provides no real weather protection and becomes extremely brittle with age. Very rarely found now.

Unreinforced bitumastic felt No longer in common use, this tears easily and becomes brittle with age. Occasionally found now.

Reinforced bitumastic felt In common use in the second half of the 20th century, this is considerably stronger than the unreinforced variety, although it will still tear if mishandled or hit. The bitumen content becomes brittle and increasingly susceptible to damage with age. Still occasionally found in current use.

A common problem of the bitumastic type of felt is its tendency to become brittle and disintegrate at eaves and other vulnerable positions, such as valley gutters, after a relatively short period. A short piece of plastic DPC or vapour permeable plastic felt installed at the bottom of each slope will avoid this problem.

Vapour permeable plastic and synthetic membranes These appeared on the market in the 1980s and allow low levels of trapped vapour (which would otherwise condense within a roof void) to escape through micro-pores contained within the material. The pores are too small to allow water droplets (eg from rain/snow) through from the outside. The plastic materials are tear proof and rot proof so, if laid in accordance with the manufacturer's instructions, should perform properly. These membranes are now the norm in domestic pitched roof construction. They received BBA Certification for warm roof construction in 1982 and Certification for cold unventilated pitched roof construction in 1999.

Some designers and contractors erroneously believe that this type of 'breathable' felt is a full substitute for proper roof ventilation as required by the Building Regulations (eg eaves and ridge ventilation). However, this can lead to condensation problems. There can also be a problem of premature aging of these membranes which can be caused by over-exposure to UV light during installation (ie due to not being covered quickly enough) or by temperature peaks throughout a roof's lifetime. These factors may lead to brittleness or material breakdown. Slope orientation needs to be considered carefully when such problems are experienced. In normal conditions, life expectancy should be comparable with that of a traditional felt underlay.

Valley and parapet gutters

The more complex the roof design, the greater the likelihood of valley gutters to drain abutting roof slopes. Often, the design will call for the roof to be 'hidden' behind a parapet wall, creating the need for a parapet gutter to drain any adjoining roof slope to a suitable discharge point.

Any gutter should be designed to drain the maximum anticipated water run-off from ALL surfaces within the roof catchment area. Account must be taken in the design for both normal rainfall patterns and short duration exceptional rainstorms, which in the past have happened rarely but are now occurring much more frequently. If the design is inadequate, or the construction not properly carried out, the gutter will not be capable of efficiently receiving and carrying away the water run-off. Unfortunately, recent climatic change means that many existing gutters that were appropriately designed at the time of installation for the worst anticipated circumstances likely will no longer fully cope with the increased intensity of current and future rainfall.

A variety of cover materials are used in the construction of valley and parapet gutters. The problems associated with each type of material are commented upon as follows:
- shaped clay/concrete tiles

  see the appropriate material in this chapter
- sheet materials, eg lead, zinc, mineral felt

  see the appropriate material in Chapter 8.

Wind related damage

The annual cost of wind-related damage to buildings in the UK runs into many tens of millions of pounds. A high proportion of this damage is to roofs. Wind may adversely affect both older and newer roofs; the former because both design and construction almost certainly made insufficient allowance for its effect. Increasing knowledge of the effects of the wind has led to better designed roofs (in particular, coverings and fixings) and to more carefully controlled construction. However, future climatic change may result in existing roofs suffering greater failure as wind speeds are tending to increase and weather patterns change.

The effect of the wind is influenced by a number of general factors.
- **Exposure** of the building to the prevailing wind.
- **Height** of the building above sea level.
- **Wind speed** – in general, this varies, depending upon the region, but tends to be greater in coastal areas than in inland districts, eg the south coast experiences winds that are up to 25% greater than those at a similar altitude in the South Midlands. It also increases the further north the building is situated within the UK; the greatest winds are experienced in North West Scotland.
- **Gusting** – air movement is subject to turbulence that causes rapid changes in its speed and direction. Such increases/decreases in wind speed are infinitely variable. However, the effect of wind gusts increases as the size of the element or component it crosses decreases. Thus, a small

roof is more affected than a larger roof and, similarly, a small component, such as a tile or slate, suffers more from gusting than a roof covered with sheeting material. The type and quality of fixings used (or not) need to be assessed when investigating a roof adversely affected by the wind, as they play an important role in how the covering performs.

- **Local topography and adjoining buildings** – valleys, hills, ridges, escarpments and ridges all have an influence on air movement and may directly influence wind behaviour. Similarly, the size, shape or height of an adjoining building or group of buildings may affect wind performance and patterns.

Wind speed increases as it passes over a roof. Wind also creates both positive and negative pressures (suction) as it moves over a building. Roof pitches of 30 degrees or lower are always subject to suction forces. However, although leeward slopes are always subject to suction, which increases as the roof pitch decreases, windward slopes are subjected to positive pressure once the pitch exceeds 35 degrees.

Normally, suction is a greater problem to a roof than positive pressure. The latter may result in great force being applied to the roof surface but, as long as the roof structure is performing adequately, wind usually only becomes a danger when it is able to get underneath the actual roof covering and lift it. More common is the damage caused by the effect of suction, which may lift, or completely remove, the tiles or slates on both the windward and leeward slopes of a pitched roof, particularly those situated along ridges, eaves and verges, where higher suction forces are experienced than on the more central areas of a roof slope. It is for this reason that it is normal practice to mechanically fix roof tiles on these 'exposed' areas of a roof. Similarly, all tiles on the entire roof in an extremely exposed location should be mechanically fixed. It is the failure to follow these practices (which are part of the manufacturer's recommendations with new tiles) that often leads to tile displacement. All slates should, in any case, be mechanically fixed, but even the best installed slate or tiled roof may suffer wind damage in extreme conditions.

Damage from aircraft wake vortices

This phenomenon affects the roofs of properties situated beneath, or close to, the flight path of a busy airport by dislodging the tiles or slates. All aircraft trail powerful air currents, or 'vortices', behind them and, in general, these are at their strongest when the aircraft is heavy and flying slowly – exactly the conditions when a modern passenger plane is taking off or landing. The likelihood of damage is greater when an aircraft is landing, as it is closer to the ground for a longer period than when taking off. Damage occurs immediately beneath the aircraft, and also to each side, as the vortices tend to drift downwards and then sideways, just above ground level.

The highest risk property is one situated within a 2km long area immediately before the runway. There is a medium risk area between 2 and 4km from the runway and a low risk area 4–6km away, where the likelihood of damage is slight but still a possibility. The footprint of the affected area is cone-shaped, widening at an angle of 10 degrees either side of the centreline the further away from the runway that an aircraft is flying.

Damage tends to be towards the centre of any affected roof slope rather than at, or near to, ridges or hips, and has been found to be in one of two distinct patterns:

- one, or more, small, localised area of loose or dislodged tiles
- longer, often diagonal, band(s) of dislodged tiles across the slope of the roof, commencing just below the ridge.

It is not uncommon for at least some of the dislodged tiles to remain on the roof surface. This residual effect is different from that of wind damage where there is a tendency for any loosened tiles to become blown off the roof surface.

CONCLUSION

Good standards of design and construction allied to regular inspection and maintenance are essential if a roof is to perform properly. Unfortunately, many building owners and occupiers (and some property managers) fail to meet even the most perfunctory standards of post occupation inspection and maintenance. All roofs should be given a full external and internal annual inspection as a matter of routine and, where problems are observed, immediate remedy undertaken. Such a policy will protect both the roof and the building beneath.

Lack of maintenance together with general long-term neglect has allowed moss to spread over the whole roof. This has impeded drainage and resulted in extensive frost attack of the tiles.

Finally, it would be impossible to predict how any building would perform in the conditions experienced below. In fact, the main structure of the house has survived surprisingly intact. Unfortunately, there is very little left of the roof.

19th December 1946, a Dakota plane crashes onto the roof of a house in Angus Drive, near Northholt Aerodrome in Middlesex.

PA Photos

CHAPTER 8

Flat Roofs

BACKGROUND

INTRODUCTION

A flat roof is a roof with a pitch of less than 10 degrees. It has been a common alternative to the pitched roof for the last 60 years or so. The decision to use a flat roof is often based on its cheaper initial capital cost. This is as a result of its easier construction and use of less material than a pitched roof of comparable plan size. However, a flat roof may not be the cheaper option when long-term costs are taken into account. Some of the covering materials used on flat roofs are not as durable as those used on pitched roofs and the extremely low pitch means that a flat roof is less efficient at draining water from its surface. Flat roofs also tend to be much more prone to leakage and water-related problems than pitched roofs, although recent advances in material technology have addressed some of these concerns.

A number of materials are used to form the structural element of a flat roof but, in the case of dwellings, although concrete is not uncommon the majority are probably timber. The waterproof covering can be any one of a number of materials – asphalt, single-ply sheet, multi-layer mineral felt, metals such as lead, zinc, copper or aluminium, or even glass reinforced plastic.

The failure of a flat roof, as with a pitched roof, may have a serious effect on other building elements and therefore any assessment of a flat roof should consider both its current condition and its future performance. Any inspection of a roof should include the adjoining elements, as it may be in these that the first signs of failure within the roof become apparent. If those first signs (in roof or adjoining element) are ignored or misread, the longer term effect may be an increased likelihood of serious failure. Inaction may also lead to 'knock on' effects, for example fungal attack leading to structural failure.

Function of the roof

This is discussed in greater detail in Chapter 7. The primary function is to provide weather protection in a form that is:

- strong
- stable
- durable
- impervious
- thermally efficient.

Flat roofs – a brief history

Both economic and skill factors have played an important part in the increased use of flat roof construction in the last 60 years or so, although, in some instances, such roofs were installed on purely aesthetic grounds. In the UK, dwelling houses constructed in the several hundred years up to the 1930s

generally had pitched roofs. There were exceptions; a considerable number of grand houses of the Georgian and Regency periods were constructed with large areas of lead flat roofing and many artisan houses built in the 18th and 19th centuries had ancillary parts formed with flat roofs over such elements as bay windows and small extensions.

In the 1920s and 1930s the influence of the 'Modern Movement' resulted in an increase of the popularity of flat roofs. This influence continued to some extent into, at least, the 1950s and 1960s.

This small extension to a Victorian houses has a lead flat roof.

(Top left) The 'purist' ideals of modernism influenced the design of these houses from the 1930s that were built with flat roofs. (Top right) An in situ concrete house from the early 1950s and (left) a 'prefab' from the period just after the Second World War.

(Left) Many Victorian houses had a ground floor extension added in the 1970s and 1980s. The roof would often be a flat one with a bitumen felt waterproof layer. Note the World War II air raid shelter in the foreground. (Far left) The original small pitched roofs over bay windows are quite often replaced with flat roofs as shown in this example from the Victorian period.

DEFECTS IN FLAT ROOFS

POOR DESIGN

A flat roof is generally considered to be an unsophisticated and simple approach to providing a building with a roof. However, flat roofs are more prone to failure than pitched roofs. This is frequently a result of poor design, particularly as simplicity of form does not mean that efficient performance is simple to achieve in technical terms – for example while a pitched roof sheds water quickly, a flat roof does not. Another aspect of poor design, and indeed specification, is the failure to properly understand how materials (especially new covering materials) perform in the extremes of climate experienced in the UK. A further important factor is a lack of thought given to the proper design of the mass of small detail that needs to be carefully worked out if the roof is to perform properly.

Examples of poor design of flat roofs include:
- inadequately sized structural members (timber, steel, reinforced concrete)
- insufficient structural members (too widely spaced)
- overloading of the structure (often when service plant is placed on top of a roof, or by replacing the roof covering with a heavier material, or by allowing general access – such as creating a roof garden – when the roof was designed for maintenance access only)
- selection of an inappropriate substrate (eg chipboard decking which disintegrates when wet)
- incorrect selection of the covering (eg zinc in contact with copper – on account of the significant risk of electrolytic corrosion of the zinc)
- poor specification (often because a product and its performance is not properly understood).

The sub-section on the poor design of pitched roofs included in Chapter 7 sets out some examples of pitched roof problems that may also be experienced by flat roofs.

THE FORM OF FLAT ROOFS

There are a number of different forms of flat roof and they are defined essentially by the position of the thermal insulation. They are referred to as either a cold roof or a warm roof – with warm roofs being divided into two categories: sandwich and upside down.

Both warm and cold roofs should be designed to avoid interstitial condensation. Interstitial condensation occurs when warm, moist internal air from the building passes through the roof structure (because its pressure is higher than the cold external air). On meeting any colder air, or colder structure within the roof, the warm air condenses. *(See Chapter 13 which discusses the principles of condensation in more detail.)*

Cold roof

This was the normal form of design until, perhaps, 40 years ago. The insulation is placed immediately above the ceiling and between the joists. The void above the insulation needs to be fully ventilated to prevent moisture condensing within the space. It is also essential to install a vapour check on the warm side of the insulation.

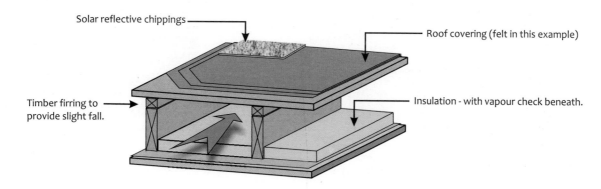

Solar reflective chippings

Roof covering (felt in this example)

Timber firring to provide slight fall.

Insulation - with vapour check beneath.

Because vapour checks are often breached by service fittings, a cold roof relies upon adequate ventilation to remove moisture vapour which may enter the roof space. In practice effective ventilation is often difficult to achieve particularly where the roof is sheltered and/or the provision of cross ventilation is not possible.

Unlike a warm roof the deck of a cold roof is exposed to the heat of the sun (warm roofs have thermal insulation above the deck). This may increase the likelihood of thermal movement.

Warm roof – sandwich type

In warm roof – sandwich construction, the insulation is placed above the decking and beneath the weatherproof roof covering. This results in the roof construction remaining at (more or less) room temperature and, as it is warm rather than cold, this reduces the likelihood of condensation occurring within the structure. Once again, a vapour check must be provided on the warm side of the insulation but the roof should not be ventilated as this will draw in cold air which will reduce the temperature of the roof construction and therefore increase the likelihood of condensation occurring.

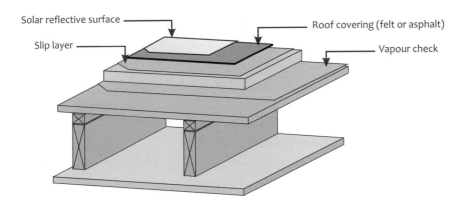

It is important to ensure that the thermal insulation layer does not cause any damage to the waterproof layer through thermal movement. To counter this possibility an isolating membrane (slip layer) can be placed between the insulation and the waterproof layer. An alternative approach is to rely on the selection of a relatively stable insulation material.

Another potential problem with this roof type is that the waterproof layer is exposed to a greater range of temperatures than other roof designs, as it cannot lose heat to the structure because the insulation is immediately below it – and it receives no heat from the building for the same reason.

Warm roof – inverted (upside down) type

With this type of flat roof the insulation is installed above the waterproof roof covering which is placed on the decking. No vapour check is required as the roof covering performs that function as well. The insulation is held in place by ballast or pavings positioned above it. The photo below shows ballast in the form of large pebbles.

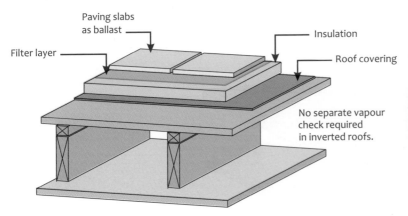

It is important that the insulation selected for this roof type is able to resist frost but its positioning also means that the structure and the waterproof covering are subjected to less thermal movement than with the other flat roof designs.

Condensation

It can sometimes be difficult to clearly differentiate between condensation and damp as the cause of moisture-related problems within a flat roof. In both cases, there may be high levels of moisture and decayed material, eg rotten timber, saturated plasterboard or concrete. The investigation should include an assessment of the potential causes of interstitial condensation as shown in the diagrams below.

Potential causes of interstitial condensation

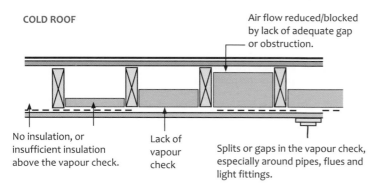

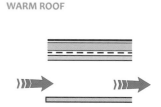

COLD ROOF — Air flow reduced/blocked by lack of adequate gap or obstruction.

WARM ROOF

No insulation, or insufficient insulation above the vapour check.

Lack of vapour check

Splits or gaps in the vapour check, especially around pipes, flues and light fittings.

Any ventilation/flow of air through the roof void will lower the temperature and can lead to interstitial condensation.

Consideration must also be given to evidence that the cause of the dampness may be condensation, such as:

- black mould – which is indicative of condensation
- the type of human activity in the area beneath the roof, eg bathing, cooking or clothes washing
- the difference in temperature between the inside and outside of the roof. The higher the internal temperature, the greater the potential for condensation within the roof element, especially if it is accompanied by high internal moisture levels and incorrect roof construction
- internal dampness without accompanying evidence of water penetration of the roof covering or adjoining structure (ie the cause is NOT penetrating damp).

POOR CONSTRUCTION

Many of the problems experienced by flat roofs would have been avoided if correct construction standards had been achieved. Even the best-designed roof may fail if the installation is poorly carried out. *(Reference is also made to poor construction in Chapter 7).*

Reasons for construction-related failure include:

- poor workmanship and/or supervision
- lack of co-operation/co-ordination between the parties involved (designer/contractor, contractor/specialist sub-contractor)
- inappropriate selection/combination of materials, including fixings (eg copper nails with zinc sheeting)
- unfamiliarity with products and processes (often when they are new on the market)
- following trades/works causing damage to the newly installed flat roof (eg painters, scaffolders)
- working in unsuitable weather conditions (eg wet weather leading to trapped moisture/blistering of the covering or a lack of adhesion to the substrate; frosty weather causing cracking of the material).

Entrapped moisture

This is a problem suffered by flat roofs that are wholly, or partially formed with insitu-cast concrete or topped with a sand cement screed. These construction processes require considerable amounts of water and the slab/screed should be allowed to become fully dry before being covered over. Alternatively, the concrete or screed may be drained through the deck, via drain tubes, or ventilated through the roof covering, by means of breather vents, so that the evaporation process can be properly completed.

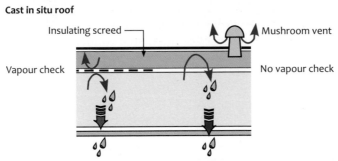

Cast in situ roof

Insulating screed

Mushroom vent

Vapour check

No vapour check

Migration of trapped moisture may take many years. Initial appearance is often around light fittings, vertical pipes or low points in the roof.

GENERAL PROBLEMS

There are a number of defects that are not strictly related to a particular form of flat roof construction, nor the actual covering. These general problems include the following.

Poor falls

Rainwater and melted snow should drain efficiently from all areas of a roof. A 'flat' roof should always be constructed with a slight fall (or slope) to ensure the proper flow of any water that collects on its surface – it should be self-draining. The fall is created by installing an upper substrate with a slope – this may be achieved by sloping the entire substrate (eg decking or insulation laid on sloping timber fillets) or by sloping the top surface of the substrate (eg shaped insulation or screed). The minimum fall for any flat roof should be 1:40 (ie a fall of 25mm in every 1.0 metre). The fall can be greater than this, and indeed it is normal practice to consider a roof to be 'flat' at any slope of less than 1:10 (ie a fall of 100mm in every 1.0 metre). Some guidance advises a fall of 1:60 or even 1:80. The danger of such a shallow pitched roof is that it will tend to drain more slowly. Also the real fall may be even less than intended if construction is not carried out extremely carefully. Such a shallow fall is also more likely to experience ponding problems as the slightest unevenness will have an adverse effect upon the flow of water across its surface (see opposite).

The fall of the roof should be towards the drainage outfall. This may be a valley/parapet gutter or an eaves gutter draining the section of roof concerned. Valley, parapet and other gutters should also be laid to an appropriate fall for the material being used. Alternatively, the roof may be drained towards a single opening (or scupper) in a parapet wall, or a sump or gully positioned above an internal rainwater pipe. The slope(s) should be shaped to direct water to the outlet.

Ponding

Ponding is an extremely common occurrence in flat roofs. Any unevenness in the covering, or the substrate beneath, may prevent free drainage occurring properly and can lead to ponding, ie trapped water collects in a puddle on the surface of the roof covering The depth of a pond can vary immensely, from a millimetre or two to many millimetres (75–100mm are not unknown), depending upon the amount of unevenness in the structure or covering. Gradually, there will be a build up of dirt deposits on the surface of the covering at the bottom of the pond. This will be apparent even after a dry spell which has caused the water to evaporate.

A flat roof should have an even fall. This should be not less than 1:40 to ensure that it is self-draining and to reduce the incidence of ponding.

Ponding.

Ponding is normally due to unevenness of the covering or sagging of the substrate/decking. A roof badly affected by ponding is much more likely to suffer from long term failure. Other causes include blocked or raised outlets, or the fall being too shallow.

The effect of the ponding is to set up irregular stresses in the covering at the edge of the water – for example in the surface temperature of the covering because the water in the pond will absorb heat. Over a period of time, this can lead to cracking, especially of built up felt. Ponding may also attract insects and larvae which in turn may attract birds whose pecking has been known to damage fabric-based coverings. A further potential hazard is that a leak in the 'pond' area, whatever the reason, will result in a greater amount of water penetrating into the building than from a properly drained roof area.

Out-fall problems

The correct size and positioning of the rainwater outlet is critical; if it is too small for the surcharge of water, or incorrectly placed to properly drain the whole of the required roof area this will cause problems. The effect of a blocked or inadequate outlet may be water building up on the roof gutter and eventually leaking into the roof void, or flooding over the ineffective outlet and down over walls.

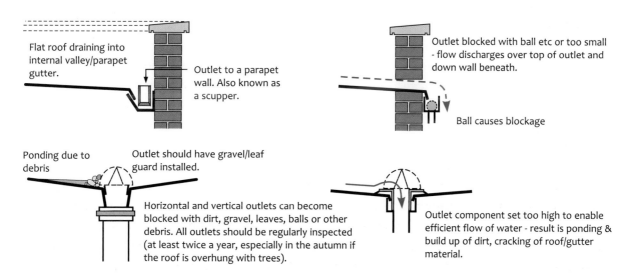

Flat roof draining into internal valley/parapet gutter.

Outlet to a parapet wall. Also known as a scupper.

Outlet blocked with ball etc or too small - flow discharges over top of outlet and down wall beneath.

Ball causes blockage

Ponding due to debris

Outlet should have gravel/leaf guard installed.

Horizontal and vertical outlets can become blocked with dirt, gravel, leaves, balls or other debris. All outlets should be regularly inspected (at least twice a year, especially in the autumn if the roof is overhung with trees).

Outlet component set too high to enable efficient flow of water - result is ponding & build up of dirt, cracking of roof/gutter material.

169

Damaged rainwater hoppers result in water being shed down face of building. (Right) Ponding and leaves around outfall.

Damage by foot traffic and/or human activity

Most flat roofs are not designed to be walked upon, except for maintenance purposes. Where access is needed for maintenance and repair, it may be necessary for crawling boards to be used. Examples of damage include gouges, rips, tears and indentations caused by human movement itself, or the placing or dropping of heavy objects on the surface. A roof covering that is already blistered or rucked, or brittle through age, will be particularly prone to damage of this nature.

Organic growth

Lichens, mosses and algae may appear on the surface of the covering, especially where it is overhung by trees or is in permanent shade. A heavy build-up of organic growth may indicate inefficient run-off of rainwater. It may also be indicative of a lack of regular inspection and maintenance of the roof.

A build up of moss is a common problem on older, mineral felt roofs, especially those that have a poor fall and are finished with chippings. Not only is it unsightly, but its presence indicates a level of retained moisture on the felt roofing which may cause problems.

Lead roofing and guttering can be corroded by the acidic run-off from an adjoining roof that is covered in lichen. The resulting acidic attack may be in the form of narrow, clean-cut grooves on the surface of the lead, or there may be round spots beneath the adjoining roof's drip-off points. Similar corrosion problems can be experienced by zinc roofs and gutters from the run-off from both mosses and lichens, as well as from timber shingles. In each case, the strongly acidic content of the run-off can rapidly corrode the metal.

Thermal and moisture movement

Failure to take account of movement is a common cause of flat roofing problems. Problems occur where a cycle of expansion and contraction are set up through either changes in moisture content or changes in temperature. Many of the common materials used for flat roof structures and coverings will undergo thermal expansion and contraction. The amount of thermal movement will depend upon the coefficient of expansion of the materials affected, and the external temperatures (sometimes the internal temperatures) involved. Thermal movement (expansion and contraction) may occur throughout the day as, for instance, the sun shines, is then obscured by clouds, then reappears etc. Sudden outbursts of rain on a sunny day will also cause a drop in the temperature of the roof. This cycle of heating and cooling, leading to expansion and contraction, can damage the roof covering.

The tendency to be affected by moisture movement is related essentially to the porosity of the material. In comparison with thermal movement, moisture movement tends to occur in less frequent cycles, except perhaps where the source of the moisture is the result of a significant amount of condensation generated from intense activities within the building.

The movement cycle in the covering material or the underlying structure, may subject the waterproof covering to damaging stresses. Often, the initial effect is to ruck up or ripple the covering. Over a period of time, the continual expansion and contraction will weaken the material and eventually lead to it splitting. When the substrate suffers from thermal movement, it invariably results in the covering being adversely affected as well.

Thermal Movement

① Expansion

Covering
Supporting substrate
Structural element

Contraction

All components of a flat roof are subject to thermal movement. In a flat roof the effect of such movement tends to be greater than in a pitched roof.

The covering will normally suffer more readily than the other components as it is fully exposed to seasonal or diurnal changes in temperature.

② moisture

③ moisture

④ Secured end of covering is loosened over a period of time.

The covering may ruck or ripple. It will eventually split. Alternatively, it will merely split as it is slowly weakened by movement.

The covering splits as the substrate moves.

Sometimes expansion detailing/jointing is inadequate to deal with the stresses involved (or the components are improperly fitted together).

(Left) Movement in the felt or perhaps the substrate below has lead to rucking (rippling) which may eventually lead to splits in the felt. (Right) Here, the rucking has created a gap at the side lap through which water can penetrate. In both examples the undulations in the felt will impede the flow of water from the roof.

Differential movement

A flat roof may be affected by differential rates of movement between the roof itself and an adjoining element of construction. The element may be at the same level as the roof (eg a parapet wall or another area of roof, perhaps of a separate construction) or below it (eg an internal wall). The effect upon the flat roof will be horizontal and/or vertical movement stress which can manifest itself in a number of defects such as:

- bucking or rippling
- stepping, unevenness or sagging
- splitting or cracking.

As with thermal and moisture movement the problem may affect only the roof covering but it is more likely to have an effect upon the substrate as well.

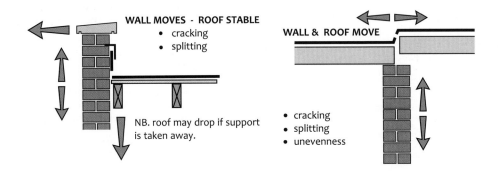

Defects at upstands and skirtings

The junction of a roof with adjoining walls and with penetrations such as pipes, flues, access hatches and rooflights, is a common point at which failure occurs particularly as the roof covering is required to be turned vertically through 90 degrees in order to form an appropriate upstand/skirting that is watertight.

Failure can occur for a number of reasons.

- Splitting along the 90 degrees turn in the covering, as it is weakened by the initial formation of the bend in the material, or by subsequent differential movement between the roof and the adjoining element. A tilting fillet should be used to reduce the angle of change (see diagram opposite).
- The height of the upstand. The waterproof covering should be dressed up the wall or component for a minimum distance of 150mm. If the distance is less than 150 mm the upstand may be surmounted by standing water left on the roof/gutter as the result of a blocked outlet or exceptional rain-burst. This height is also considered sufficient to prevent the effects of splashing from rainwater as it hits the roof surface.

A cover flashing, ideally in lead, should be used where felt is dressed up a wall or other vertical component. The felt should not be attached to the wall in order to avoid the consequences of differential movement between the roof structure and the wall. Asphalt, however, if laid on a concrete base can usually be bonded to the wall and dressed into a suitable chase. A cover flashing is not normally necessary. However, if asphalt is laid on a timber deck, or there is a risk of differential movement between deck and wall, the asphalt should be supported by a freestanding upstand and should not be bonded to the wall; a separate cover flashing is then necessary.

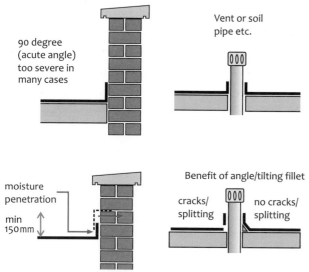

90 degree
(acute angle)
too severe in
many cases

Vent or soil
pipe etc.

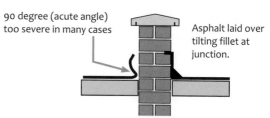

Where the roof meets a wall or pipe flue, hatch or rooflight, the junction to the roof needs to be properly formed with an upstand. Often, turning the roof covering through 90 degrees creates unacceptable stress in the material and it cracks. (Cause may be weakness induced by installation or thermal movement).

90 degree (acute angle)
too severe in many cases

Asphalt laid over
tilting fillet at
junction.

moisture
penetration

min
150mm

Benefit of angle/tilting fillet

cracks/
splitting

no cracks/
splitting

Top of upstand (min 150mm) should be protected with cover flashing - with asphalt & mineral felt this is often omitted - leading to potential water ingress problems.

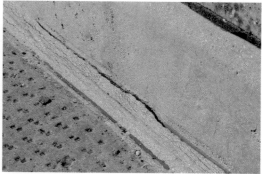

Asphalt is often dressed into a chase in the background. It has a tendency to slump where chase is not provided or is inadequate.

Mineral felt should not be chased in because it is vulnerable to tearing through differential movement. It is better to cover it with a separate lead flashing.

Upstands around outlets (scuppers) in parapet walls are frequent sources of water penetration. Often this is because they are poorly formed as it can be difficult to properly shape or dress the roof/gutter material through the various vertical and horizontal planes.

(Top left) The lead flashing has been omitted from a section of the the upstand detail. It is possible that the builder assumed the coping would provide sufficient protection. As a consequence, the felt has had to be attached to the upstand; this may cause problems due to differential movement. (Top right) Cracks such as this in the asphalt, at the junction of the decking and the upstand, could be caused by differential movement or the lack of a tilting fillet. (Left)The asphalt has slumped because it has not been properly attached to the upstand.

Decking problems

In domestic flat roof construction, the structure is normally formed with a reinforced concrete slab or by timber joists. The former will usually be finished with a sand – cement screed topping laid to a fall, or insulation boarding with an integral sloping top. The latter will normally be covered with board or sheet decking, with the roof slope formed by either installing the decking on firring pieces fixed to the joists, or by shaped insulation boarding.

UNDERSTANDING HOUSING DEFECTS

There are a number of materials that may be used for the decking but the following are the most common.

Softwood timber boards These are the traditional decking above a timber structure. They should be at least 19mm thick, tongued and grooved, and fixed with two nails to each joist. Straight edged boarding is often used, on the grounds of financial economy, but it is more prone to warping, either due to initial drying out or the effect of subsequent water penetration of the covering. Both types may shrink as a result of initial drying out.

Excessive warping or shrinkage may result in the roof membrane being affected. Warping will produce ridges in the covering, while shrinkage results in dips or mini-valleys between the individual boards. In a badly affected roof, either may be a substantial problem because of the considerable number of board joints that may be involved.

Plywood External quality plywood or marine plywood sheets should be used for decking, as the glues in these are water resistant. Plywoods of internal quality are adversely affected by moisture and may expand, disintegrate or develop fungal infection (wet or dry rot) if the roof leaks.

Chipboard This is manufactured in a number of grades, including moisture resistant which is intended for use in roofs. However, some leading authorities have expressed concern over the use of any grade of chipboard in a roof, as they are all capable of absorbing moisture under certain conditions, particularly those where high levels of humidity are present. If this occurs, the chipboard loses its structural integrity, will sag under its own moisture-sodden weight and will, eventually, disintegrate.

Wind related damage

There are a number of potential problems that may be experienced by flat roofs because of the effect of the wind. They are discussed below. *(The general background to this subject is discussed in Chapter 7 in the section entitled 'Wind related damage'.)*

The effect of suction As with pitched roofs, negative pressure (suction) is a greater problem than positive pressure and affects the edge of the roof (the eaves and verges) more than the central area of a flat roof. This is because maximum suction occurs at the edge, although flat roof coverings that are particularly light in weight may be prone to lifting by suction across their whole surface area. Any weakness in the (edge) fixing is exploited by the wind and may lead to the covering being lifted. Multi-layered roofs can experience successive removal of the individual layers of the roof, ie the covering, insulation, decking, sub-structure. At its worst, it is not unknown for a complete flat roof to be physically removed as an integral unit from the top of a building and deposited some distance away.

Scouring action on protective dressings The wind can have a scouring action on any gravel covering used as a protective dressing on a flat roof. Loose ballast of less than say 20mm diameter on inverted roofs may be particularly affected, leading to damage of the underlying substrate. The movement of the dressing may also result in the exposed roofing material being degraded by ultra-violet light, drainage out-falls becoming blocked, injury caused to pedestrians or damage to adjoining glazed areas, especially where there is a lack of any raised parapet walling.

Pressure on laps, joints and steps A wind speed of 26 metres per second is not unusual in the UK and is often exceeded (this wind speed is described as 'storm' on the Beaufort Scale, which indicates wind force). Such a wind speed, which is becoming more frequent due to climatic change, will support a 50mm column of water. Therefore, any welted or folded joint or seam, any upstand or step, that is less than this height must be viewed with suspicion as there is every possibility that it will leak under high wind conditions.

PROBLEMS EXPERIENCED BY SPECIFIC ROOF COVERINGS

A wide range of materials can be used to cover flat roofs. The following discussion concentrates on those that are most commonly found on domestic flat roofs.

Built-up mineral felt

This roof covering usually consists of two or, preferably, three layers of felt, bonded with hot bitumen. The exact number of layers depends on a number of factors including whether a partially bonded first layer is needed to accommodate substrate movement; the type of reinforcement in the individual layers – polyester felt membranes normally consist of two layers, glass and other fibre membranes of three. The top layer is normally finished with stone chippings or, sometimes, reflective paint to provide protection against solar radiation (which can rapidly degrade the material).

Mineral felt has been in use as a flat roof covering (and, occasionally, as a gutter lining) since before the Second World War. It became very popular in the 1960s and 1970s. This was because of its relative cheapness and ease of installation. It has since declined in popularity, because of its apparent lack of durability. In the 1960s and 70s, the life expectancy of mineral felt roofing was thought to be 15–25 years, but it rarely achieved this, often suffering failure in less than 10 years. This was due to a mixture of poor installation and inadequate materials (eg a core of rag fibres, asbestos fibres or glass fibres coated with oxidised bitumen).

Manufacturers of felt now claim that advances in material content (eg the introduction of polyester fleece cores – especially when coated with polymer modified bitumen to form a 'high performance felt') and better quality installation techniques should ensure a longer life.

Apart from general faults covered in the previous sub-section, built-up mineral felt can have the following problems.

Loss of surface protection An even spread of solar chippings, or the use of solar reflective paint, over the complete roof area is essential. This is to prevent the bitumen content oxidising and prematurely ageing the covering. The cause is the damaging effect of ultra-violet light from exposure to the sun. The result of the loss of any surface protection will be the general deterioration of the felt, usually accompanied by embrittlement.

Protective ballast (small chippings or grit) should be dressed over the entire surface of the felt to a depth of 10–15mm, but they can become displaced for a number of reasons including the fall being too steep, allowing the chippings to be swept by rainwater to the bottom of the 'slope', or sagging of the substrate.

Chippings can be moved around by wind or rainwater or simply brushed to one side during repair works and then not swept back in place.

The use of solar reflective paint as an alternative to chippings will provide protection against the damage caused by ultra-violet light. Chippings however can also work to some extent as a 'heat sink' absorbing the heat of the sun and, in so doing, potentially reducing thermal movement – paint will not provide this benefit.

175

Solar reflective paint is applied to a roof surface that is free of any debris or encumbrance, such as chippings or grease. It requires frequent renewal at approximately 2–5 year intervals, depending upon weathering and exposure, to ensure that its integrity is maintained and the felt is fully protected from the damaging effect of ultra-violet light.

Blistering Blisters are quite common in felt and asphalt roofs. The blisters are usually formed through the expansion of trapped air and moisture underneath the covering (and/or between sheets in the case of felt). The blisters usually form as a result of the trapped, moisture laden air expanding as it is heated by the sun. The blisters will tend to expand and contract as the trapped air is heated and then cooled (a process some-times referred to as 'thermal cycling'). The expansion and contraction cycle can result in the felt being permanently stretched and not returning to its original state. The stresses can cause a split – although often the damage is caused through a mechanical impact on the blister.

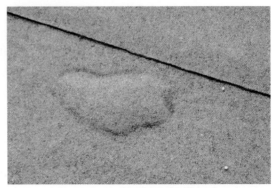

Blisters in flat roofs. In this example the roof is above a kitchen which is likely to be the source of the moisture laden air. Solar reflective chippings help in reducing the risk of blisters forming because they absorb some of the heat from the sun.

Problems with jointing of the top layer The joints in the lower layers of a mineral felt roof are formed by butting together the individual sheets/pieces, ensuring that the joint in each layer is staggered to avoid a through joint with the one above. For the top (or weathering) layer, the adjoining sheets are overlapped and then bonded together with hot bitumen.

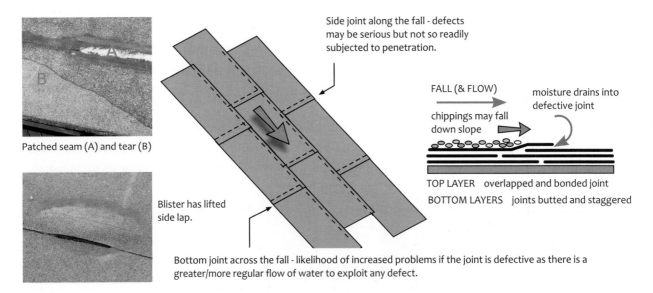

Patched seam (A) and tear (B)

Blister has lifted side lap.

Side joint along the fall - defects may be serious but not so readily subjected to penetration.

FALL (& FLOW)

moisture drains into defective joint

chippings may fall down slope

TOP LAYER   overlapped and bonded joint
BOTTOM LAYERS   joints butted and staggered

Bottom joint across the fall - likelihood of increased problems if the joint is defective as there is a greater/more regular flow of water to exploit any defect.

A joint may fail because it was, initially, poorly bonded, or it may pull apart subsequently, perhaps as a result of blistering. The failure of any joint in the top layer is a serious matter, but it is particularly so when it is across the fall of the roof or a gutter, when any weakness will tend to be rapidly exploited by water flowing across the surface.

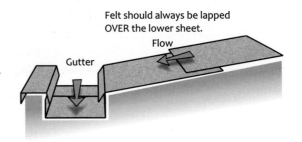

Felt should always be lapped OVER the lower sheet.

Flow

Gutter

Eaves and verges It is normal practice to finish the mineral felt by dressing it down over the face of the eaves to form a drip. The same detail may be formed at a verge, although it is better to create a weathered edge, or kerb, to direct water away from the verge (and avoid it running down the face of the wall below). It is also common practice to incorporate an extruded aluminium edge trim along a verge (but not the eaves) to provide a mechanical fixing that is resistant to wind uplift.

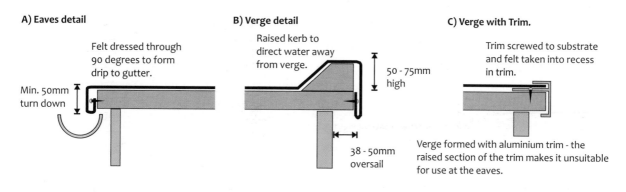

A) Eaves detail

Felt dressed through 90 degrees to form drip to gutter.

Min. 50mm turn down

B) Verge detail

Raised kerb to direct water away from verge.

50 - 75mm high

38 - 50mm oversail

C) Verge with Trim.

Trim screwed to substrate and felt taken into recess in trim.

Verge formed with aluminium trim - the raised section of the trim makes it unsuitable for use at the eaves.

Problems that may occur at the eaves include:

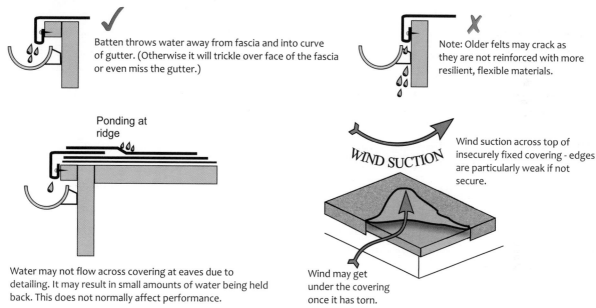

Batten throws water away from fascia and into curve of gutter. (Otherwise it will trickle over face of the fascia or even miss the gutter.)

Note: Older felts may crack as they are not reinforced with more resilient, flexible materials.

Ponding at ridge

Water may not flow across covering at eaves due to detailing. It may result in small amounts of water being held back. This does not normally affect performance.

WIND SUCTION

Wind suction across top of insecurely fixed covering - edges are particularly weak if not secure.

Wind may get under the covering once it has torn.

Problems that may occur at the verge include:

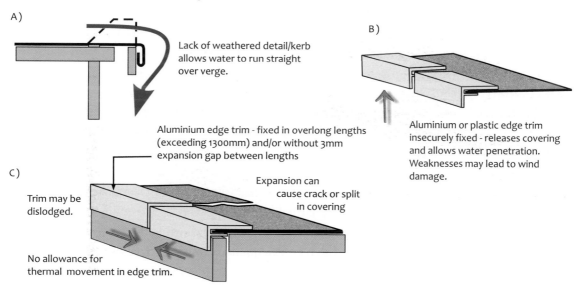

A)

Lack of weathered detail/kerb allows water to run straight over verge.

B)

Aluminium or plastic edge trim insecurely fixed - releases covering and allows water penetration. Weaknesses may lead to wind damage.

Aluminium edge trim - fixed in overlong lengths (exceeding 1300mm) and/or without 3mm expansion gap between lengths

Expansion can cause crack or split in covering

C)

Trim may be dislodged.

No allowance for thermal movement in edge trim.

Single-ply membrane

Recent years have seen the introduction of single-ply roofing membranes from Europe and the USA. The introduction of these membranes into the UK took place in the late 1960s, but they had not been fully developed and did not perform successfully. In part, this was due to the thinness of some of the membranes – less than 1mm thick. Subsequent research and development has resulted in materials of much greater integrity and durability.

Generally, they have been installed over commercial premises but there have been some examples of their use on domestic roofs. They are manufactured from various plastics or from synthetic rubber-based materials. Sheets of the former can be softened and welded together by heat or solvent, whereas the latter must be joined by adhesive. Site welding is thought to produce a more satisfactory joint than gluing, although it is a skilled operation and requires carefully controlled site conditions.

The membrane should be mechanically fixed to a warm deck roof by screw fasteners and disc washers. Each fastening should be set into a seam, or be beneath a spot patch on a larger roof. On an inverted warm roof, the membrane is loose laid and covered with a layer of ballast.

Problems include the following.

- Some manufacturers claim that their product may be laid on an absolutely flat roof without any drainage fall. The result is standing water on the covering and an increased likelihood of water penetration if the membrane suffers from a defect.
- Certain single-ply membranes may be loose laid with only an edge fixing. These are more susceptible to the effect of wind lift if the manufacturer's installation requirements are not precisely followed.
- For those materials requiring a mechanical fixing, the size of the roof and its degree of exposure may dictate closer spacings than normal for the fixing points, otherwise the roof may suffer from wind-lift.
- Careful selection of adjoining materials is important. Some plastics/rubbers are not suitable for use beneath the extruded expanded polystyrene sheeting commonly used on inverted warm roofs.

Mastic asphalt

Mastic asphalt is an extremely durable material that has been used as a waterproof covering for flat roofs since about the beginning of the 20th century. Mastic asphalt consists of fine and course aggregates which are mixed with bitumen. It is laid over a separating layer, normally of black sheathing felt, that isolates it from any movement in the substrate.

Mastic asphalt comes in the form of solidified blocks that are reheated on site until it becomes soft enough to be applied by trowel. This is usually carried out in two layers, to a combined thickness of 20mm. The work is carried out in bays of around 2.5 to 3 metres and, while the proper jointing of each bay needs to be properly carried out, the fact that an essentially seamless surface can be produced is partly why mastic asphalt roofs can be so durable.

The action of applying the mastic asphalt by trowelling commonly results in a bitumen rich skin to the top coat. This skin is susceptible to surface crazing in cold weather. To avoid the formation of this skin, a light dressing of sand, is rubbed in to the final coat after it has been applied. The surface should then be protected by solar reflective paint or a 10–15mm thick dressing of small chippings.

Asphalt has a tendency to slump when incorrectly laid on a slope or vertical upstand. It, also, tends to crack when formed into a sharp angle such as occurs at an upstand, and so should be laid over an angled fillet *(see earlier)*.

An older asphalt roof (40 years +) will tend to deteriorate by developing cracks as elasticity is lost. This is a result of the effect of continual changes in temperature. Newer roofs may be less prone to these stresses, due to the introduction of polymer-modified asphalts.

Apart from those defects that were discussed in the earlier sub-section dealing with general flat roof problems, the following problems may also occur.

Surface crazing Surface crazing (see photo below left) together with wrinkling, pimpling and crocodiling are normally the result of solar radiation, and are often seen on older roofs. They may also be caused by ponding. These cracks are not considered serious as they are usually of little depth. However, those cracks associated with ponding can, over a period of time, develop into a deeper and more serious problem.

Blistering Blistering (right-hand photo below) in asphalt has the same cause as that in felt *(see earlier explanation)*. Because of the inherent durability of asphalt, blisters tend to cause less problems than they do with mineral felt roofs.

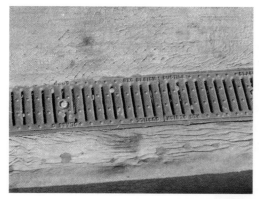
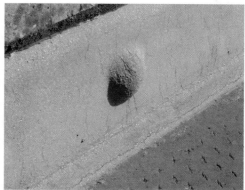

Adverse effect of the substrate Mastic asphalt is probably more dependent upon the condition of the substrate than any other covering material. Any movement or unevenness in the substrate tends to be transferred to the asphalt and will frequently induce major cracking/splitting. A multi-jointed decking, such as timber boarding, can be particularly uneven if the boards are insufficiently fixed and/or suffer curling. Sheet decking or insulation board may have uneven or widely spaced joints. All forms of decking may be poorly supported by the structure, resulting in dipping and sagging. Any movement in the substrate will be readily transferred to an asphalt roof unless adequate precautions (such as thermal or differential joints/details) have been incorporated.

Lead roofing

Lead sheet is commonly used as both a pitched and a flat roof covering. Lead is a well-proven and long-lasting material, which is extremely malleable and easily dressed into shape. As it ages, the surface dulls as a coating of lead carbonate is formed due to contact with the atmosphere. This coating protects the lead from further deterioration. The traditional method of manufacture, cast lead sheet, which is formed by pouring molten lead over a bed of sand, has been found to last over 400 years. A small quantity of this type of sheet lead is still produced. It is used for conservation and other high quality work.

The majority of the lead used in roofing is known as milled lead sheet. It is thinner than cast lead and has a shorter life (at 100+ years). Milled lead sheet is made by running a slab of lead through a rolling mill several times, until a sheet of the correct thickness is achieved. There is a range of six thicknesses, each being given a British Standard Code – Code 3 (the thinnest and lightest) to Code 8 (the thickest and heaviest).

The type of use will dictate the thickness selected. Large exposed areas of lead (such as are found on flat roofs and in valley/parapet gutters) are subject to considerable thermal expansion, caused by exposure to the sun. The thicker/heavier sheets (Codes 6-8) are normally selected for such positions, as they are not so easily, and so adversely, affected by extremes of temperature. They can also be installed in larger panels or bays. The thinner grades (Codes 3-5) are used where there is a lower level of exposure involved, eg for flashings and soakers, and for smaller gutters such as those found on pitched roofs. Selection will also be influenced by the life-span sought and the shape and/or profile of the area to be covered.

Many of the problems associated with lead roofs are caused by inadequate design and/or construction. Such problems include the following.

Excessive bay size Lead is extremely sensitive to changes in temperature. In summer, leadwork that is continuously exposed to the sun, may be subjected to daily changes in temperature of up to 40°C. It is important that the individual pieces of lead sheet or guttering do not exceed maximum recommended sizes, as there is an increased likelihood of failure when they do. This is due to the expansion/contraction cycles involved and the effect will be rippling and cracking of the lead. Overlarge bay size is the most common cause of failure of lead roofing. Guidance on the maximum recommended dimensions for lead covered roofs and gutters is given in the 'Lead Sheet Manuals' published by the Lead Sheet Association.

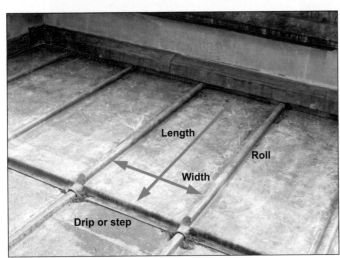

The maximum size of each bay is determined by the thickness (code) of the lead. For flat roofs Code 6 lead (2.65mm thick) is suitable for bays up to 2250 x 675mm. Code 4 lead (1.80mm thick) is suitable for bays up to 2000 x 500mm. When using Code 4 or 5 lead bay length may need to be a bit shorter depending on exposure to sun.

To avoid premature failure the lead must be free to move – nowadays a geo-textile membrane is often specified below the lead.

Junction problems – rolls The junction along the length of each bay is normally formed with a roll, and sometimes with a welted upstand (see later). The roll may be hollow or formed over a solid softwood core. A hollow roll is only suitable for use on a pitched roof, as on a flat roof, foot traffic can easily crush it.

The lead from each bay should be freely dressed over the timber roll to form a junction that allows thermal movement. Only the lower sheet should be nailed, if absolutely necessary (and at one side only). Inappropriate fixing will result in buckling and cracking of the sheet, often along the roll.

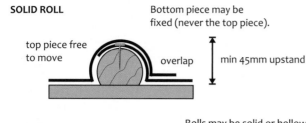

SOLID ROLL

top piece free to move

Bottom piece may be fixed (never the top piece).

overlap

min 45mm upstand

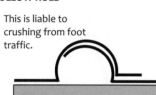

HOLLOW ROLL

This is liable to crushing from foot traffic.

Rolls may be solid or hollow. Lead is dressed over solid core or metal spring (subsequently withdrawn) to form a detail that will allow thermal movement, prevent capillary action and avoid penetration of any normal level of standing water.

Boards are laid diagonally to minimise effects of warping (ie causing ridges in the lead).

Junction problems – drips A drip is used to form the longitudinal junction between adjoining bays of a flat roof, or of a gutter where the fall is less than 15 degrees. It is a stepped detail that should be both watertight and able to allow thermal movement to properly occur. Its installation requires care, as any deficiency in design or construction may lead to moisture problems, usually as a result of capillary action. Failure may occur for a number of reasons including those illustrated below.

DRIP PROBLEMS

Too small a splashlap may lead to moisture penetration. A minimum lap of 40mm is required to prevent capillary problems. The horizontal tail of the splashlap stiffens the free edge and helps keep it in position.

- Undercloak should be minimum 25mm long. A ridge/step will be formed if the deck is not rebated (ponding may result).
- Undercloak should be nailed at 50mm centres (not less than 25mm from end).

min 25mm

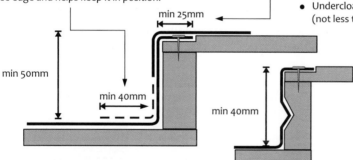

min 50mm

min 40mm

min 40mm

Minimum height of drip should be 50mm to prevent water being drawn up between the sheets by capillary action and/or wind pressure. This can be reduced to 40mm if an anti-capillary groove is incorporated. Even then, a high flow may penetrate the detail.

Rebate formed in edge of boarding

Other junction problems – overlaps and welts A welted joint may be used on the side junctions of small flat roof areas (eg a door canopy). This is only acceptable where the flow of water is unlikely to exceed the height of the welt. A welt should never be used for lead sheet jointing on a large flat roof area, nor in any low-pitch gutter. A roll should be used in both situations. (Note: Pre-clad lead panelling on flat roofs and other areas involves a standing seam joint at the side junctions. Similar in principle to a welt, this type of joint usually requires mechanical site assembly.)

Simple overlapping of two adjoining sheets on flat roofs and gutters is incorrect, however wide the amount of overlap. Moisture will penetrate such a detail and get into the roof substrate and structure beneath. However, the simple overlapping of bottom joints and the use of welted side-joints are frequently found on steeply pitched roofs and gutters. Both are only acceptable practices on surfaces where rainwater is drained away quickly.

181

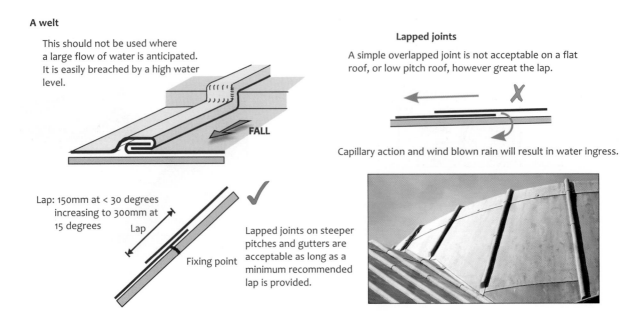

A welt

This should not be used where a large flow of water is anticipated. It is easily breached by a high water level.

FALL

Lap: 150mm at < 30 degrees increasing to 300mm at 15 degrees

Lap

Fixing point

✓

Lapped joints on steeper pitches and gutters are acceptable as long as a minimum recommended lap is provided.

Lapped joints

A simple overlapped joint is not acceptable on a flat roof, or low pitch roof, however great the lap.

✗

Capillary action and wind blown rain will result in water ingress.

Incorrect underlay Building paper or felt underlay have both been used as the separating layer between lead and substrate for some considerable time. The bitumen or resin content of the felt may leach out under solar heat, leading to adhesion between lead and substrate. This will restrict thermal movement. A further problem is that even slight condensation may cause the organic fibres of the felt to rot. To avoid these problems a needled, non-woven polyester textile underlay should be installed, although building paper may still be used where there is no risk of condensation forming under the lead.

Incorrect fixing Correct fixing of leadwork is essential in order to avoid premature failure. The fixing nails should have a life equal to that of the lead. They should be sufficient in number, and fixed in the correct pattern, to properly support and secure the lead, while permitting thermal movement. Fixing nails should also be of the correct type – either copper or stainless steel ring-shanked clout nails. Other materials, including aluminium or galvanised steel, are now considered to be inappropriate. Common fixing defects are shown below.

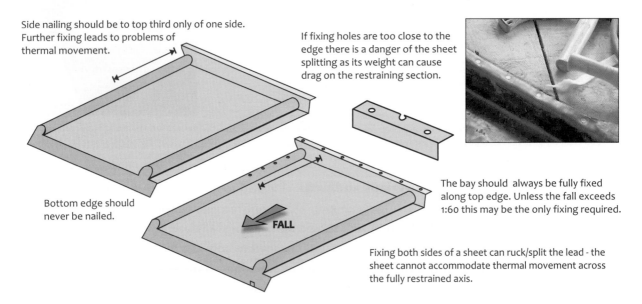

Side nailing should be to top third only of one side. Further fixing leads to problems of thermal movement.

If fixing holes are too close to the edge there is a danger of the sheet splitting as its weight can cause drag on the restraining section.

Bottom edge should never be nailed.

FALL

The bay should always be fully fixed along top edge. Unless the fall exceeds 1:60 this may be the only fixing required.

Fixing both sides of a sheet can ruck/split the lead - the sheet cannot accommodate thermal movement across the fully restrained axis.

Incorrect flashings Lead flashings are to be found on all types of flat and pitched roofs, not just those that are covered with lead. This is because of the ability of lead to be easily beaten into any shape, together with its durability and its compatibility with a wide range of other materials.

It is normal to use Code 4 lead for flashings and a properly installed flashing will perform for

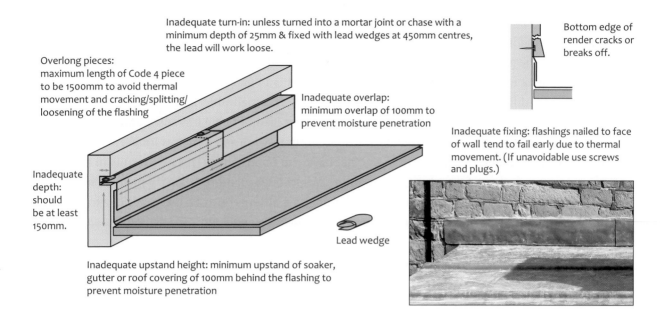

Inadequate turn-in: unless turned into a mortar joint or chase with a minimum depth of 25mm & fixed with lead wedges at 450mm centres, the lead will work loose.

Bottom edge of render cracks or breaks off.

Overlong pieces: maximum length of Code 4 piece to be 1500mm to avoid thermal movement and cracking/splitting/loosening of the flashing

Inadequate overlap: minimum overlap of 100mm to prevent moisture penetration

Inadequate fixing: flashings nailed to face of wall tend to fail early due to thermal movement. (If unavoidable use screws and plugs.)

Inadequate depth: should be at least 150mm.

Lead wedge

Inadequate upstand height: minimum upstand of soaker, gutter or roof covering of 100mm behind the flashing to prevent moisture penetration

80–100 years, often longer, without any attention. Occasionally, Code 3 lead flashings may be found. These are installed for cheapness but do not perform as well as the higher code. Most problems arise from poor installation, some of which are shown above.

Staining of adjoining surfaces Newly installed lead sheet produces an uneven white carbonate on its surface as a result of wet or damp conditions. Unless treated with a patination oil immediately after installation, any run-off from the affected surface may create stains on adjoining surfaces. The staining on the lead surface will eventually disappear as the surface of the lead acquires an overall patina of lead carbonate. As mentioned earlier this patina of lead carbonate is a key factor in ensuring leads' durability.

Corrosion The design and construction of the roof should ensure that there is no problem with interstitial condensation. It has long been known that condensation may affect the underside of leadwork (where it is not protected by the lead carbonate patina that forms on its topside). This can lead to corrosion because the condensate is distilled water (which slowly dissolves lead).

A more recently recognised problem is sometimes called 'summer condensation'. This was identified in the late 1980s and affects the underside of a lead covering or gutter above a warm roof construction. Its cause is condensation created by the following, previously unknown, phenomenon in warm roofs.

In summer, rainwater falling on a hot roof covering, and rapidly cooling it, results in sub-atmospheric pressure being created in the zone between the vapour barrier and the covering (ie within the insulation). Rainwater is then drawn into this area by the lower pressure through the expansion and other joints in the roof covering. The trapped water subsequently vaporises under further hot weather conditions, migrates through the substrate and condenses on the underside of the lead as further reductions in temperature take place, leading to corrosion.

This phenomenon also occurs with other roof coverings such as zinc, copper, aluminium, asphalt and mineral felt, although the effect may be different.

Zinc roofing

This became popular as a flat roof covering, gutter lining and for flashings in Victorian times, because of its cheapness when compared with lead. It is also considerably lighter, but is not as durable. In recent years, this has led to a decline in popularity, even though it is now manufactured as an alloy – zinc/titanium for sheeting or zinc/lead for flashings – to give better performance. It is more popular in the rest of Europe, especially France.

Zinc can, possibly, achieve a life of up to 60 years in reasonably clean urban areas and possibly longer in rural areas, but 25 years is more likely, especially on flat roofs. In heavily polluted industrial areas, or even some inner city areas, it can fail in an even shorter period. Its durability depends, to some extent, on the gauge (thickness) of the zinc used. The five thicknesses most likely to be found on roofing are from 12 to 16 gauge – the normal thickness used is 14 gauge (0.8mm thick).

Zinc is a more difficult material to work than lead, being less malleable, and more prone to splitting where it is turned or shaped. It does have the advantage over lead of being less susceptible to thermal movement, so can be used in larger sized bays – up to 600mm wide and 5 metres long. For this reason, extremely long zinc gutters, having fewer individual bays, can be installed. This also results in fewer expansion joints, which are formed with drips and rolls, although standing seams, which are more popular elsewhere in Europe, may sometimes be found. Care still needs to be taken in not exceeding maximum recommended bay lengths and girths, or else buckling and splitting may occur.

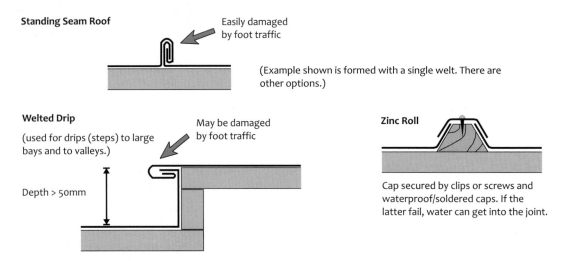

The zinc sheeting should be given full support by the substrate and be laid over a suitable separating layer. This is to prevent adhesion and possible thermal movement problems. As with lead, careful material selection is needed for this layer.

The problems associated with zinc include the following.

Generally A number of the problems experienced with zinc are similar to those suffered by lead, eg failure of flashings, excessively large bays/overlong gutters affected by thermal movement.

Brittleness As zinc ages, its surface is increasingly affected by pitting and it also takes on a crusty appearance. This is zinc carbonate, which develops as a result of the zinc's exposure to air and water. The carbonate does not (as with lead carbonate) develop into a skin of sufficient thickness or density to provide a protective coat to the zinc. The metal gradually deteriorates as it ages, becoming extremely brittle, and prone to cracking and splitting. This process is accelerated in a polluted atmosphere.

Soldered joint failure It is a difficult metal to bend and shape, so it is normal practice to cut the material and solder the pieces into the shape required. This may result in many such joints in a flat roof or gutter. Thermal movement of the metal tends to place soldered joints under great stress. They are common points of failure in zincwork. Great care is needed in the formation to ensure that the cutting of the metal and the soldering are properly carried out.

Joints and welts It is possible to form changes of direction in the metal – the zinc can be folded with hand tools (although the factory pre-forming of shapes is often undertaken to ensure straightness of the folds). The folding of the metal tends to weaken it and this may be accentuated when the work is clumsily undertaken, resulting in early failure due to cracking.

Foot traffic may damage the joints, especially standing seams. Welts may also suffer from human clumsiness.

Restrictions of height often mean that parapet and valley gutters will not tolerate a stepped lining.

The result is that zinc is used for the gutter lining, as it can be laid in relatively long lengths without the need for too many drips. Sometimes, sections of zinc are soldered together to form an extremely long gutter without any drips (steps), in order to deal with a severely restricted height. The lack of an expansion detail (drip) may place considerable stress on both the zinc and, in particular, on the soldered joint(s). As a result, these may fail, allowing water into the roof construction beneath.

Bi-metallic corrosion Zinc is subject to electrolytic action when placed in close proximity to metals that are more 'noble' (eg copper). Inappropriate use of adjoining materials, especially fixings, may lead to corrosion of the zinc. Care should be taken to avoid run-off from these materials. Its use next to lead roofing and components is acceptable. Fixings should be of heavily galvanised steel screws or nails, and clips formed in zinc.

Corrosion Zinc, like lead, is adversely affected by both interstitial condensation and 'summer condensation' (for an explanation refer to lead corrosion). In these conditions, water vapour in the roof condenses on the underside of the zinc, and reacts with the metal to create zinc hydroxide on its surface. This is known as 'white rust', because of its appearance.

Copper roofing

Copper is a generally durable roofing material. There are examples of it being used for flat roofs but it has tended to be more common on pitched roofs (albeit often on quite a shallow pitch). On exposure to the atmosphere copper produces a number of chemical reactions which changes its colour from almost 'pink' through shades of brown until it eventually becomes the light green colour that is most commonly associated with it. This natural process adds to the durability of copper.

The light green colouring of these copper roofs is produced by the patination of copper as it reacts with the atmosphere. The undulations in the copper roof (right) are caused by condensation which has adversely affected the decking.

CHAPTER 9

External Rendering

INTRODUCTION

This chapter summarises 'good practice' in external rendering and explains a number of defects which commonly occur.

In the early 19th century it became fashionable to render houses to give them an appearance of stone. The renders, usually made from lime, sand or other bulk fillers, were often incised with fine lines to imitate the courses of stonework. In the Victorian period modern cement was invented. It was called Ordinary Portland cement because, when it hardened, it looked not dissimilar to natural Portland Stone (or so its makers claimed). During the latter part of the Victorian period hundreds of thousands of speculatively built houses were erected. Many of these were rendered, partly to hide the poor quality brickwork used in their construction, and partly to improve their weather protection (cavity walls were very rare until the early 1900s). From the 1900s to the present day rendering has maintained its popularity as a walling finish in both new build and rehabilitation work.

The first picture, top left, shows wattle and daub. Vertical staves of oak with an interwoven hazel wattle usually formed the base for the daub. The daub itself could comprise a variety of materials including lime, straw, mud, sand and dung. The top right hand photograph shows a traditional lime render on a masonry background.

The bottom left hand photograph shows a render which has been incised with fine lines to look like stone work. It can be very convincing.

The bottom right picture shows the main shopping street in Monmouth. By the late 19th century it was quite common to find whole streets of properties rendered, particularly in Wales and the West, and in seaside towns.

Rendering is applied to buildings for a number of reasons. These include:
- weather protection
- hiding materials such as blockwork or common brickwork
- providing a decorative surface
- providing a smooth surface to take a painted finish.

In recent years renders have also been used to protect and hide external insulation systems.

Cement based renders are normally applied in two or three coats (three coats are recommended in areas of high exposure) and may be finished in a number of ways. Perhaps the most difficult finish, in terms of workmanship, is smooth finished render. Smooth renders can be painted or may include special cements, sands and pigments to give the render a particular self-finished appearance. Textured finishes are a common alternative to smooth renders. The most common is pebble dash or dry dash. In this finish tiny pebbles or other coloured aggregates are 'flicked' onto the wet top

Some renders are quite unusual – pebble dash is common enough but not using pebbles of this size.

coat. Its appearance is unmistakable. Another textured finish is wet dash, sometimes known as rough cast. In this approach the stone or pebble aggregate is mixed-in with the final coat and spread onto the base coat, basically in the form of a weak concrete. It can be a very durable finish. There are also a number of modern proprietary finishes which, from a distance, look similar to rough cast. These are usually based on factory mixed, pre-bagged mixtures of cement, lime, aggregates and pigments which are applied 'wet' to a dry top coat using special hand held machines. 'Tyrolean' is probably the most common.

Unfortunately, render defects are all too common. Their causes are attributable to a number of factors, some of which are related to structural movement, an area which has been explained in earlier chapters. This chapter is more concerned with defects in render specification, inadequate background preparation and poor site practice. Before considering specific defects it is important to understand some simple principles relating to render specification.

RENDER SPECIFICATION

The strength and density of a render is related to its cement content. Strong or cement rich renders are mixes of about 1 part cement to 3 parts sand. This mix produces a dense but brittle render which has excellent weatherproofing qualities as long as cracking can be avoided. The render is dense because the cement fills all the voids in the sand aggregate. A given volume of sand contains approximately 25–33% air voids. By replacing the air with the cement binder a void free material is produced. However, renders containing such a high proportion of cement are likely to shrink on drying. As the render dries, there is consequential shrinkage which can cause the render to crack. Some strong back-grounds with a good mechanical key can resist this shrinkage because of the good bond formed with the render. However, most backgrounds do not form a strong key and cannot resist the ensuing shrinkage forces created by the drying render. The cracks which may appear form a path for rain penetration. Water entering the cracks cannot escape (because the render is dense) and may, therefore, possibly increase the incidence of damp penetration as water can only evaporate through the internal face of the wall.

Weaker mixes, which contain less cement, tend to be more durable, but a mix of say 1 part cement to 5 or 6 parts sand can be very difficult to 'work'. This problem can be resolved by adding lime to the mix or by using chemical additives. Chemical additives, or plasticisers as they are more commonly known, act as air entraining agents; they reduce the internal friction of the render making it more 'plastic', with consequential improvements in its 'workability'. Lime fulfils a similar function although in a slightly different way. Lime improves the workability and the cohesive nature of the mix. It also reduces the amount of drying shrinkage and the risk of cracking. The cement and lime are usually batched to ensure that the two materials together form approximately 1:3 of the volume of the sand. This ensures that the voids in the sand are filled. If the voids are not filled rain penetration is more likely.

Typical mixes:

	Cement/Sand	Cement/Lime/Sand	Cement/Sand+Plasticiser
Weak	1:7 (difficult to work)	1:2:9	1:7
Medium	1:5 (difficult to work)	1:1:6	1:5
Strong	1:3	1:0.5:4.5	1:3

Although weaker mixes are less dense, in practice they are more effective at resisting rain penetration because they are less likely to suffer drying shrinkage and cracking. Although the render will absorb some rainwater it will soon evaporate when the weather changes. In addition, weaker renders are more able to accommodate the stresses of thermal and moisture expansion which may occur in the render itself or in the backing material. Generally, it is accepted that weaker render mixes result in fewer overall problems than strong, cement-rich ones.

The sand used in the mix should be clean, sharp sand (coarse sand of a sharp angular nature) and should be well graded. In a well graded sand there is a mix of particle sizes; this helps reduce the voids to a minimum. Very fine sands require a lot of water to achieve a workable mix and this will subsequently evaporate resulting in high shrinkage and increased risk of cracking.

Some renders, by their very nature, are only suitable for backgrounds with a good strong key. Roughcast, for example, is essentially a mix of concrete trowelled onto a base coat. It is therefore a relatively strong mix and will tend to shrink as it dries. To maintain durability the backing coat also needs to be strong to help resist this shrinkage movement. The base or backing coat, in turn, needs to be applied to a wall with a good mechanical key if shrinkage is to be avoided. Rough cast is therefore rarely successful when applied to backgrounds such as old friable brickwork or aerated concrete blocks.

Thickness and number of coats

On most surfaces two coat work is acceptable although in areas of severe exposure three coats will be necessary. The thickness of the first coat should not exceed 15mm. Coats thicker than this will tend to sag under their own weight. In addition, when applying a thick render coat to a wall it is more difficult to squeeze out any air pockets; trapped pockets of air will prevent a good bond being formed with the background.

When the undercoat has begun to stiffen it should be combed or scratched to form a good key for subsequent coats. The surface scratching also provides stress relief points which help to prevent large cracks appearing as the render sets. The base coat should be left for a few days before applying further coats. Subsequent coats should be thinner and preferably slightly weaker than base coats to minimise the risk of shrinkage. A strong render coat should never be applied to a weaker base coat.

A common error on site is incorrect scratching of the base coat. Some plasterers assume this is just to provide a bond for subsequent coats. In fact, it is also to provide stress relief points for the drying (and shrinking) render. A properly scratched base coat will enable the drying render to spread these stresses across its face thus preventing wide cracks from appearing. Good practice is shown on the left; poor practice on the right.

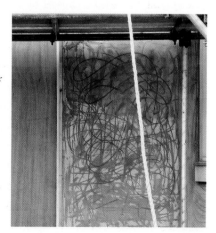

COMMON DEFECTS DURING CONSTRUCTION PROCESS

POTENTIAL PROBLEM AREAS	EXAMPLES	IMPLICATIONS/MANIFESTATION
Background/bond/surface preparation	Insufficient suction on smooth brickwork. Porous bricks provide too much suction. Flush mortar joints prevent key	Lack of bond, loose and hollow render
Render mix and number of coats	Single coats rarely satisfactory. They are not weatherproof and cause 'grinning'. Render mix too strong	Damp penetration caused by insufficient protection. Uneven drying after rain
	Sand too fine, requires more cement and more water to get workable mix	Shrinkage cracking – lack of weather protection
Application	Thick coats will sag. Dubbing out therefore required on uneven surfaces	Uneven surface, usually with some loss of bond
	Base coats must be property scratched to provide key for top coats	Top coat partly bonded. May detach from base coat
	Over trowelling	Surface crazing
Insufficient drying time	Top coats should not be applied until base coats have completed shrinkage. Fast dry-out	Shrinkage cracking of base coat affects top coats
Chemical agents	Lime bloom. In situ slaking of quicktime nodules causing cone shaped depressions in surface	Surface finish affected
	Incorrect angle beads, bedding beads	Rusting of beads, sulfate attack in gypsum plaster
Quality of finish	Day joints show at scaffold lifts etc, most common in smooth renders	Aesthetic mostly
	Bare or uneven dash finishes caused by top coat drying too quickly	
	Tyrolean finish not confined to render (gutters downpipes etc!). Tyrolean too wet or too dry (see text)	

DEFECTS – MANIFESTATION AND CAUSES

Hollow renders

If a render sounds hollow when tapped the cause may be lack of adhesion to the background. Hollow spots will often be accompanied by cracking (see later section) but this is not always the case. Hollow renders can occur through lack of adhesion of the base coat to the wall itself or lack of adhesion in the subsequent coats. It can be caused by incorrect mix (usually too strong), incorrect thickness of coats or inadequate surface preparation.

Occasionally a hollow sounding render may not actually be defective, eg renders supported on metal lathing which are not designed to bond directly with the wall.

Hollow base coats

In order for the base coat to bond successfully to the wall there should be a good mechanical key and the right degree of suction (absorbency). In addition the render strength should be designed to match the nature of the background. Dense concrete blocks (or bricks) with well raked out joints provide a good key and the right level of suction and may even be suitable for quite strong mixes, say 1:4, cement:sand.

Aerated blocks have higher suction. Their relatively porous nature can 'suck' water out of the render mix before the process of hydration (setting) has taken place. To reduce this risk the manufacturers recommend that dry blockwork is dampened before rendering takes place. Excess wetting should be avoided as aerated blocks will expand slightly as their moisture content increases. In addition, aerated blocks have inadequate key to resist the shrinkage of dense renders and mixes should be no stronger than 1:1:6, cement:lime:sand, or its equivalent. On smooth surfaces such as in situ concrete or ashlar stonework,

some form of surface treatment is necessary before applying the render. Surface treatments usually take the form of a spatterdash coat (a slurry of cement and coarse sand thrown onto the wall to provide a key – *see photograph right*) or various water based adhesives. To bond successfully the render must be applied while the adhesive is still 'tacky'. If the render remains wet for long periods or the background is damp these adhesives can break down after a number of years resulting in hollow spots in the render.

Soft, porous stone or friable brickwork is rarely suitable as a base for rendering and metal lathing is usually required to provide separate support for the rendering.

Surface preparation of any background should include complete removal of moss, paint, loose material and soot. Hollow rendering can also be caused by applying the render in extremes of hot or cold weather. In hot weather the water may evaporate before hydration occurs. In cold weather the render may freeze.

Right – a very dense roughcast render, almost a fine concrete. Unfortunately, there is virtually no bond with the flush-jointed background. Shrinkage and rain penetration are almost inevitable. Below left – rendering on stone work is rarely successful particularly if strong mixes are used. Below right – the crude marks around the window are not some form of provincial art but an attempt to provide a mechanical key for render (long since removed).

Hollow subsequent coats

If the hollow spots are in the second or third coat the likely cause is one or more of the following:
- inadequate scratching of base coat
- incorrect thickness; second and third coats should normally be between 6 and 10mm
- inappropriate mix – too strong a mix will shrink, too weak a mix may be difficult to apply, may not bond successfully and, in extreme cases, may not harden adequately
- render applied in weather conditions which are either too hot or too cold.

Cracking – movement of the background

In many cases renders will crack as a result of structural movement in the background. Even minor background movement can cause cracking in renders, particularly in stronger mixes. The pattern of cracking will give some clues to the cause; these have already been illustrated elsewhere but a few common examples are included below as an aid to diagnosis.

If chimneys lean like this, always suspect sulfate attack.

Cracking above a window suggests lintel failure

Cracking which follows the brick joints is likely to be caused by structural movement.

Cracking along every course, especially in exposed positions, suggests sulfate attack.

Cracking at 450mm or so intervals suggests wall tie failure.

Note: Rendering can increase the risk of sulfate attack; there is more cement to react with the sulfates.

NB. Weak renders are more tolerant of minor movement than strong ones, but render cannot cope with the stresses caused by sudden structural movement.

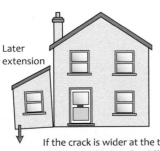

Later extension

If the crack is wider at the top the cause is likely to be differential structural movement.

Cracks at high level may be caused by lateral thrust from 'cut' roofs or the lifting of modern trussed rafters which are not strapped down.

These examples all show failure of the wall itself, not just the render. The top left hand photo shows a 1930s semi detached house with a failed segmental arch over the window. The gate post and the single storey extension are both suffering from sulfate attack. The bottom right picture shows that the cracks in the extension run right through the extension wall.

Most of the examples above often result in fairly pronounced cracks. Finer cracks are often caused by differential movement of the walling material. This can occur in a number of situations. Concrete framed buildings with block infill panels will suffer from differential movement. The infill blocks will contract and expand at different rates from the load bearing concrete frame; this is mostly caused by thermal movement.

Where fine cracks run horizontally and vertically, and coincide with the joints between frame and infill block differential movement is the likely cause. At the design stage this could have been prevented by fixing a galvanised or stainless steel mesh across the columns and beams secured to the blockwork either side. A layer of building paper should be sandwiched between the mesh and the in situ concrete to prevent the render bonding to the frame.

Cracking due to differential movement of frame and blockwork

Special precautions are necessary to prevent renderings from cracking - see text.

Textured renders may hide cracking for a short period.

In modern construction where extensions are added, but are built from different materials, similar movement can occur. Cracking of a render finish can be prevented by fixing a flexible movement bead at the junction of the two different walling materials.

Where terraces are constructed without provision for movement joints cracking can occur in the walls and in the render. A building constructed with a weak mortar and covered with a weak render may be able to accommodate thermal movement; strong mixes are more likely to crack.

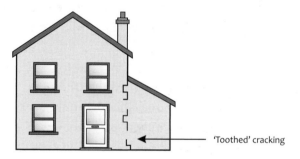

'Toothed' cracking

If the crack is of even width and vertical, the cause is likely to be differential thermal and/or moisture movement. The nature of the masonry and the strength of the mortar affect the shape of the crack. It may follow a fairly straight line or be 'toothed'.

Rendered brick or stone terrace

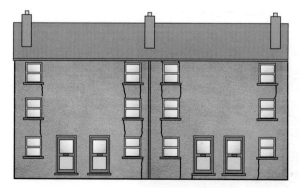

If a building is constructed with fairly weak mortar cracking is less likely to occur in the wall itself. However, if it is rendered with a strong mortar, cracks may still appear between windows if there is thermal movement.

Terraces built in a strong cement mortar and with inadequate provision for horizontal movement are likely to crack due to long-term thermal movement.

The cracks will appear at the weakest points, usually between windows. With weaker mortars any cracks in the masonry itself are likely to follow the joints; with strong renders the cracks can run through the stone or brick. In either case the cracks in the render tend to be fairly straight.

CRACKING – PROBLEMS RELATED TO RENDER SPECIFICATION AND WORKMANSHIP

If the foundations and walls appear stable, problems of cracking are probably due to inadequate specification or workmanship in the render itself.

Shrinkage cracking

One of the most common problems is shrinkage cracking, usually as a result of using a render which is too strong for the background. In many cases drying shrinkage will also result in hollow spots. Drying shrinkage is normally fairly easy to identify. It is characterised by a pattern of cracking known as 'map cracking'. If dense renders are used on weak backgrounds shrinkage cracking is almost inevitable. Rain penetrates the render but cannot escape through evaporation due to the render density.

Cracking along edges of angle beads

Vertical cracks at quoins or at window jambs are often caused by problems with angle beads. In traditional work angle beads were rare but, nowadays, their use is widespread. It is, unfortunately, not uncommon to find angle beads bedded in gypsum plaster. This material sets quickly and is often seen as an ideal material in which to bed the beads. In wet conditions a chemical reaction between the cement and the gypsum (sulfate attack) causes expansion of the render and this will result in cracking.

In addition unsightly staining can be caused by the use of galvanised angle beads which are easily damaged by the operative's razor edged float. The sharp float edge can scrape off the zinc protection leaving the unprotected steel free to rust. This can be avoided by using plastic coated, stainless steel or non-ferrous angle beads. In 'traditional' rendering, angles were formed by working to a batten fixed proud of the wall at the quoin. This method is still occasionally used today and prevents the problems above from occurring.

Angle beads should never be spot bedded in gypsum plaster. In the long term it will lead to localised sulfate attack. Galvanised beads are easily damaged and will soon rust – particularly if they are in very exposed locations or near the sea.

Rendering over horizontal DPCs

Render should terminate just above the DPC, usually about 150mm above external ground level. Timber battens or proprietary stop beads can form a neat render 'stop'. On many houses, particularly those built in the 1920s and 1930s, the render runs down to ground level. Thermal and moisture movement will cause the wall above and below the DPC to move at different rates; the DPC providing a slip layer. This can result in cracking at DPC level. Apart from its unsightly appearance the crack encourages rain penetration; this is often confused with DPC failure. Even if cracking is not apparent damp problems can occur internally as ground water can bridge the DPC through capillary action.

The render on the left runs down to ground level and bridges any DPC. The example on the right has cracked, probably at DPC level.

Surface defects

There are a number of defects related to workmanship or specification which may not necessarily affect the render's long term performance but which may affect its appearance. Surface crazing is a common form of cracking. Surface crazing occurs when the render surface (smooth renders) is over trowelled. Over trowelling results in a mix of water and cement rising to the face of the render. As this dries it forms a pattern of very fine cracks. It does not necessarily affect the performance of the render itself although it can be very difficult to disguise.

Pebble dash can be quite a durable finish. However, it is quite common to find areas of pebble dash which are almost 'bald'. This is usually caused by inadequate workmanship and there are two aspects of good practice which need to be considered. First, it is usually necessary to add an approved waterproofer to the base coat, not to improve the weather protection of the wall, but to reduce its suction so that the top coat (butter coat) will dry slowly. While the top coat is still soft the selected aggregate can be thrown against the wall with a scoop. Second, once applied, the aggregate should be lightly tamped into the butter coat to ensure a good bond.

When adding waterproofers to renders caution needs to be applied. If the renders shrink and crack they can allow rainwater penetration; the waterproofer restricts subsequent evaporation.

When repairing or patching pebble dash it is also very difficult to get a seamless, well blended finish. The example below eventually blended in – but the second photo was taken 15 years after the first one!

Tyrolean failures

The quality of a Tyrolean finish will depend on the suction of the coat to which it is applied. If the suction is too high the finish will not bond properly and, if the suction is too low, the Tyrolean finish may sag or run down the wall ruining its appearance. To be successful it must be applied in one complete 'run'.

Scaffold lift marks

Smooth renders, particularly if they are not painted, require considerable skill if they are to provide a consistent, smooth finish. The normal practice is to start the top coat at the highest scaffold lift. The render is trowelled onto the wall working left to right (for a right handed plasterer) and top to bottom. Before dropping to the next scaffold lift the render is rubbed with a plastic float and possibly finished with a sponge. This exercise is then repeated on the lift below. Achieving a good joint at the junction of the lifts requires skill and fast work. It also requires the right weather conditions. If the weather is too hot dry-out will be too fast and it will be impossible to achieve a 'seamless' joint.

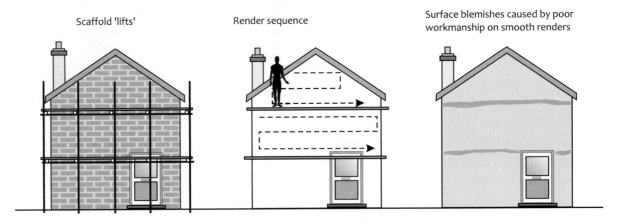

Scaffold 'lifts'   Render sequence   Surface blemishes caused by poor workmanship on smooth renders

Surface popping

This problem is most noticeable on smooth renders which contain lime. If the lime contains small particles of quicklime *(see Chapter 10: Plastering)* there is the risk of hydration (slaking) occurring after the render has been applied to the wall. The resultant expansion causes the render to 'blow' leaving small conical holes in the render surface. The first evidence of this is small circular cracks in the render about 50–75mm in diameter. After a few weeks or months a small cone of render is forced out by the expanding lime. It is often possible to see a small spot or blob of white slaked lime at the base of the cone.

The use of good quality hydrated lime should avoid this problem. If the lime is in powdered form the best results are achieved if it is soaked in water overnight before being added to the mixer.

Patching popping and blowing render is rarely successful – often the wall will need to be painted.

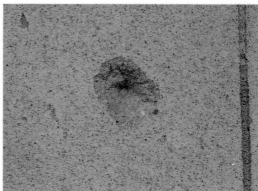

Grinning

If a faint pattern of brick or block joints can sometimes be seen on the surface of the render 'grinning' may be occurring. Grinning usually occurs after rainfall and is often caused by using mortars of different strength, and therefore varying suction, from the bricks or blocks around them. The phenomenon is exacerbated by thin renders. The differences in suction, and therefore rainwater absorption, will result in the wall drying out at different rates.

In some cases grinning is caused by 'cutting corners'. One example is the practice of applying Tyrolean direct to fair faced blockwork without any render base coats. However thick the Tyrolean, it is difficult to hide the minor imperfections in the blockwork and, thus, the joints 'grin' through.

It is particularly noticeable on sunny days becuse the imperfections create small shadows which emphasise the defect.

Lime bloom

Lime bloom manifests itself as patches of a pale, misty white powder on the render surface. It is caused by calcium hydroxide (formed during hydration) coming to the surface of a render during drying and slowly absorbing carbon dioxide to form calcium carbonate. Calcium carbonate is water soluble and usually disappears after long periods of heavy rainfall.

PROBLEMS WITH PAINTED FINISHES AND OTHER SURFACE TREATMENTS

Paint

Smooth renders and rough cast can be painted. If the correct paints are used and if due attention is paid to surface preparation, painted finishes can be durable. In addition, they will help to reduce the tendency of a wall to absorb rainwater. In practice, unfortunately, there are a number of problems which can, and do arise.

Paints should be vapour permeable. External grade emulsions and proprietary 'stone' paints are acceptable although conventional oil based paints should be avoided. Apart from their impermeability (see bitumen paint below), they are not compatible with the alkaline nature of the cement in the render. They are likely to blister and flake after relatively short service.

Even if the correct paints are used, life expectancy can be quite short if the render surface is not properly prepared. Cracks should be cut out and filled, preferably with a flexible filler, and loose surface material and organic growth should be brushed, or washed off, the wall. Special 'stabilising' solutions are usually required on older renders to overcome the adverse effect of loose flecks of old paint, microscopic organic growth or other undesirable surface materials.

In some cases the painted finish is less than satisfactory in terms of appearance. This often occurs where the wall is 'marked out' into small manageable 'bays' as the work progresses. This effectively means that some parts of the wall have more coats than others. The pattern of the marking out is often visible and the only solution is to provide one, or more, extra coats.

'Marking out' can cause an uneven painted finish. Additional coat(s) will be required to hide the blemishes.

Do not confuse with 'grinning' or shrinkage cracking.

Bitumen paint

Completely impermeable coatings such as bitumen paint can cause problems more serious than the ones they are meant to eradicate. They prevent moisture escaping from a wall and can, in freezing conditions, lead to frost attack in the render. In many cases this type of coating is often specified in an attempt to resolve perceived problems of penetrating damp. However, incorrect diagnosis of the causes of dampness is common; surface condensation is often mistaken for penetrating damp. Where this is the case external surface coatings will have no effect on the condensation. In some cases, for example where interstitial condensation is occurring, (see Chapter 13) impermeable surface coatings will prevent moisture evaporation from the wall and will be counter productive.

Silicon

In an attempt to reduce the incidence of damp penetration it is not uncommon to find renders coated with various clear silicon solutions. These can be effective in the short to medium term although they will be counter productive if the render is cracked. Although these solutions are, to a degree, vapour permeable they do limit evaporation from a wall. Rainwater will penetrate the cracks and its impeded evaporation can lead to damp patches on the wall's inner surface. Where DPCs are non-existent or unreliable, silicon treatment of the render can lead to problems of damp penetration and efflorescence on the internal plaster just above DPC level.

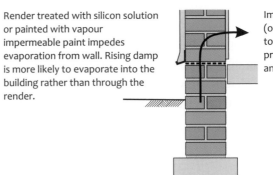

Render treated with silicon solution or painted with vapour impermeable paint impedes evaporation from wall. Rising damp is more likely to evaporate into the building rather than through the render.

Imperfections in DPC (or no DPC) can lead to increased problems of damp and efflorescence.

CHAPTER 10

Plastering and Plasterboard

INTRODUCTION

This chapter describes the various types of internal wall finish found in housing and explains a number of common defects; defects, in the main, caused by inadequate specification, incorrect use of materials, poor workmanship, building movement and chemical attack.

Lime plaster

Lime plaster was common until the 1950s, although its use nowadays is largely confined to conservation work. Lime is produced by heating ground limestone or chalk (calcium carbonate) to a high temperature to remove carbon dioxide. The material thus formed is known as quicklime (calcium oxide). This material was traditionally slaked with water in large pits to produce slaked lime (calcium hydroxide). The lime pits could contain lime putty (slaked lime for later mixing with fine aggregate) or 'coarse stuff' (slaked lime and fine aggregate already mixed). The longer the material was left the better the plaster it produced. In Roman times it was not uncommon for lime to be left three years before use. Nowadays, there are still some specialist suppliers who slake lime in the traditional way, although most slaking takes place in factories where exactly the right amount of water is added to enable the material to break down into a fine white powder. This bagged material, known as hydrated lime, can be tipped straight into a mixer, although better results are obtained if it is mixed with water and left overnight before use.

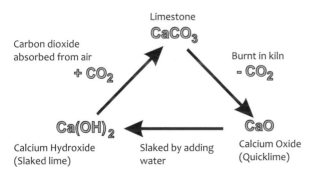

Limestone
$$CaCO_3$$

Carbon dioxide absorbed from air
$$+ CO_2$$

Burnt in kiln
$$- CO_2$$

$$Ca(OH)_2$$
Calcium Hydroxide (Slaked lime)

Slaked by adding water

$$CaO$$
Calcium Oxide (Quicklime)

Slaked lime can be used in three main forms:

- lime putty
- 'coarse stuff' - mixed with sand
- hydrated lime (powdered form, sold in bags)

Limestone burnt in open kiln to form quicklime.

Slaked lime in the form of lime putty can be stored indefinitely in the right conditions.

Lime plaster was usually applied in three coats; render, float and set. The render coat roughly levelled out the wall, the floating coat provided a true, even surface with uniform suction and the setting, or finish, coat provided a smooth finish for decoration. The base coats contained sand, or other fine aggregate, and were normally mixed in the proportions of 1 part lime to 3 parts sand. They were usually reinforced with animal hair in an attempt to resist the shrinkage action of lime as it set. Each coat would, or should, have been left for several weeks before applying subsequent coats. The setting or finish coat usually contained equal parts of lime and fine sand.

A thick base coat of lime plaster

Lime mortars are relatively flexible and surprisingly able to adjust to seasonal movement and subsidence but their major drawback is their slow set. Lime plaster sets by slowly absorbing carbon dioxide from the atmosphere and returning to its initial chemical form, calcium carbonate.

The process of absorption is very slow and the plaster often forms a thin skin on its outer surface, as this is the part which carbonates first. It was quite common to leave these walls several months before applying paint or wallpaper. This delay would be unacceptable today. Carbonation is helped by the presence of moisture; plasters which have dried out too quickly will have little strength.

Typical late Victorian specification

Lime plaster was traditionally applied in three coats; Render, float and set.

The first two coats were normally reinforced with animal hair to improve tensile strength and reduce the risk of cracking.

Typical 1930s specification

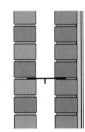

Two coat work sometimes replaced three coat work. Cement was sometimes added to the mix to provide strength and increase the speed of set.

If the binder was cement or cement/lime the hair was usually omitted.

In both periods the sand was sometimes 'bulked out' with ash - a cheaper aggregate.

In conservation work an understanding of lime plaster is essential. Here a 16th century ceiling from a Scottish castle is being re-created for Historic Scotland.

Lime plastering on timber lathing

Until the 1940s ceilings were usually formed from timber lath finished with lime plaster. The best practice used riven laths (split by hand) although, from the Victorian period onwards, the majority of housing used sawn softwood. The thin strips of timber were nailed to the joists. Small gaps in between the laths provided the key for the plaster.

Three coats of lime plaster, sometimes gauged with a little cement, were used to form the finish itself. The first coat of plaster which was squeezed in between the laths was known as the 'pricking up' coat. A similar technique could be used on walls where the laths were fixed to a series of vertical battens fixed to the walls or a free standing stud framework.

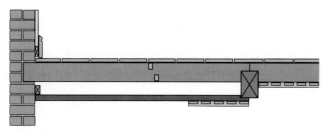

Most ceilings are attached to the floor joists - in some larger properties, particularly those with ornate ceilings, there may be separate ceiling joists

Fir lath (nailed) and three coats of plaster.

Lime gauged plasters

Lime plasters, in their later years, were often gauged with a small proportion of cement or gypsum plaster to impart early strength.

The three coat work to the wall dates from the late 1940s. The two base coats both contain cement, lime, sand and ash. The ceiling plaster 'pricking-up' coat is a similar material although from the early 1930s. These plasters only contain small amounts of cement and should not be regarded as brittle.

Cement plasters

Although the Romans discovered materials with properties similar to modern cement, the material used today has only become widespread since the early part of the 20th century. Cement differs from lime in a number of ways. Unlike pure lime, it sets by hydration rather than carbonation, in other words, the chemical reaction occurs when in contact with water, not carbon dioxide. Again, unlike lime, it cannot be 'knocked back' once the set has started. It is strong and brittle, virtually impermeable if mixed properly, and it develops initial strength rapidly. Dense or cement rich plasters (say a mix of 1 part cement to 3 parts sand) have high drying shrinkage and their brittle nature means that they are unable to accommodate minor thermal and moisture movement; where these occur, cracking is common. They must therefore, be used with caution, particularly where strong mixes are specified (strong or dense mixes are sometimes specified to resist damp penetration). Increasing the amount of sand in the mix (say 1 part cement to 5 or 6 parts sand) will produce a weaker, more tolerant material mix but it is difficult to work and not favoured by plasterers.

By adding lime to the mix greater plasticity can be obtained; the material is easier to work and because it hardens more slowly than cement/sand there is less risk of shrinkage cracking. The cement and lime together (the binder) should be about one third of the sand volume to ensure that the air voids in the sand are filled. A moderate strength mix would be 1:1:6 (cement:lime:sand), a weaker mix 1:2:9. Sand cement plasters are usually applied in two coats (render and float) with a finish or setting coat of gypsum plaster. Sand/cement plasters cannot be trowelled to such a fine, smooth finish and are rarely used as a setting coat.

Another method of producing a workable, plastic mix is to add a plasticiser. This is added as a liquid to cement/sand mixes and usually acts as an air entraining agent. It surrounds the tiny particles of sand and cement with air, thus reducing the internal friction of the mix and making it easier to work.

Sand/cement plasters are most likely to be found on the inside face of solid walls or where DPCs have been added or renewed. They are also common in basements, where the addition of chemical waterproofers can help reduce problems of damp penetration.

Most sand/cement plasters are batched on site although some manufacturers produce pre-bagged cement/lime/lightweight-aggregate plasters suitable for use on most backgrounds (except plasterboard). Probably the most common of these is 'Limelite'. A number of grades of Limelite are manufactured to suit a variety of backing materials. They can also be used to replaster walls after installation of new or replacement DPCs *(see Chapter 14: Damp)*. Unlike gypsum plasters they are not suitable for application to plasterboard.

Gypsum plasters

The mineral gypsum (hydrated calcium sulfate: $CaSO_4.2H_2O$) was formed by the deposition of salts in inland lakes during the course of many geological periods. As these lakes dried, and as the surface formation of the region altered, the mineral so formed was buried to varying depths. Pure gypsum is white in colour, but small proportions of various impurities cause the colour to vary to shades of grey, brown and pink. The use of gypsum plaster spread to this country from France in the 13th century. Plaster of Paris (hemi-hydrated plaster), formed by heating ground gypsum and driving off part of the

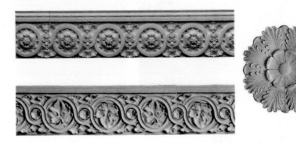

water, was renowned for its white nature and its ability to be moulded into complex shapes. It was, however, very expensive and was reserved for the 'top end' of the market. Plaster of Paris sets very quickly and is not therefore suitable as an in situ wall finish. It was, and still is, used to form precast plaster mouldings found in decorative ceiling work.

By adding various substances to Plaster of Paris, its setting time can be delayed, and by adding bulk filler, such as sand, this material is ideal for plastering internal walls. Keratin is the chemical most commonly used to retard the set (thus giving retarded hemi-hydrated plaster); keratin is a natural protein found in the hair, horns and hooves of cattle. Much of it is imported from South America.

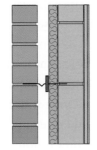

A cavity wall will normally be finished with two coat work; Float and Set. 11mm floating coat and 2mm setting coat.

Gypsum plasters are the most common although cement/lime based pre-mixed plasters are also available.

Cement/lime/sand or cement/sand/plasticer base coats are fairly rare but quite acceptable.

Unlike lime, gypsum plasters set quite quickly and they do not shrink on drying; in fact they are accompanied by a very slight expansion. The plaster sets by combining with water to form a mass of needle-like crystals which interlock and provide a set material of considerable strength. Because of their fast set, gypsum plasters can be successfully laid on a wall in thicker coats than lime plaster; two coat work (float and set) is normal with an overall thickness of about 13mm. Its fast set and reduced number of coats made it an obvious choice in the post-war re-construction of Britain. A typical float coat would contain 3 parts sand to 1 part gypsum (batched on site) and the set or finish coat would be neat gypsum plaster, possibly with the addition of some lime to improve its working characteristics. Another advantage of gypsum plaster is that it can be applied to plasterboard, a material which first became popular during the Second World War where it was extensively used to help patch up bomb damage.

Gypsum plaster should never be mixed with cement. In wet conditions sulfate attack can occur, caused by a reaction between the calcium sulfate and a by-product in the cement.

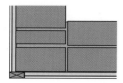

In traditional lime plastering external angles were often formed with long strips of timber...

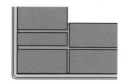

...or, in some cases, rounding the plaster arris and/or using a harder grade of plaster, possibly a hydraulic lime.

Nowadays galvanised steel, stainless steel, or even plastic angle beads are used.

Lightweight gypsum plasters

In the last 30 years, or so, pre-mixed lightweight gypsum plasters have replaced the sanded mixes of earlier years. Nowadays, there are a number of grades available to suit almost any background. The pre-bagged plasters usually contain a lightweight bulk filler (aggregate) and only require the addition of water on site. This provides a material with reliable quality control that is both light and easy to use, and a material which, at the same time, provides improvements in thermal insulation when compared with cement based plasters. Lightweight plasters can also provide substantial fire protection to vulnerable materials, such as steel beams and columns. The lightweight aggregates are normally formed from exfoliated vermiculite (by expanding a mica-like mineral at high temperatures) or perlite. Perlite is a volcanic rock which is chemically combined with water. When heated to high temperatures the combined water turns to steam, thus expanding the perlite to many times its original volume.

COMMON SITE ERRORS DURING CONSTRUCTION PROCESS

PROCESS/STAGE	TYPICAL SITE ERRORS	FUTURE IMPLICATIONS
Background etc	Smooth backgrounds such as dense concrete not primed. Mould oil left on face of concrete	De-bonding
	High suction backgrounds may suck water out of plaster	Plaster will dry before it has set properly – powdery finish with poor bond
	Solid, possibly damp walls, are not generally suitable for most gypsum plasters	Efflorescence, hygroscopic salts, damp penetration
	Plaster board is not suitable for most undercoat plasters	De-bonding
	Metal lath is not suitable for some plasters	Rust staining
Plaster choice (other than items above)	Incorrect grade of plaster. For example, where high impact resistance is needed, 'hardwall' plasters should be used	Surface damage to plaster
Plaster storage etc	Bags stored in damp conditions. Use of old plaster. Use of plaster straight from works	Plaster will not achieve its full strength
Mixing water	Water from butt may be contaminated with chemicals such as plasticisers and cement	Plaster will not achieve its full strength, sulfate attack
Undercoat plasters	Excessive thickness	De-bonding – it is difficult to exclude air when laying on thick coats
	No key for finish coats – light scratching is preferred	De-bonding
Finish plasters on plasterboard	Excess thickness, over 2mm or so an appropriate undercoat plaster should be applied first	

For failure in cement sand plasters at construction stage see Chapter 9. For failure in plasterboard see end of chapter.

DURABILITY AND DEFECTS

PLASTER

The durability of any plaster depends on a number of factors. These include:

- background key
- background suction
- background movement
- strength of the mix
- number and thickness of coats
- dampness
- chemical attack
- defective workmanship.

Defects which occur through the above problems manifest themselves in a number of ways. There may be cracking, either in the plaster itself or in the walling material as well. There may be surface deterioration accompanied by loss of bond, and in extreme cases, partial or complete detachment of one or all the plaster coats. Finally, there may be staining, mould growth or efflorescence which, while not necessarily affecting the durability of the plaster, may damage decorations. In many cases, the cause of the defect can be diagnosed without detailed investigation; shrinkage of sand/cement undercoats, for example, forms a characteristic and unique pattern. In other cases, thorough chemical or physical analysis may be necessary to pinpoint the exact cause of the problem. The first problem, shown below left, is caused by shrinkage of the sand/cement undercoat, the second problem, below right, is caused by minor structural movement.

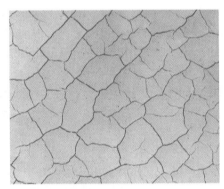 

Specification and site practice

However good the mechanical key and suction of the background, bond failure, surface deterioration and cracking can still occur through poor specification of the plaster and inadequate site practice. There are a number of common problems.

Mix Gypsum plasters, nowadays, are nearly always factory 'bagged'. This ensures that the mix is of the correct proportions. As long as no materials other than water are added on site, the mix should not cause any problems. With cement-based plasters site batching is often a fairly crude process. Excess water in the mix, or a badly graded sand, will lead to problems of shrinkage cracking. In addition, it is common to find mixes containing too much cement. These will shrink as they dry and, although some backgrounds such as dense blockwork can resist this, most will not. As the plaster dries and shrinks, it forms the characteristic map pattern.

Water Water containing contaminants can affect the ability of the plaster to set properly. This can affect both cement based and gypsum plasters. Water should be drawn from the mains, not from butts or drums which have been used to clean tools etc. Water required in the setting process can, as already mentioned, be removed by high suction backgrounds. It can also be removed by evaporation. This is most likely to occur where walls have been 'forced' to dry by using hot air blowers. A similar problem can occur where very thin plasters have been applied in very hot weather. The resultant finish, in both cases, is soft and powdery.

Additives To improve the workability of site batched cement based plasters a plasticiser can be used. This should be gauged strictly in accordance with the manufacturer's recommendations. Most plasticisers act as air entraining agents. They surround the tiny particles in air, thus reducing the friction in the mortar. Too much plasticiser, or plaster too long in the mixer, may contain too much air which will ultimately reduce the strength and bonding properties of the plaster. Washing up liquid has also been used as a plasticiser. This material produces a very workable mix but chemical additives in the liquid will ultimately affect the strength of the plaster.

It is not unknown for milk or urine to be added to gypsum plasters in an attempt to slow their set and keep them workable for longer periods. These prevent a good finish from being achieved and may ultimately affect the bond of the plaster.

Sand For best results, sand should be well graded and clean. Sea sand, if not washed properly, can affect the set of plasters, can cause efflorescence (because of the presence of chlorides) and recurrent problems of dampness (they are hygroscopic). Chlorides can also affect paint finishes and will corrode metals such as angle beads bedded in the plaster.

Pit sand often contains a small proportion of clay. Although this provides a very workable, loamy mix and is, therefore, favoured by many plasterers, the plaster will not achieve its intended strength.

Thickness of coats On uneven walls 'dubbing out' (filling in hollow spots) will be necessary before applying the render or floating coat. Trying to compensate for the uneven nature of the wall by applying an extra thick render coat is rarely successful. Thick plaster coats need considerable pressure if they are to bond to the wall and ensure that any air pockets are squeezed out. At the same time, thick coats are likely to sag under their own weight. Thick coats are a particular problem with heavy sand/cement plasters.

If cement and sand render coats are too thin, they will dry out too quickly. This will inevitably result in shrinkage and loss of bond.

'Dead rats' This is a problem sometimes found in old lime plaster. The 'dead rats' are bundles of reinforcing hair which have not been distributed correctly in the mixer. It can be a problem where hair has been added to the setting coat because it affects the smoothness of the surface finish.

Cracking

The nature and pattern of cracks will help to diagnose their cause. Perhaps the first stage in any diagnosis is to identify whether the cracking is caused by movement in the background, ie the wall, or is confined to the plaster itself. Some cracks are usually attributable to a fault in the building itself. These can take many forms (see Chapters 1 and 2). Diagonal cracks, either stepped or in fairly straight lines, are usually caused by building movement. In new buildings these may be fairly minor and may be caused by initial building settlement. Cracks which occur in later years, particularly cracks over 2 or 3mm wide, are likely to be caused by differential structural movement. Careful monitoring is required before taking any remedial action.

In older properties, horizontal cracking at junctions of ceilings and walls, or vertical cracks in the corners of rooms, may be the first signs of progressive vertical or lateral movement. In some cases, further monitoring and exploration will be required, although in others it may be due to minor differential movement caused by a combination of shallow foundations (possibly at differing depths) and a long hot summer. As the weather changes and the ground absorbs water, the cracks will often close.

In older houses with solid walls, horizontal and vertical cracking of the brick joints can be caused by sulfate attack. Cavity wall tie failure can also cause horizontal cracking of the joints (unlike sulfate attack these cracks will normally be at 450 mm centres – 2 courses of blockwork or 6 of brick). This problem only occurs in cavity walls and will not usually manifest itself until a property is at least

20 years or more old. In many cases wall tie failure will only be noticeable in the external leaf. *This phenomenon is explained in more detail in Chapter 4: Brickwork and Stone.*

Horizontal hairline cracks at junctions of walls and ceilings are usually due, in new buildings, to drying shrinkage. A wider crack at the junction of the upper floor ceiling and wall may be caused by lack of roof strapping. If the crack is horizontal and about 50–70mm below the upper floor ceilings, it can be caused by applying plaster directly to timber wall plates. As the plate shrinks slightly, a hairline crack is formed.

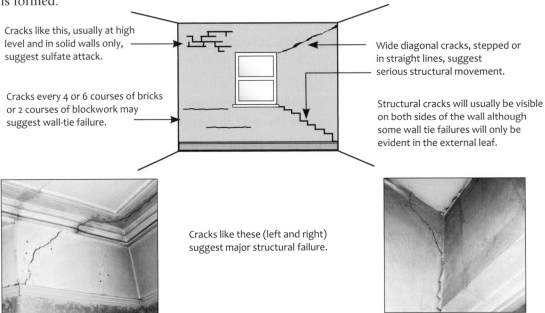

Cracks like this, usually at high level and in solid walls only, suggest sulfate attack.

Cracks every 4 or 6 courses of bricks or 2 courses of blockwork may suggest wall-tie failure.

Wide diagonal cracks, stepped or in straight lines, suggest serious structural movement.

Structural cracks will usually be visible on both sides of the wall although some wall tie failures will only be evident in the external leaf.

Cracks like these (left and right) suggest major structural failure.

Cracks which follow mortar joints (internal leaf of cavity walls only) and which occur in the early stages of a building's life are probably due to shrinkage of wet blockwork or blocks laid in too strong a mortar.

As mentioned earlier, map cracking is most likely to occur where gypsum finishing plasters have been applied to a sand/cement floating coat. The sand/cement coat will shrink as it dries, particularly if it contains a high proportion of cement. If the gypsum setting coat is applied before shrinkage of the floating coat is complete, the continuing shrinkage will crack the finish coat. This type of failure is most likely to occur where houses with solid walls have been replastered. Sand/cement render and floating coats are rarely used in new housing.

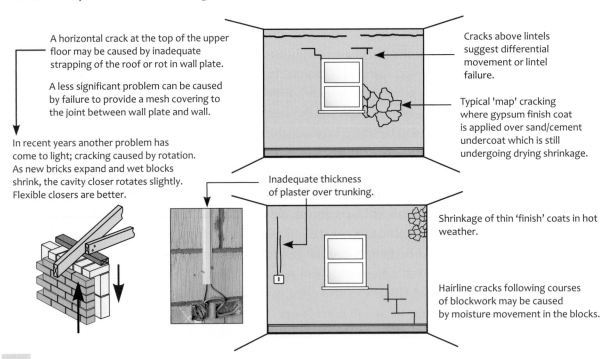

A horizontal crack at the top of the upper floor may be caused by inadequate strapping of the roof or rot in wall plate.

A less significant problem can be caused by failure to provide a mesh covering to the joint between wall plate and wall.

In recent years another problem has come to light; cracking caused by rotation. As new bricks expand and wet blocks shrink, the cavity closer rotates slightly. Flexible closers are better.

Cracks above lintels suggest differential movement or lintel failure.

Typical 'map' cracking where gypsum finish coat is applied over sand/cement undercoat which is still undergoing drying shrinkage.

Inadequate thickness of plaster over trunking.

Shrinkage of thin 'finish' coats in hot weather.

Hairline cracks following courses of blockwork may be caused by moisture movement in the blocks.

In two coat gypsum work, fine vertical cracks above switches are usually caused by inadequate thickness of the finish coat where it hides plastic trunking containing electricity cables.

Cracking commonly appears where two different backgrounds meet. Even if blockwork and brickwork are 'toothed' together, cracking may occur because of the differing rates of thermal and moisture movement. In housing, a common example of two different materials causing cracking is where extensions have been added. In framed blocks of flats, the problem can occur where concrete blockwork abuts concrete columns.

Bond failure

If a plaster surface appears to be in good condition but sounds hollow when tapped, this usually indicates a failure to bond properly with the background. This may be due to top coats not adhering to base coats or base coats not adhering to the wall. Localised hollow spots do not necessarily mean that the plaster is loose and may not warrant further action. However, where hollow spots are extensive, or where there is associated cracking, hacking off and replastering may be the only option.

Failure of a base coat to bond successfully to a wall can be caused by a number of factors. Two of the most significant are the mechanical key and the degree of suction offered by the walling material. Bond failure caused by chemical attack is included in a later section.

Mechanical key and suction

Hollow or loose plaster may indicate that there is insufficient key with the background. This may be due to inadequate surface preparation or because of the nature of the walling material. Problems with the former include failure to remove loose material, paint, dust, soot and organic growth. The nature of the walling material can vary considerably. Dense concrete blocks, for example, with well raked out joints provide a good key for all plaster work. However, some materials such as glazed bricks, in situ concrete, old friable brickwork, stone with wide lime mortar joints and composite surfaces, such as timber and brick, will rarely provide a suitable background. In existing properties where plaster has been removed, much of the key will be lost because it is virtually impossible to remove all the plaster from the walls. The tiny pockets and blemishes on the blocks and bricks form an important part of the key. These will usually be filled with the previous plaster and are difficult to remove. There are a number of techniques used to help deal with surfaces such as these including the use of spatter dash coats *(see Chapter 9: External Rendering)*, metal lath and proprietary bonding agents. Metal lath is often used on weak backgrounds but there are a number of ways in which it can fail. These include:

- inadequate fixing (screwing is better than nailing)
- flat rather than ribbed meshes fixed flush against the wall thus preventing the 'pricking up' coat being forced round the back of the mesh to form a good key
- no overlaps in sheets.

Three coat work is normally required on metal lath, the first coat, known as the 'pricking up coat', should squeeze through the mesh and form a strong mechanical key. Sand/cement plasters, because of their high alkalinity, will help prevent the lath from rusting. If gypsum plasters are used, a grade should be used which includes a rust inhibitor.

Adhesives and spatter dash are more likely to be applied to smooth surfaces where there is little key (Thistle Bonding is one of the few plasters which will adhere well to in situ concrete). However, they need to be applied with care if problems are to be avoided. Water based adhesives such as PVA (polyvinyl acetate) have a tendency to emulsify in wet conditions. They are also ineffective if the plaster is applied after the PVA has dried. Silicon and bitumen based adhesives are often preferred.

Clean common brickwork with raked-out joints provides a good key for plaster.

This cement based render coat has been firmly scratched to form a key for the floating coat.

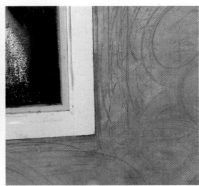

The floating coat is lightly scratched to form a key for the gypsum setting coat.

The suction, or absorbency of the background, can also lead to problems. Backgrounds with moderate suction are suitable for gypsum plasters and sand/cement plasters with a moderate or low cement content. Low suction surfaces are also suitable as long as there is a good mechanical key. Where suction is high, even if there is a good mechanical key, problems can occur through incorrect choice of plaster. Some aerated blocks, for example, have very high suction due to their open, porous nature. Aerated blocks, because of their high thermal insulation, are the normal specification for the internal leaf of modern cavity walls. British Gypsum produce a plaster that is suitable for aerated blocks; it contains a water retaining agent which prevents water (required for the chemical reaction which binds the plaster together) being drawn into the blocks. If other gypsum or strong sand/cement base coats are used, the high suction of the aerated blocks will draw water out of the plaster and reduce its strength, bond, and the quality of its surface finish. On site, plasterers sometimes try and overcome the problem of high suction by dampening or wetting the background. If done with care and in moderation, this can be successful. If the blocks are saturated, they will expand; the resulting shrinkage will form cracking which tends to follow the block joints. This can be confused with problems in the initial construction. Lightweight blocks, for example, laid in a dense mortar, can result in cracking of the bed and perpend joints.

Loss of bond in gypsum finishing coats

In many cases, plaster failure will be a result of failure between coats rather than failure of the coat in contact with the background material. Most vulnerable is the thin setting coat which, in most cases, will be formed from gypsum plaster.

The floating coat should be lightly scratched to form a key for the setting coat. Inadequate key will result in the finish coat 'shelling off'. Where plasterboard is finished with a setting coat, the correct choice of plaster is vital. Plasterboard's relatively smooth surface does not provide a strong key for the setting coat and most manufacturers produce a finish plaster designed specifically for plasterboard. It has reduced expansion on setting to reduce the stresses which may occur at the interface of board and plaster.

An inadequate key can also occur if the backing coat (sand/cement) is over trowelled. This will bring the cement to the surface and can result in the finish de-bonding.

If a sand/cement backing coat is still wet when the finish is applied, the excess moisture can prevent the crystallisation growth which bonds the finish plaster to the backing coat. In addition, sand/cement base coats which are wet will not have undergone their full shrinkage.

In warm weather, or when using finish plaster that has started to set, there is a temptation to add water. This can result in a soft friable finish. A similar problem occurs when using plaster that is beyond its shelf life. Plaster should normally be used within two months of manufacture. Old plaster has a 'flash set' which results in a rough powdery finish.

Chemical attack and staining

There are a number of problems caused by chemical action which may result in surface deterioration or, in some cases, disintegration or de-bonding of the plaster.

Efflorescence Soluble salts, usually present in the wall itself (or some plasters), can be deposited on the surface of the plaster during drying. These salts are usually carbonate and sulfate ions. They are normally harmless and can be brushed off once the plaster is dry. However, if the efflorescence continues it may suggest a problem of damp penetration that warrants further investigation *(see Chapter 14)*. More serious are problems caused by nitrates and chlorides. The former are carried up the wall in solution and indicate a problem of rising damp. The latter may be present in the ground or in some building materials (unwashed sea sand for example). These salts tend to be deposited at the peak of the rising damp. They are hygroscopic (they absorb water) and this prevents them from crystallising. Chapter 14 on Damp explores this phenomenon in more detail.

In some cases efflorescence can be caused by chemicals in the plaster itself. Lime which contains a small proportion of clay (hydraulic lime) may contain the soluble alkalis of potash and soda. These can effloresce in the form of sulfates or carbonates. Some gypsum plasters are also affected. The photo below (left) shows efflorescence which is derived from chemicals in the bricks not the plaster. This is often a temporary phenomenon in new, drying brickwork. It should not be confused with the staining cause by long-term rising or penetrating dampness (right).

Example of efflorescence caused by problems in the brick – not the plaster.

Example of efflorescence caused by long-term penetrating dampness through a solid wall.

Magnesium sulfate attack This type of failure only occurs in cement or lime based plasters (ie alkaline). Gypsum plasters are not affected. It causes a loss of bond due to the formation of needle like crystals of magnesium sulfate below the surface of the plaster. It occurs because magnesium sulfate, which is present in some bricks, is dissolved by water in the wet plaster and reacts with slaked lime (calcium hydroxide), forming a deposit of magnesium hydroxide. In gypsum plasters crystallisation of magnesium sulfate occurs on the surface of the plaster and appears as efflorescence.

Sulfate attack Chapter 4: Brickwork and Stonework, explains the phenomenon of sulfate attack. Internally it is comparatively rare, although it can occur where cement and gypsum are in contact and remain wet for long periods. There is a temptation for jobbing builders and, perhaps, DIY enthusiasts to add gypsum plaster to a sand/cement mix to accelerate its set. However, as already explained, this risks future problems of sulfate attack.

A similar problem occurs where, on walls intended for sand/cement, angle beads are bedded in gypsum plaster because it sets quickly. In damp conditions, sulfate attack can occur and leads to expansion of the plaster and its eventual failure.

Sulfate attack can also occur in cement-based undercoat plasters if they contain ash (sulfur) and if they remain wet for long periods. The plaster will expand slightly, forcing away the finish coat – whether it be gypsum-based or lime-based (see example right).

Staining and damp penetration Rust stains along angle beads can occur in rooms of high humidity, either because of condensation or because of damp penetration. They are more likely to occur where gypsum plasters are used because cement based plasters and lime are alkaline materials which prevent rust from occurring. The galvanising can also be scraped off the beads when plasterers run their sharp trowels along the bead's edge.

Staining of plaster is a common problem on chimney breasts. It is often assumed that this is due to water running down the stack itself, or possibly due to flashing failure where the stack runs through the roof. Although this can be the cause of staining, there are other possibilities. Ammonia and sulfur dioxides (from solid fuel or gas or oil) form salt deposits within the stack or flue. These salts are hygroscopic and are often accompanied by rusty looking stains. Condensation (or rain water) occurring in a flue can absorb these soluble salts and deposit them on the plaster surface when the water eventually evaporates.

Flashing failure

Rainwater or condensation can dissolve chemicals deposited inside the stack.

Damp patches on the chimney breast may be due to condensation as well as penetrating damp.
As the damp evaporates it leaves behind hygroscopic salts which can cause staining.

Both the images shown right were caused by long-term saturation and both are at ground level.

Other surface defects

This final section includes a number of common problems which, although unsightly, will not necessarily affect the durability of the plaster.

Popping and pitting These occur in plasters which contain lime. They are caused by small particles of unslaked lime which slowly continue to slake after the plaster has been applied. As the plaster slakes, it expands. The small particles form small conical holes about 50mm in diameter. At the base of the cone, a small white lump of slaked lime can often be seen.

Lime bloom This can occur in plasters and renders which contain lime. Lime bloom is a fine film or mist of calcium carbonate which forms on the face of the work. It is caused by hydrated or slaked lime dissolving in water and rising to the face of the plasterwork on drying. There it reacts with carbon dioxide to form calcium carbonate. It can also occur in cement sand mixes which do not contain lime as calcium hydroxide is formed during the hydration process.

Pattern staining   Pattern staining, where the faint outline of the joists or battens can be seen on the surface of plasterboard, is caused by dust in the air which is deposited to a varying degree on the materials it comes into contact with. Cold surfaces are more likely to attract dust than warm ones. Thus, a plaster ceiling fixed to timber joists is more likely to find dust settling on the underside of the plaster in between the joists than directly under the joists themselves. The plaster immediately below the joists is better insulated and therefore attracts less dust.

Mould growth   In conditions of high humidity mould growth can occur in plastered walls. Gypsum walls are particularly at risk. The phenomenon is explained in more detail in the chapter on Condensation. It is most likely to occur on cold walls or where ventilation is limited (in bathrooms, kitchens, behind wardrobes etc). Well heated and well ventilated buildings are not usually at risk.

Mould growth (but not the underlying problem of condensation) can be prevented to some extent by adding a small proportion of lime to a gypsum finishing plaster; mould growth is less likely in an alkaline environment. However, adding lime will result in a slightly softer surface finish.

Grinning   This usually occurs where thin or insufficient plaster coats are applied. The joints of the block or brick wall or the scratch marks of the previous coat appear to 'grin' through the top coat. *See Chapter 9: Render, for an example.*

Lack of drying time between coats   Gypsum plasters normally set within a few hours. If a setting coat is applied before the base coat is firm, it's very difficult to achieve a smooth, even finish. The setting coat (right) was applied too soon (within a couple of hours). Extensive sanding was required, and lining paper, to provide a finish suitable for painting.

Paint defects   Insufficient thickness of finish plaster can also result in problems with painted finishes. The finish coat, where it is thin, is unlikely to provide even suction. This affects the painted finish because there is an uneven build up of paint film. Trying to use single coats of emulsion on plaster finishes of the correct thickness can result in a similar appearance. Most paint manufacturers recommend a 'mist coat' followed by two full coats.

Problems can also occur where plasters, which are still wet, are painted with impervious paints. The trapped moisture may cause subsequent blistering or bubbling of the paint work, and possibly, breakdown of the plaster itself.

Plaster failure following DPC installation and problems of dampness
See Chapter 14: Damp, page 280.

Failure of lath and plaster

Timber lath ceilings (and partitions) can fail for a number of reasons. A common problem occurs where the laths are fixed too close together. This prevents the first coat of plaster from squeezing between the laths and forming a good key. Another problem is fixing laths to wide joists. This, again, prevents the plaster from forming a key. Counter battening prevents this problem from occurring but its use is rare in all but the best quality work. Other problems with lath and plaster include:

- excess weight from items stored directly on the lath in a roof space
- rusting of nails – usually caused by damp penetration or old age
- dampness causes rotting of lath – causing bulging or collapse of ceiling plaster
- vibration from traffic and human activity breaks the mechanical key between lath and plaster – this again can result in bulging or detachment of the ceiling plaster
- the use of oak laths (rare) – tannic action on the nails causes corrosion.

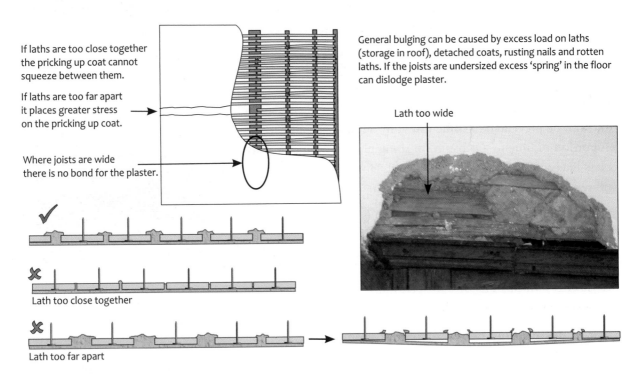

If laths are too close together the pricking up coat cannot squeeze between them.

If laths are too far apart it places greater stress on the pricking up coat.

Where joists are wide there is no bond for the plaster.

General bulging can be caused by excess load on laths (storage in roof), detached coats, rusting nails and rotten laths. If the joists are undersized excess 'spring' in the floor can dislodge plaster.

Lath too wide

Lath too close together

Lath too far apart

PLASTERBOARD

The use of plasterboard first became common in the 1930s when it slowly but steadily replaced plaster and timber lath in ceiling construction. It was also used extensively during the Second World War to help patch bomb-damaged buildings. Since then, its popularity has grown steadily and in modern buildings it is also used to line internal walls and in partition systems.

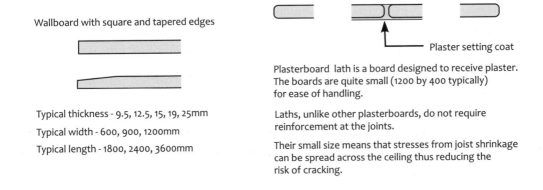

Wallboard with square and tapered edges

Typical thickness - 9.5, 12.5, 15, 19, 25mm

Typical width - 600, 900, 1200mm

Typical length - 1800, 2400, 3600mm

Plaster setting coat

Plasterboard lath is a board designed to receive plaster. The boards are quite small (1200 by 400 typically) for ease of handling.

Laths, unlike other plasterboards, do not require reinforcement at the joints.

Their small size means that stresses from joist shrinkage can be spread across the ceiling thus reducing the risk of cracking.

212

Plasterboard is available in a range of sizes and thicknesses. Plasterboards comprise a gypsum plaster core covered on both sides (and along the long edges) with a heavy paper. The long edges can either be square, rounded or tapered. Tapered edges are designed for taping and direct decoration. There is no need to provide a plaster skim to the boards.

As a dry lining plasterboard can be fixed in a number of ways. Probably the most common is to apply dabs of adhesive to a block or brick background. Another option which is becoming common is to fix the boards to a system of steel channels.

Plasterboard is also used as a finish to studding; it's also invaluable for loft conversions and the like. Traditional plasterboard was nailed to timber supports – nowadays it's usually screwed.

Plasterboard can also be used below existing lath and plaster ceilings, for example to improve fire protection or to hide minor defects.

Specialist boards

A number of specialist boards are available. These include plasterboards backed with a metallised polyester film to provide water vapour resistance, and plasterboards bonded to expanded polystyrene or polyurethane for better thermal insulation.

Finishing the boards

When the boards have been fixed, they are usually taped and jointed, as shown in the photographs on the right. Additional compound is feathered across the tape thus disguising the joint.

When the joint treatment has dried, a slurry of special sealer or similar product should be applied to the board surface using a roller. This provides an even texture and suction for the painted finish. It also provides a sealed surface so that any wallpaper can be 'wet stripped' before repapering. A second coat of sealer can be applied over the first to provide a vapour control layer.

An alternative approach often used for ceilings or for stud partitions is to apply one or more coats of plaster to the boards. Where a setting coat alone is applied to the boards, they can be finished by filling in the joints between the boards with plaster (a grade suitable for boards) and then pressing joint tape or a cotton scrim into the plaster. The tape or scrim should then be covered again with a thin layer of plaster. When the joint treatment has stiffened, the plaster should be laid on the boards with firm pressure, to give an overall thickness of about 2mm. Consult manufacturers for specific details.

If a thicker ceiling finish is required (for improved fire resistance or to hide any imperfections in the ceiling line), the boards can be given a floating coat of Thistle Bonding plaster (about 5mm thick) followed by a skim of Thistle Finish (2mm).

Defects

Plasterboard is an inert material and should be maintenance free. It can be affected by dampness, impact damage or by incorrect loading. Shelves, for example, on dry lined walls should be fixed to the wall itself not just the plasterboard although light loads such as ornaments or pictures can normally be supported using special fixings designed for plasterboard.

Some defects are largely cosmetic; these boards (right) were not taped before being plastered.

Most defects result from incorrect nailing, inadequate background support or incorrect preparation prior to applying a plaster finish. The diagrams below show typical problems which can occur if correct installation procedure is not followed.

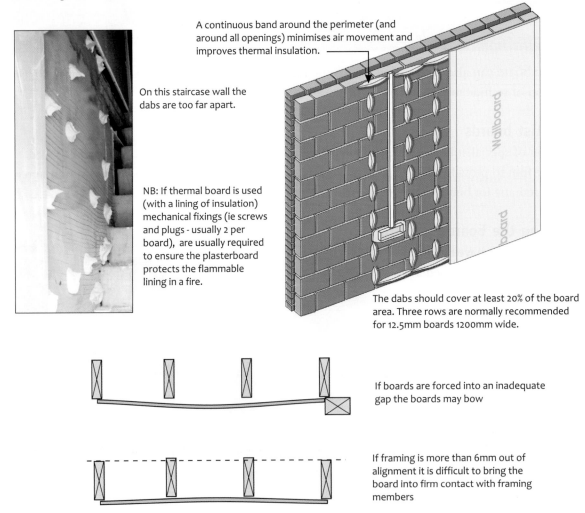

A continuous band around the perimeter (and around all openings) minimises air movement and improves thermal insulation.

On this staircase wall the dabs are too far apart.

NB: If thermal board is used (with a lining of insulation) mechanical fixings (ie screws and plugs - usually 2 per board), are usually required to ensure the plasterboard protects the flammable lining in a fire.

Wallboard

The dabs should cover at least 20% of the board area. Three rows are normally recommended for 12.5mm boards 1200mm wide.

If boards are forced into an inadequate gap the boards may bow

If framing is more than 6mm out of alignment it is difficult to bring the board into firm contact with framing members

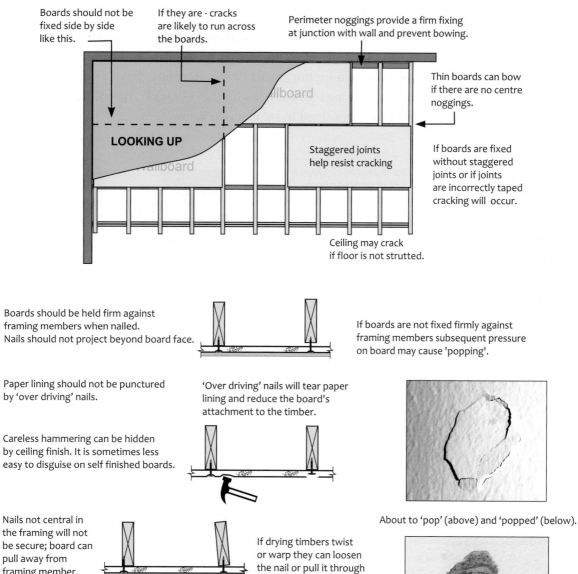

Boards should not be fixed side by side like this.

If they are - cracks are likely to run across the boards.

Perimeter noggings provide a firm fixing at junction with wall and prevent bowing.

Thin boards can bow if there are no centre noggings.

LOOKING UP

Staggered joints help resist cracking

If boards are fixed without staggered joints or if joints are incorrectly taped cracking will occur.

Ceiling may crack if floor is not strutted.

Boards should be held firm against framing members when nailed. Nails should not project beyond board face.

If boards are not fixed firmly against framing members subsequent pressure on board may cause 'popping'.

Paper lining should not be punctured by 'over driving' nails.

'Over driving' nails will tear paper lining and reduce the board's attachment to the timber.

Careless hammering can be hidden by ceiling finish. It is sometimes less easy to disguise on self finished boards.

Nails not central in the framing will not be secure; board can pull away from framing member.

If drying timbers twist or warp they can loosen the nail or pull it through the board.

About to 'pop' (above) and 'popped' (below).

Battens stored externally may contain high levels of moisture.

When the battens shrink they can cause popping.

Ridging can occur if battens shrink. As the boards are squeezed together the compressive forces form a ridge along the joint.

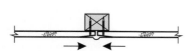

Ridging can also occur if battens twist.

Plasterboard in damp conditions

In damp conditions, plasterboard will quickly deteriorate. The plaster can swell and the paper linings can become detached from the inner plaster core. Perhaps, the most common cause of dampness is applying plasterboard to solid walls which are prone to penetrating damp. In many cases, uninformed builders and DIY enthusiasts see this as a solution rather than a potential problem. Plasterboard used on ceilings can also fail if it is subject to constant wetting from leaks. Ceilings below bathrooms (especially showers) and leaks in ducts lined with plasterboard are particularly at risk.

Vapour checks (vapour control layers)

Even if solid walls are dry, there is a risk of dampness, caused not by penetrating damp but by interstitial condensation. Where external solid walls are drylined, the risk of interstitial condensation requires a vapour check, sometimes referred to as a vapour barrier or vapour control layer. The purpose of the vapour check is to prevent moist air coming into contact with the cold wall. The wall is cold because the drylining acts as insulation. This is explained in the diagram below.

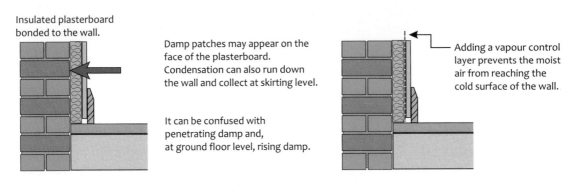

Insulated plasterboard bonded to the wall.

Damp patches may appear on the face of the plasterboard. Condensation can also run down the wall and collect at skirting level.

It can be confused with penetrating damp and, at ground floor level, rising damp.

Adding a vapour control layer prevents the moist air from reaching the cold surface of the wall.

Asbestos boarding and fibreboard

The use of asbestos products has been closely controlled for many years. However, there are still many houses where asbestos boarding has been used to dryline walls or finish ceilings. It was a very common material in the post Second World War reconstruction of Britain, particularly in non-traditional housing. Some grades of plasterboard were suitable for painting, others received a plaster finish. Where asbestos boarding is suspected specialist advice is necessary.

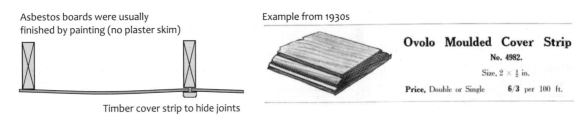

Asbestos boards were usually finished by painting (no plaster skim)

Timber cover strip to hide joints

Example from 1930s

Ovolo Moulded Cover Strip
No. 4982.
Size, 2 × ½ in.
Price, Double or Single   6/3 per 100 ft.

Thin boards can twist and bow, especially if the joist centres are 'stretched'. There are post-War examples of joist centres as wide as 900mm.

Many house dating from the 1940s and 50s have ceilings (and walls in many non-traditional houses) lined with fibreboard or hardboard. These can be quite durable but will suffer if they get wet: they can also crack through impact damage or through incorrect fixing of shelving etc. They also tend to sag and bow, causing an uneven wall or ceiling finish. Fibreboard and hardboard, unlike asbestos or plasterboard, will burn and are therefore potential fire risks. The advert right is from the late 1930s when various boards made by companies such as Essex, Celotex, Sundeala and Masonite were growing in popularity.

ESSEX Board can be useful in many ways

Not only in big jobs such as lining houses, but in many of the minor repairs that constantly crop up. It is light and very easy to handle—you can saw it more easily than wood—and fixing is simple. When it comes to decoration, a better surface could hardly be imagined for paint or distemper and very effective panelled designs can be secured.

Saves TIME, LABOUR, MONEY.

Another thing about "ESSEX" Board is that it won't crack and it laughs at vibration—which makes it ideal for ceilings. Not only is the Board pleasing when fixed, but getting it fixed is quick work.

ESSEX BOARD
MADE IN BRITAIN

CHAPTER 11

Internal Walls

INTRODUCTION

This chapter briefly describes the construction of load-bearing and non load-bearing walls (partitions) and explains the most common defects which can occur. Many of the most common defects are, in fact, caused by incorrect plastering or incorrect fixing of plasterboard. These aspects have been covered in Chapter 10: Plastering and Plasterboard.

LOAD-BEARING OR NON LOAD-BEARING?

Partitions divide space within a building and they can be load-bearing or non load-bearing. Load-bearing partitions usually provide support to upper floor joists and, in buildings with 'cut', or traditional roofs, support for part of the roof structure. They may also provide restraint to the external walls. Non load-bearing partitions divide space and should not carry any loads from floors or roof structure. Depending on their construction, they may also provide some restraint to external walls.

Load-bearing and non load-bearing partitions can be made from brick, concrete blocks, clay blocks and timber studding. Non load-bearing partitions can also be made from compressed straw boards and proprietary plasterboard systems.

In practice, it can be quite difficult to determine whether or not a partition is load-bearing. Although the construction of the partition (where it is known) may give some clues (straw boards and proprietary plasterboard systems can never carry loads), caution is required before reaching any firm conclusions. Timber studding is a common form of load-bearing partition, particularly in Georgian and Victorian construction where they were often used to form internal walls prior to the mass production of common bricks. In some Georgian and Victorian houses timber partitions were trussed; in other words, they were constructed to span unaided across an open space, possibly, at the same time, providing support to the joists immediately beneath them.

The age of a building will give some clue as to the nature of its partitions. Modern houses, with roof structures constructed with trussed rafters, are unlikely to have load-bearing partitions on the upper floor. Modern trussed rafters are designed to span from external wall to external wall. Modern houses are, however, likely to have one or more load-bearing walls on the ground floor (and the first floor, if the house is three storeys) to support the floor joists. Only in the very smallest of houses will the joists span from external wall to external wall without intermediate support.

Houses built before the 1960s are likely to have traditional or 'cut' roofs. In modest houses an internal load-bearing wall is required to support the upper floor ceiling joists. In larger houses, where there are strutted purlins, the wall will also be supporting part of the roof structure.

To identify whether or not a partition is load-bearing requires, therefore, a broad knowledge of building together with an understanding of the evolution and development of construction methods. A roof inspection will fairly quickly identify the nature of its structure. To assess whether or not partitions

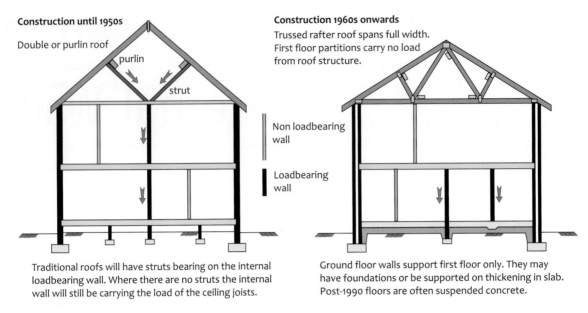

Construction until 1950s

Double or purlin roof

purlin

strut

Construction 1960s onwards
Trussed rafter roof spans full width.
First floor partitions carry no load
from roof structure.

Non loadbearing
wall

Loadbearing
wall

Traditional roofs will have struts bearing on the internal
loadbearing wall. Where there are no struts the internal
wall will still be carrying the load of the ceiling joists.

Ground floor walls support first floor only. They may
have foundations or be supported on thickening in slab.
Post-1990 floors are often suspended concrete.

are supporting floor joists is not quite so easy and may require lifting of floor boards. Where joists run across internal walls and bear upon them directly, it is likely that the walls are load-bearing although it is possible that the joists are merely providing the wall with lateral restraint. If joists meet, and lap, on internal walls there is no doubt that they are carrying load.

CONSTRUCTION SUMMARY

Brick and block partitions

Load-bearing partitions in modern housing are usually formed in blockwork and are most likely to be found at ground floor level where, besides dividing space, they support the upper-floor joists. They will usually be built on a foundation although, in some areas, they may be built on a thickening in the concrete slab. The partitions can be built from dense or lightweight blocks and are usually 100mm thick. In very small houses, where the joists can span from external wall to external wall, load-bearing partitions are unnecessary. Load-bearing partitions should always be fixed securely to the external walls as they provide mutual restraint. 'Toothing-in' is the best structural method although precautions will be necessary to ensure a cold bridge is not created (*see next section on defects*). An alternative is to bond the two walls together with wall ties or mesh.

A typical Victorian house. The centre wall is carrying part of the upper floor, ceiling joists and roof. The internal wall has a foundation although it is often smaller and shallower than the foundations under the external walls.

Non load-bearing partitions formed from lightweight blocks are rare in modern construction but were quite common from the 1930s to the 1960s. The blocks can be made from a variety of lightweight aggregates, most of which are no longer available. These include breeze, coke and clinker.

The blocks are usually 50 to 75mm thick and are often built directly off concrete and timber floors. The thinner blocks often require storey-height door linings to maintain stability. A storey-height lining has jambs (sides) which run up, beyond the head of the door, and are fixed to the joists. Although these blocks do not support loads, they may provide restraint to the external walls and certainly depend on the external walls for their own stability.

Brick partitions can be load-bearing or non load-bearing. Despite the economic advantages of blocks, brick partitions were quite common until the 1960s. They are usually a half brick thick (112mm) although in larger Victorian houses one brick thick load-bearing partitions sometimes formed the spine wall which often supported part of the roof structure.

Hollow pot partitions are no longer manufactured but their use was quite common in the 1930s to 1950s. They were made from concrete or clay and were usually non load-bearing. The thinner examples required storey height door linings to prevent deformation.

Timber partitions

Timber partitions have been used in housing for several hundred years. Since the beginning of the Georgian period in the early 18th century imported softwood has been the most common framing material. Timber partitions can be load-bearing or non load-bearing, the former usually constructed from timbers of larger cross-section (100 x 50mm is typical). Timber studding comprises a series of vertical studs, at centres ranging from 350 to 600mm. The studs are fixed to horizontal bottom and top plates and will normally be stiffened with centre noggings to prevent deflection and twisting. Early studding was covered with timber laths and covered with three coats of lime plaster.

In Victorian and Edwardian housing the studs may have an infill of bricks on edge, ie laid on their stretcher face, to provide improved sound insulation.

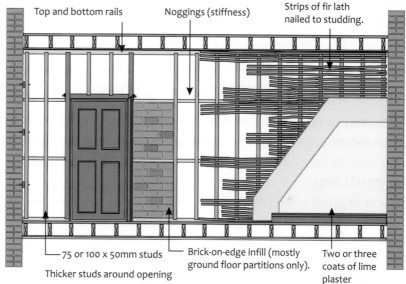

Top and bottom rails   Noggings (stiffness)   Strips of fir lath nailed to studding.

75 or 100 x 50mm studs   Brick-on-edge infill (mostly ground floor partitions only).   Two or three coats of lime plaster

Thicker studs around opening

In modern construction, studding is normally covered with plasterboard or, possibly, with a metal lath finished with in situ gypsum plaster. Stud partitions are rarely found in new housing (apart from timber framing) although their use is more common in rehabilitation work.

As previously mentioned, in large Victorian houses trussed partitions were sometimes used when a load-bearing partition was required over a void. This might be necessary where rooms in an upper floor were situated over a larger room on a lower floor.

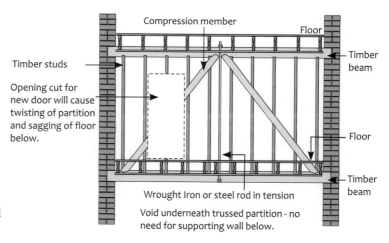

Compression member   Floor
Timber beam
Timber studs
Opening cut for new door will cause twisting of partition and sagging of floor below.
Floor
Wrought Iron or steel rod in tension
Timber beam
Void underneath trussed partition - no need for supporting wall below.

Trussed partitions usually comprised two or more angled compression members to spread the load from above into the supporting walls and a tension member (often an iron or steel rod) to prevent the lower beam from deflecting. Trussed partitions are very effective but problems may arise if subsequent building alterations cut through the compression or tension members.

Proprietary partitions

In modern housing non load-bearing partitions are likely to be formed from laminated plasterboard, cellular core plasterboard, studding or metal framing – both with a 12.5 or 15mm plasterboard covering. Laminated and cellular core partitions have fallen out of favour in recent years due to their relatively poor sound insulation.

Cellular core partition (far left) and metal stud partition (left). A quilt in between the studs or thick plasterboard improves sound insulation.

Other partition types

In the 1950s and 1960s, 'Stramit' partitions were quite common. 'Stramit' partitions were made from large slabs of compressed strawboard covered with a thick paper facing. Storey height panels approximately 50mm thick and 1200mm wide were fixed to a timber or metal framing. The side joints were covered with a jute scrim and finished with a skim of gypsum plaster.

SUMMARY OF CONSTRUCTION ERRORS

This chapter describes the main generic causes of defects in partitions. It should be recognised, however, that there are also a number of common construction errors related to specific partition types.

Stud partitions

- Studs at stretched centres or studs of inadequate size. As a rule of thumb 100 x 50mm studs are suitable for load-bearing partitions, 75 x 50mm for non load-bearing.
- Inadequate fixings; especially to external walls.
- Lack of noggings to stiffen partition.
- Lack of additional noggings to support shelving loads.
- Poor sound insulation due to inadequate thickness of plasterboard finish.
- Poor fire protection due to undersized timbers, lack of noggings or inadequate thickness of plasterboard.
- Inadequate nailing.

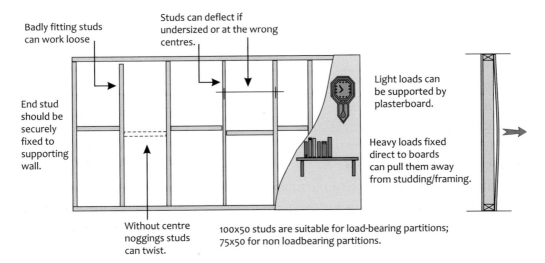

Badly fitting studs can work loose

Studs can deflect if undersized or at the wrong centres.

End stud should be securely fixed to supporting wall.

Light loads can be supported by plasterboard.

Heavy loads fixed direct to boards can pull them away from studding/framing.

Without centre noggings studs can twist.

100x50 studs are suitable for load-bearing partitions; 75x50 for non loadbearing partitions.

Load-bearing partition (far left – timber framed house) and non loadbearing partition (left). The loadbearing partition contains a raking brace because the partition is helping to keep the timber frame rigid.

Block partitions

- Poor fixing to external walls; partitions can be bonded to pockets in the internal leaf or secured by mesh/wall ties.
- Incorrect choice of mortar; a dense (cement rich) mortar may shrink with consequential loss of bond and cracking.
- Inadequate thickness of blocks; 75mm blocks are generally suitable for non load-bearing partitions; 100mm for load-bearing partitions.
- Use of wet blocks and subsequent problems of shrinkage.
- Blocks laid on ground floor slabs without DPC below.
- Lack of bond with other partitions.
- Use of thin blocks without storey height linings.

Plasterboard systems

- Inadequate fixings at top, sides and bottom.
- Use of incorrect door linings.
- Hanging heavy doors on thin partitions.
- Reduced fire protection and sound insulation caused by addition of electrical sockets or pipework penetrating the partition.
- Excess loading caused by heavy shelves, etc.

DEFECTS

There are a number of common defects in partitions. These can occur through:
- imposed loading
- lack of support
- poor connections to external walls and other partitions
- inadequate thickness
- poor sound insulation
- poor fire protection
- thermal/moisture movement, condensation and damp penetration
- poor initial construction (see above)
- overloading (from shelving etc).

Imposed loads

In modern construction one of the most common forms of failure is caused by erecting non load-bearing partitions on the upper floor before tiling the roof. The weight of the coverings (3-4 tonnes) will inevitably cause slight deformation in the truss. If a truss compression member is over, or near, a partition, it acts as a strut transferring load from the roof into the

Non load-bearing walls

Compression member

Partitions under the centre section may distort truss.

Under load the bottom chord of the truss will drop. Lightweight partitions under joints may buckle as they are compressed.

partition. This may cause buckling in the partition and may also cause deformation in the floor immediately under the partition. Partitions which are not in close proximity to the compression members may cause upward deflection in the bottom chord of the truss.

221

Problems of a not dissimilar nature occur in traditional construction. Settlement of foundations (often of inadequate depth) below internal walls will manifest itself in sloping floors and out of square door openings. Where roofs are recovered with heavier materials, the same effect will occur. If the wall is central under the ridge, the force from the struts will be downwards. If the wall is offset slightly, the uneven forces may cause lateral movement as well.

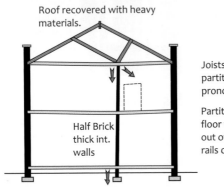

Roof recovered with heavy materials.

Half Brick thick int. walls

Joists bearing on partition develop pronounced sag.

Partitions bearing on sagging floor will have openings out of square and/or picture rails out of level.

Differential settlement will depend on size and depth of original foundation.

In modern and traditional roofs excess loads in the roof space can also cause buckling in non load-bearing partitions. A cold water storage tank, for example, can impose a considerable load and, in modern roofs, may also cause damage to the truss unless correctly supported.

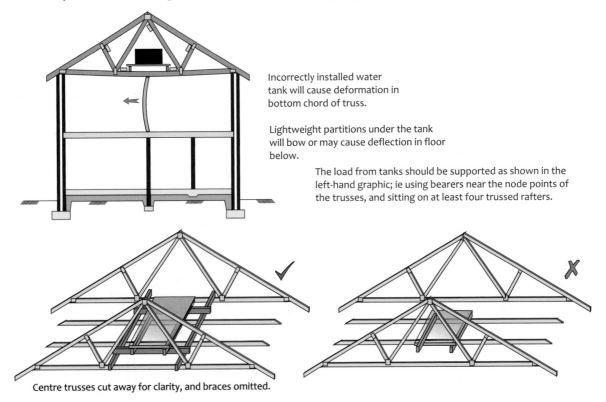

Incorrectly installed water tank will cause deformation in bottom chord of truss.

Lightweight partitions under the tank will bow or may cause deflection in floor below.

The load from tanks should be supported as shown in the left-hand graphic; ie using bearers near the node points of the trusses, and sitting on at least four trussed rafters.

Centre trusses cut away for clarity, and braces omitted.

Changes of use which require the addition of extra partitions (converting large houses into flats) may also cause deflection in a floor, and subsequent overloading of partitions on the storey below. *This is covered in more detail in Chapter 6: Upper Floors.*

Lack of support

The problem of internal load-bearing walls without adequate foundations has already been mentioned. Non load-bearing walls can be built directly on concrete or timber floor although, in the latter, additional joists may be necessary if the partitions are made from blockwork and the partition runs parallel to the floor joists. Blockwork supported by floorboarding alone will result in board deflection. In addition, it is virtually impossibly to remove the boards at a later date.

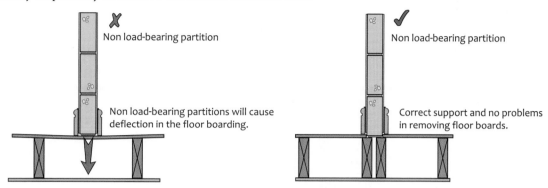

✗ Non load-bearing partition

Non load-bearing partitions will cause deflection in the floor boarding.

✓ Non load-bearing partition

Correct support and no problems in removing floor boards.

Lack of side fixings

Most partitions, load-bearing or non load-bearing, are quite thin. A secure connection to the external walls and other partitions is vital because they may be providing mutual support. The restraint from the internal walls is sometimes vital to ensure the structural integrity of the external wall. This is particularly important where the external walls receive little or no restraint from the floor joists. *This mode of failure has been covered in more detail in Chapters 2 and 6.* Brick and block partitions should be bonded or 'toothed-in' to the wall; an alternative is to use wall ties or expanded metal. Timber partitions can be plugged and screwed to the external walls or, in earlier construction, nailed to timber pads built into the brickwork. The latter method was common before the introduction of cavity walls and is not good practice; dampness in the solid wall can lead to rot in the pads and possibly the partition itself.

Some forms of roof construction may impose a lateral thrust on the external walls which can overcome the bond between partitions and wall. The right hand drawing (below) shows how this can occur. A vertical crack (often wider at the top) indicates a problem of lateral movement.

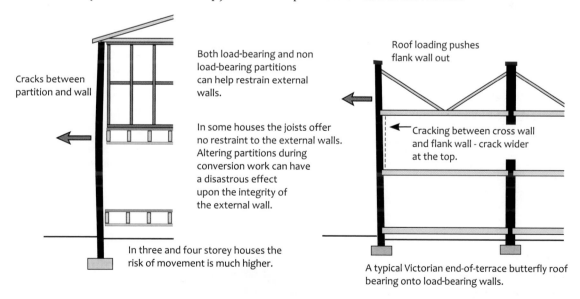

Cracks between partition and wall

Both load-bearing and non load-bearing partitions can help restrain external walls.

In some houses the joists offer no restraint to the external walls. Altering partitions during conversion work can have a disastrous effect upon the integrity of the external wall.

In three and four storey houses the risk of movement is much higher.

Roof loading pushes flank wall out

Cracking between cross wall and flank wall - crack wider at the top.

A typical Victorian end-of-terrace butterfly roof bearing onto load-bearing walls.

Where partitions are not necessary to the structural integrity of the external wall a good connection is still important because thin partitions rely on this connection for their horizontal stability. Thin partitions, whether built from bricks, blocks, studding or plasterboard, can twist or bend if this support is not provided.

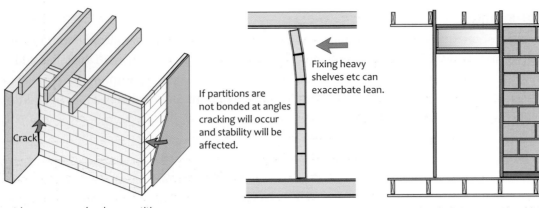

A long, even crack, where partition meets external wall suggests poor bond in original construction.

If partitions are not bonded at angles cracking will occur and stability will be affected.

Fixing heavy shelves etc can exacerbate lean.

50mm and, possibly, 75mm block partitions may lean or buckle if not wedged or fixed to joists above.

Storey height linings provide additional stiffness.

In these two photos you can see some of the stresses cause by lateral movement of external walls. In both examples movement of the left-hand wall (an end of terrace wall) has cause tearing at the junction with the internal wall

Dampness, condensation and thermal/moisture movement

Penetrating and rising damp can affect partitions in a number of ways. Where partitions are built into solid walls or into defective cavity walls, damp may penetrate the edge of the partition. In studding and proprietary plasterboard partitions this can cause rot in the framing materials and deterioration of the plasterboard. Rising damp can affect partitions at ground floor level. This can be due to failure (or non-existence) of DPCs. Where partitions are built directly on concrete ground floors with a membrane below the slab, a separate DPC is still necessary. This ensures that any construction water, or rainwater trapped in the slab, cannot migrate up the internal wall.

Condensation can occur on the face of internal block partitions, particularly if they are built from dense concrete. It occurs in two common situations. The first is where partitions are built next to open or partly ventilated spaces; a typical situation is a wall separating a garage from the main body of the house. The cold wall will allow condensation to form on its inner face. The second problem can occur in modern construction where dense blocks are built into the inner leaf of a well insulated cavity wall. The dense blocks form

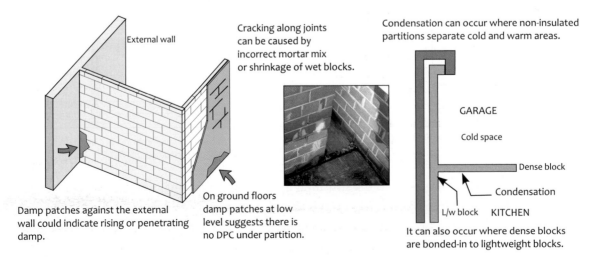

External wall

Cracking along joints can be caused by incorrect mortar mix or shrinkage of wet blocks.

Condensation can occur where non-insulated partitions separate cold and warm areas.

GARAGE

Cold space

Dense block

Condensation

L/w block   KITCHEN

Damp patches against the external wall could indicate rising or penetrating damp.

On ground floors damp patches at low level suggests there is no DPC under partition.

It can also occur where dense blocks are bonded-in to lightweight blocks.

a cold bridge which allows condensation to form on the face of the partition adjacent to the external wall.

Finally, blocks left out on site uncovered and subsequently used in partitions will eventually shrink as the building slowly dries out. As the blocks shrink, fine cracks occur along the bed and perpend joints.

In timber stud partitions, cracking may occur if wet timbers are used. Subsequent dry-out causes cracking and usually manifests itself as fine cracks along the plasterboard joints.

In long lengths of blockwork (usually 6m or more), cracking can be caused by thermal and moisture movement. Fortunately, walls of this length are comparatively rare in housing.

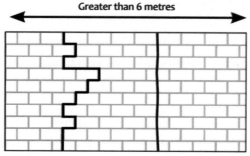

Greater than 6 metres

On lengths of 6 metres or more cracks may appear if movement joints are not provided. Depending on the strength of the mortar these can appear as vertical lines through the blocks or cracked joints.

If partitions are altered through improvement works or alterations blocks of the same type should be used. Different blocks have different characteristics and cracking at the interface of the two materials is a possibility.

In most houses block walls are less than 6 metres long; movement joints are not usually required .

Other cracking

In new houses, where partitions abut ceilings, cracks often appear within a few months of construction. This is usually caused by shrinkage of timber joists as their moisture content slowly settles into equilibrium with its surroundings. A more serious problem occurs in older housing as shown in the photograph right. This cracking appeared after a long hot summer and is probably caused by minor differential settlement or temporary changes in moisture content of the roofing or ceiling timbers.

Cracking can also be caused by excess vibration (traffic or machinery) and heavy doors which are continuously slammed.

Cracks like these may sometimes close slightly when weather conditions change. They are rarely of structural significance and are best dealt with by raking out and filling prior to decoration. Modern silicon based flexible fillers are probably more durable than traditional plaster based ones.

Overloading

Heavy shelving should not be fixed to lightweight or thin partitions. There are two potential problems, failure of the fixings and buckling of the partitions.

SOUND INSULATION AND FIRE PROTECTION

Sound insulation and fire protection are complex topics; far too complex to be tackled in any detail in a book of this nature. This last section summarises some of the most common problems related to partitions and separating walls. Where there is any doubt about the provision of acceptable sound insulation and fire protection, appropriate specialist advice should be sought. This is particularly important in multi-storey housing and buildings in multiple occupation (flats, hostels etc).

Sound

In houses, the occupants should have some degree of control over levels of noise. In flats, the situation is very different. Airborne sound can be very stressful and is a particular problem in properties which have been converted from a previous use. Walls separating dwellings can allow unacceptable levels of sound transmission if they are of inadequate thickness and mass, if the bed and perp joints are not filled, and if joists or services penetrate the separating wall. Separating walls should also be built up to the underside of the roof.

In converted properties, problems of sound insulation often occur where living accommodation is next to landings, corridors and other dwellings. Stud partitions, which are commonly used in this type of conversion, often provide inadequate sound insulation. They may be too thin so that they readily flex and reverberate; they may have inadequate thickness of plasterboard either side; or they may not include insulation mats. Also, it should be remembered that lightweight partitions are inherently less effective than dense, solid partitions with regard to reducing airborne sound. Walls, unlike floors, are not required to provide resistance to impact sound.

Even within flats, partitions may not provide adequate sound insulation. Although most forms of block partition are usually adequate, stud partitions can sometimes be found with hardboard as opposed to plasterboard linings. In stud partitions and cellular core partitions the addition of extra power sockets (recessed not flush fitting) necessitates cutting part of the plasterboard away. This forms a path for the transfer of sound.

Fire

Lack of adequate fire protection is another potential hazard, particularly in properties which have been converted from a previous use. Nowadays, all conversion work is subject to scrutiny by Local Authorities but it should be remembered that not all alteration works are 'approved', and that older conversions may have been subject to less stringent scrutiny. Walls separating dwellings and walls lining routes of escape are particularly at risk. Although block and brick partitions provide good fire protection, timber studding depends on the size of its framing members, and the thickness and supports of its plasterboard covering, for its resistance to fire. It should also be remembered that some timber partitions, especially in older conversions, may be constructed with a facing of hardboard, ply or fibre board. These will not provide acceptable levels of fire protection.

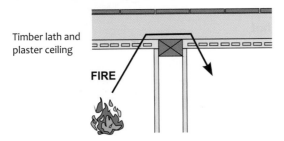

Timber lath and plaster ceiling

FIRE

Timber lath and plaster ceilings do not provide adequate levels of fire resistance.

Stud partitions often have inadequate fire resistance; problems include:

· studs of inadequate size
· no noggings behind board joints
· plasterboard too thin
· linings may be unsuitable (hardboard, fibreboard etc.)
· services passing through partition without fire stopping

Timber studding covered with lath and plaster offers very little resistance to the spread of fire. Similarly, ceilings finished with lath and plaster will readily allow the spread of fire even if a timber stud partition is correctly constructed. Fire can also spread from one dwelling, or room, to the next, through the floor void immediately below the partition.

As with sound insulation, walls separating semi-detached houses or terraced houses should be built up to the underside of the roof. Depending when the houses were first constructed, there may be more onerous requirements regarding fire-stopping to prevent fire from spreading behind the facia board or over the top of the separating wall.

The Building Regulations specify minimum periods of fire resistance for separating walls and wall adjacent to means of escape. These periods range from half hour to one and a half hours. In addition, where a partition supports a floor with a specified level of fire protection, the supporting wall should provide the same level of protection. The Building Regulations, and a number of companies such as British Gypsum, provide more detailed information on specific requirements for periods of fire resistance together with appropriate construction guidance.

In timber-frame housing, fire is obviously a major risk. More detailed guidance on the fire related defects can be found in Chapter 15: System Building.

CHAPTER 12

Timber Pests

INTRODUCTION

Timber has been used successfully in construction for many hundreds of years. Although the 17th century saw the increasing popularity, and then dominance, of stone and timber for external walls, timber has remained a significant material for house construction. Timber is valued for its functional and its aesthetic qualities. It is also a relatively durable material. Unfortunately, this quality of durability may be seriously affected if it is attacked by what are known as timber pests. In the context of buildings the term timber pests refers to those organisms which attack timber for food. These organisms can be divided into two groups:

- fungi
- wood-boring insects.

The risk of damage by timber pests for any given piece of timber in a building will be related to a number of factors including:

- the natural durability of the species of timber
- the 'food value' of the timber
- adverse environmental conditions – particularly moisture content of the timber
- protection afforded through any treatment by chemical preservatives.

The existence of timber frame buildings dating from the Medieval and Elizabethan periods is a testament to the natural durability of some timber species. But even as far back as the Elizabethan period there were problems in obtaining sufficient economical supplies of appropriately durable timber for building. This was largely due to demands from other uses, such as ships, but made worse by the poor management of timber supplies. During the Georgian and Victorian periods there was a vast increase in the use of imported softwood. This coincided with changes in design and building use which resulted in these timbers being utilised in particularly vulnerable positions (eg suspended timber ground floors). The problem of quality being reduced as demand increased has been repeated in more recent times with the use of smaller sections of timber from relatively immature trees.

Some timbers are naturally more resistant to fungal attack than others. This durability is essentially determined by the presence in the tree of naturally occurring substances which are toxic to fungi. Natural durability is primarily determined by the type, amount and distribution of these substances. Because this development of an inbuilt resistance has evolved with varying degrees of success between species, some timbers are more durable than others. In terms of timber use in buildings, knowledge of natural durability will have an effect on a number of decisions including where the timber might be used and the necessity for preservative treatment. Specifying an appropriately durable timber and reducing the likelihood of adverse, particularly damp, conditions by attention to appropriate design and construction details can reduce the risk of attack by timber pests. Although preservative chemicals can reduce the risk of fungal or insect attack, there are financial and environmental costs arising from their use.

Wood is the raw material; timber is the term for wood that has been converted for use. Wood consists of millions of cells, whose function is to provide the physical support to the tree as well as to store and circulate food. The cells consist of cellulose fibres. There is a 'crust' around these cellulose cells formed by a material called hemicellulose. A material known as lignin is the matrix between the cellulose and hemicillulose. Lignin helps to determine hardness and resistance to fungal decay. It is these materials which are attacked by decay organisms. The physical nature of the cells is also a factor in durability.

The seasonal cycle of growth of tree cells produces the rings that can be seen when the tree is cut. Sapwood is that part of the tree that carries nutrients from the ground to the branches and leaves and, in the other direction, the products of photosynthesis from the leaves to the roots. As a tree ages the older cells die. The portion of the trunk that consists of these dead cells is known as heartwood. Obviously, with age, the proportion of heartwood to sapwood increases. The proportion of sapwood to heartwood will also vary between tree species. Heartwood is in effect dead sapwood; it does not carry nutrients but it functions as a depository for waste matter and as a skeletal support to the trunk. The heartwood can often be distinguished from sapwood because of a distinct colour difference – it is usually darker due to the presence of toxic organic substances. The colour distinction is often less clear once the timber has been incorporated in a building. In living trees the combination of natural toxins and the fact that the cells are full of water means that they are relatively well protected against attack by fungi. Where attack does take place it is usually associated with damaged or dead areas. In some instances, heartwood can be relatively vulnerable due to its lower moisture content. However when a tree is felled, and then converted into timber for use in a building, the moisture content will be reduced (through natural means and by drying processes). As the timber dries the moisture content of the sapwood will fall within the optimum range that will encourage attack by fungi (and beetles).

Timber should not be used in a building until its moisture content is below the minimum necessary for fungal attack. In some cases this may not happen or, due to a defect, the moisture content may rise. In such situations the sapwood is vulnerable to decay because it contains the nutrients that timber pests feed on. Because it will contain natural toxins and does not carry nutrients the heartwood of timber in a building has a greater resistance to insect and fungal attack than sapwood.

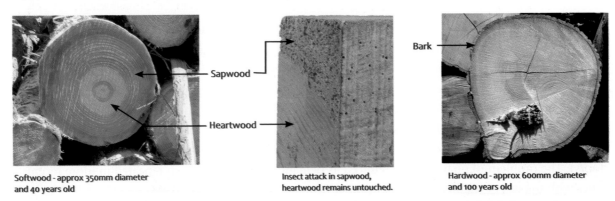

Softwood - approx 350mm diameter and 40 years old

Insect attack in sapwood, heartwood remains untouched.

Hardwood - approx 600mm diameter and 100 years old

Many older properties will contain timbers with a higher percentage of heartwood than is the case in newer properties. In the recent past commercial pressures on forestry production have meant that timber for building use tends to have a high proportion of sapwood and early wood because it is taken from younger, faster growing trees. This goes some way to explain why, for instance, relatively new timber windows, notably from the 1960s onwards, may prove to be less durable than those from the Victorian or Georgian periods.

Timber is classified as either softwood or hardwood. The terms hardwood and softwood can be confusing as they refer to botanical differences between broad-leaved trees (hardwood) and coniferous trees (softwood) rather than the actual hardness of the material. Balsawood is a much quoted example of timber which is categorised as a hardwood but is physically soft. Conversely, yew is a very hard softwood. In reality though, the majority of hardwood timber used in buildings is harder than the majority of softwoods. The terms may also mislead in assumptions of durability because, for example, some softwoods are more resistant to fungal decay than some hardwoods.

Many older buildings will have been constructed using hardwoods, although as mentioned previously, softwoods became the most common type of timber during the Georgian period.

FUNGAL ATTACK

Fungi are important organisms. They are present in great numbers in many situations and they play a number of roles. They are responsible for the breaking down of organic matter, they can cause disease both in animals and in man, they are the basis of fermentation and they help to produce chemicals and medicines (such as antibiotics). They are important factors in enhancing the growth of living trees but are responsible for the decay of timber 'in service'.

In very simple terms, fungi are a group of plants which do not possess chlorophyll and do not, therefore, have the ability to produce their own food through the process of photosynthesis. Because of this they feed by attacking a cellulose based food source such as wood. They do this by secreting acids and enzymes into the wood, which break it down, the soluble end product then being used as food by the fungi. This process has evolved so that dead trees and plants which fall to the ground, can be broken down into their essential components and re-integrated into the ecological chain. If fungi did not exist the Earth would be covered in dead trees.

Fungi are an important part of the ecological chain.

Fungi which attack timber in buildings will usually develop and thrive in certain environmental conditions. It has been observed that fungi will

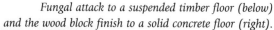

Fungal attack to a suspended timber floor (below) and the wood block finish to a solid concrete floor (right).

C Edmans

F Blampied, Bristol City Council

thrive in those parts of a building which go some way to replicate the conditions of the forest floor. The key factor is the moisture content of the timber but the amount of light, air movement and humidity are also important. If timber is attacked by fungi then the key to effective remedial treatment is to alter the environmental conditions and deal with the fault which is allowing or causing the high moisture content of the timber.

There are generally recognised to be six groups of fungi growing on timber:

1. brown rots
2. white rots
3. soft rots
4. stains
5. moulds
6. plaster fungi.

Of these the ones which will actually break down timber belong to the groups known as brown rots and white rots. Fungi belonging to the other groups will not usually cause serious damage to timber, often because they cannot break down the cell walls. Moulds and stain fungi are not a serious problem, although they can increase a timber's porosity and they are able to neutralise some of the timber's natural fungicides. Soft rots can cause some surface decay, particularly in very wet timbers, but the growth is usually not very extensive in buildings (it can be more significant with timber in contact with the soil – garden fence posts for example). However, because their formation is associated with damp conditions, the presence of these 'non-destructive' fungi will indicate that conditions are suitable for the more serious rots to occur.

Brown rots

These can cause severe decay in timber. The timber cracks into cubes during the decay process. This cuboidal cracking is a primary identifier of a brown rot attack. Brown rots can cause decay and a consequent lose of strength in the timber quite quickly. As a consequence of the attack the wood becomes darker in colour and often becomes so dry that it can be crumbled easily between the fingers.

Cuboidal cracking is caused by brown rots.

White rots

With white rots the timber develops a rather fibrous consistency which is sometimes described as being similar to lint. As a consequence of the attack wood generally becomes lighter in colour – in some cases almost as though it has been bleached. The cracking that appears tends to be along the grain, rather than along and across (ie cuboidal cracking) as is the case with brown rot.

As a generalisation white rots tend to prefer a higher moisture in timber than is the case with brown rots.

Life cycle

The life cycle of wood rotting fungi is shown in the diagram to the right.

The cycle begins when the spores (seeds) are dispersed from the fruiting body. These spores are very small, but numerous and light, and can be carried great distances by air currents. A fruiting body can produce literally millions of spores over a short period of time. Some of these spores will inevitably alight on timber. If conditions are suitable the spores will germinate and

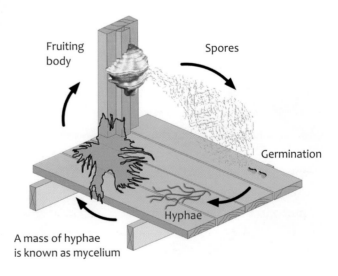

Fruiting body

Spores

Germination

Hyphae

A mass of hyphae is known as mycelium

develop microscopic root like filaments, known as hyphae, which penetrate the wood. A massing of hyphae, which usually renders them visible, is known as mycelium. Fungi use enzymes and acids to dissolve the timber cells. The nutritious products of this breakdown are absorbed and used as food.

When a fungus is well established, and conditions are suitable, it may complete its life cycle by producing a fruiting body. The fruiting body is known as a sporophore. It is the most easily identifiable part of the fungus and will produce the seeds/spores which can be distributed in order to start the cycle again.

The process described above may be interrupted if environmental conditions change.

The timber floor in this building was replaced following an attack by wood rotting fungi.

Essential conditions for wood decay by fungi

Fungi will only be able to survive and to thrive in certain conditions; these are discussed briefly below:

Food Susceptibility to fungal decay varies between the different species of timber. In general, the sapwood will be attacked because it contains the nutrients that the fungi needs (although in certain timbers the heartwood can be susceptible). Modern construction, as we have already observed, tends to use timber from relatively young trees which therefore tend to have a high proportion of sapwood.

Moisture Moisture, sufficient to allow spores to germinate and the fungus to develop, must be present for decay to occur. In a building which is correctly designed, constructed and maintained, the moisture content of the timber should remain low enough to prevent fungal attack. The moisture content of the timber is usually the most critical factor in the establishment and continuation of decay – without a high moisture content in the timber, fungal decay will not start, and if excess moisture is removed growth will stop.

Moisture content may also have an influence on the type of fungi that develops.

There is some debate about what the minimum moisture for fungal growth is. Although some sources have a lower figure, a minimum moisture content of around 22% (for dry rot) is often referred to. In practice this figure is commonly reduced to 20% to allow for margins of error in assessment. The optimum moisture content range for the various fungi will be higher than this figure.

Temperature Fungal growth can be killed off by low and high temperatures and its growth will be slowed down as temperatures are lowered beyond a certain point. Some fungi can become dormant at lower temperatures but may start to grow again if the temperature rises. The optimum temperature for growth will depend on the type of fungus.

Air Fungi require air, or more specifically, oxygen, for growth. If air is excluded then decay will not occur. For example, timber lying underwater may have no oxygen in its cells – because they are full of water – and therefore the wood will not decay.

If one or more of the above conditions are absent or unfavourable then fungal decay will not occur. Even when decay has become established a change of conditions may kill off the fungus, alter its pattern and rate of development, or cause it to enter a dormant state. If the fungi remains in a dormant state it will eventually die, but if the change of condition is only temporary the fungus may resume its evolution.

Daylight This apparently has a complex relationship with fungal growth. Daylight is not generally required, indeed, for most of the time it will be detrimental to growth. However, it is believed that at certain stages of fungus evolution, particularly in relation to the formation of the fruiting body, exposure to daylight may actually accelerate growth.

Effect of fungal attack on wood

The most important effect of an attack is the loss of strength. As decay proceeds the wood becomes weakened producing splits and cracks, and at a more advanced stage the wood may crumble at the touch. Other effects include a loss of weight and a change in colour. A further effect may be that the wood is more easily attacked by wood-boring insects because it has been 'softened up' by the fungi.

Types of fungal attack

There are a number of fungi which will attack and break down timber but in the context of buildings they are divided into two groups known as wet rot and dry rot. Dry rot is a specific type of brown rot. All other rots which attack timber – whether they are brown or white rots – are referred to as wet rots.

DRY ROT

Dry rot is often referred to by its Latin name *Serpula Lacrymans*. It is the term given to a particular identifiable brown rot. The popular name refers to its effect on wood – which becomes dry and crumbly – rather than the conditions that create it. Dry rot spores are probably present in most buildings, although some buildings are more prone to attack than others. This is because of the presence of excessive moisture, perhaps due to defects, but it can also be because once a successful attack occurs there are more spores around increasing the likelihood of further successful dry rot germination. Figures for the level of moisture content which make timber susceptible to dry rot vary. It has been suggested that it can occur in timber with a moisture content of about 17%, but figures of 20% or 22% are more usually referred to.

Dry rot prefers damp still air. An environment where timber is in contact with damp brickwork, and where ventilation is poor, is ideal for its growth. Therefore sub-floor voids or cellars with no ventilation are examples of ideal locations for its propagation.

Because of its sensitivity to environmental change the dry rot fungus is only rarely found outside buildings. Reports of specimens being found in the pine forests of central Europe or the Himalayas seem to result in expeditions being mounted with an enthusiasm similar to that involved in searching for the Yeti.

How the fungus spreads

The mycelium of dry rot can spread extensively on the surface of affected wood, and as the growth progresses it can develop various visual characteristics, often in response to differing environmental conditions such as air temperature, air speed, light, etc. For example it may look like what can be best described as a mass of cotton wool, bright lemon patches may be seen and other patches of colour, particularly lilac, may occur. In other circumstances a thin rather 'leathery' skin may appear. Dry rot produces conducting strands which develop within the mycelium. These strands (rhizomorphs) supply water and nutrients to the active growth area. These strands can move not only over wood but also over the surface of inert materials such as masonry. The 'strands' are able to penetrate some plaster and mortar joints, which effectively allows an outbreak to spread from building to building via say a masonry party wall. This is most likely to happen in older properties where the alkaline nature of lime based mortars and plastering has reduced over time. However replastering or repointing, for example, may help to restrain growth because the mortar will be relatively highly alkaline. Obviously, the strands can obtain no sustenance from masonry or plaster – although the fungi may utilise, and indeed depend upon some of the chemicals present in these inert materials to assist in its development. The important point is that the dry rot strands can travel over or through the inert materials, in search of a suitable food source. As they do this they will, in the right conditions, continue to supply food to the growing ¬ ¬ considerable period of time. In many older buildings dry rot will find it relatively easy to ¬ ¬ through the mortar joints and to grow behind older plaster particularly where it has lost its key. Many older buildings have timbers 'buried' in the brickwork behind plaster and this can provide a potential food source for the fungi (if the timber is sufficiently 'wet').

It is sometimes stated that dry rot can create its own condition for growth by producing water and therefore wetting timbers sufficiently to make them vulnerable to attack. It would appear that while the breakdown of timber can produce significant amounts of moisture this will occur over a period of time and will often tend to be removed by evaporation and other mechanisms.

Identification of dry rot

There are some distinguishing features of dry rot attack which can help with identification.

- The wood becomes dry and crumbly. It becomes light in weight and has a dull brown colour.

- Although both dry rot and the brown rot species of wet rot produce a cuboidal cracking effect in the timber under attack, in the case of dry rot the cracking is often more pronounced. In practice, however it is often difficult to make this distinction.

Cuboidal cracking caused by dry rot.

- The strands are grey or white and 2–8mm thick.

- Depending on environmental conditions the mycelium may develop some different physical attributes, for example, as noted previously, it may have a leathery, pearly white skin or it may look like cotton wood balls or silky sheets.

These photos show different stages in the development of the mycelium of dry rot.

- The spores are often the first indication of dry rot. Individually they are microscopic but 'en masse' they appear as a rusty red coloured dust.

The reddish dust like spores of dry rot.

233

- The fruiting body is usually only found indoors and is quite distinctive. It is usually either pancake or bracket like in shape. The fungus has a cratered surface, sometimes with folds, and is a reddish brown with grey-white margins.

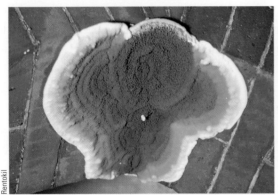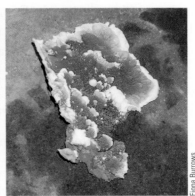

The fruiting body of dry rot.

Fruiting bodies are useful identifiers because they are unique to each species. Unfortunately they are often hidden or they may not have been produced.

In practice identification of the fungus can be difficult, and confusion can arise. A common confusion is to think that cuboidal cracking only occurs with dry rot whereas in fact it is a symptom of brown rot.

At least one specialist uses specially trained dogs ('rot hounds') to seek out (living) dry rot by scent.

Dry rot – note the fruiting body in the top left and the mycelium in the top right of the image.

Dry rot showing fruiting body and hyphal threads (as mycelium).

WET ROT

As mentioned previously there are a number of fungi which can be identified by the term 'wet rot', but it is normal to consider them under this single heading.

The table on the opposite page summarises the identifying characteristics of the common wood-rotting fungi. It is produced by the Building Research Establishment *(see BRE Digest 345)*.

Identification of wet rot

There are a number of wet rot fungi and their characteristics differ as, to a certain extent, do the optimum conditions for growth. The mycelium produced may differ between the species, as may the fruiting body and the strands. Because wet rots can be either brown rots or white rots some of the wet rots produce a cuboidal cracking reminiscent of dry rot attack, some may cause a darkening of the wood, while others may bleach it.

Wet rots tend to thrive in timber with a moisture content higher than the optimum for dry rot.

FUNGUS	MOST COMMON SITUATION	FRUIT-BODY	STRANDS	MYCELLUM	DECAY TYPE	TO DISTINGUISH FROM DRY ROT
Serpula lacrymans (dry rot)	Timber in contact with wet brickwork	Rusty red folded surface; flat or bracket shaped. Spore dust produced	White to grey, robust; brittle when dry	White to grey silky sheets, often with patches of lemon or lilac	Brown rot, commonly with large cuboidal cracking	—

BROWN ROTS

FUNGUS	MOST COMMON SITUATION	FRUIT-BODY	STRANDS	MYCELLUM	DECAY TYPE	TO DISTINGUISH FROM DRY ROT
Coniophora puteana	Very damp situations	Olive green – brown, surface irregular lumps. Flat on substrate	Dark brown, yellowish when young	Rare. Cream to brownish; off white under impervious floor coverings	Brown rot, commonly leaves a skin of sound timber	Brown strands
Coniophora marmorata	Floors, cellars	Pinkish-brown, smooth to lumpy. Flat on substrate	As *C. puteana*	As *C. puteana*	As *C. puteana*	Brown strands
Fibroporia vallantii Poria placenta Amylorporia xantha	Areas of higher temperature	White to cream to yellow; surface covered in minute spores	White to off-white (only well developed in *F. vallantii*); flexible when dry	White to off-white, woolly or fern-like	Brown rot; usually paler than dry rot	Flexible strands; fruit-body has spores
Dacrymyces stillatus	Exterior joinery	Small orange fleshy lumps	None	None visible	Brown rot	Fruit-bodies
Lentinus lepideus	Skirting boards	Pale beige mushroom with gills; commonly 'staghorns' produced	None	White with pinkish brown patches; distinct sweet organic smell	Brown rot	Mycelium and smell
Ozonium stage of *Coprinus*	Laths	None in this growth form	Ochre, thin	Ochre; can be very resilient to the touch	Brown rot; laths crumble	Colour of mycelium and strands
Paxillus panuoides	Very wet situations	Yellow to amber with gills	Yellow to amber	Yellow to amber, woolly	Brown rot; yellow coloration initially	Colour of fungal growths
Ptychogaster rubescens	Timber in contact with wet brickwork	Pinkish-brown powdery cushions	White, not common; may be brittle when dry	White, woolly	Brown rot; may leave skin of sound timber	Fruit-body

WHITE ROTS

FUNGUS	MOST COMMON SITUATION	FRUIT-BODY	STRANDS	MYCELLUM	DECAY TYPE	TO DISTINGUISH FROM DRY ROT
Donkioporia expansa	Ends of large oak beams embedded in walls	Brown or buff, covered in minute spores	None	Not always present; yellow to red-brown felted growth	White rot	Decay type, fruit body
Asvomycete decay eg *D. concentrica*	Hardwood floors particularly sports halls	Hard black rounded, concentric rings inside	None	None visible	White rot plus superficial date zone	Decay type
Asterostroma spp	Skirting boards	Cream to beige to light-tan, smooth, flat on substrate	White or light tan, sometimes across masonry for long distances	White, cream or buff sheets	White rot; often little decay	Decay type; strands flexible when dry
Phellinus contiguus	Exterior joinery	Ochre to dark brown, covered in minute pores	None	Tawny brown tufts	White rot	Decay type; colour of fungus growth
Pleurotus ostreatus	Chipboard	Grey or fawn on top, white gills underneath	None	Whitish woolly mat; strong mushroom smell	White rot; chips tend to separate	Decay type; fruit-body

Wet rot is often found where wood is repeatedly wetted, for instance as a result of faulty plumbing or leaking gutters, or where there is poor protective detailing. For the reasons mentioned above a description of every wet rot fungi is inappropriate in this publication. The brief description below illustrates the point by comparing the characteristics of just two types.

A common wet rot is cellar rot *(Coniophora Puteana)*. It is a brown rot which attacks both hardwoods and softwoods. The wood becomes darker and suffers from cuboidal cracking. The fruiting body is thin and flat and is olive coloured. The fruiting body is rarely found in buildings. The strands are yellowish coloured during their early development but become darker, reaching a blackish colour with age. Because of the cuboidal cracking this fungi is sometimes confused with dry rot and undoubtedly often treated as such.

Phellinus Contiguus is a white rot that is commonly found on external joinery. Its mycelium looks like brown tufts. The fruiting body is rare and it is an elongated shape which is coloured dark brown. As it is a white rot the wood becomes bleached and has a stringy, fibrous appearance.

A wet rot attack to the frame and threshold of an external door.

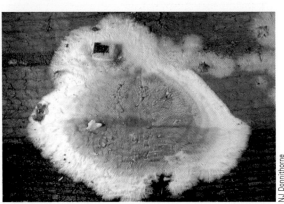

Coniophora puteana.

NJ Donnithorne

The remedial treatment of dry and wet rot

It is important that the cause as well as the symptoms of a rot outbreak is adequately dealt with. The first step in remedial treatment for rot is to determine the cause of the outbreak. This will involve finding and eliminating the source of dampness and rectifying those faults which have contributed to the problem. Fungal decay happens because conditions are created which allow for the germination and evolution of an attack and the key factor in creating and maintaining these conditions is timber with a high moisture content – without which there will be no rot. Even when wood is infected, if its moisture content is reduced to an appropriate level the fungus will stop developing and then die off. It is important to remember that the removal of damp conditions is the key treatment in dealing with fungal decay. Therefore the cause of the excessive moisture should be identified and rectified. Measures should also be undertaken to dry out the structure and fabric as quickly as possible. Clearly timber which has been unacceptably weakened structurally should be replaced. In reality it is common practice for all infected wood to be removed plus, in the case of dry rot, seemingly unaffected timber say 300–450mm beyond, to allow for the possibility of less obvious infection. There is a debate about the necessity for this approach, because drying out and/or selective 'preservative' treatment will kill the fungi (although infected timber will remain a risk while it dries out).

There may of course be situations where it is not possible to cost-effectively ensure that dampness does not reoccur. Timber which is removed can be replaced by timber which has been pre-treated with preservatives, or alternatively material which will not rot can be substituted (for example concrete or steel). It may be necessary to treat remaining timbers in situ with preservatives, particularly where damp conditions may remain for some time, or where isolation from, say, damp masonry is difficult.

Because the dry rot fungus can spread rapidly throughout a building, including through mortar joints and behind plaster, it is best if the full extent of the outbreak can be established in order to ensure that all vulnerable or infested timbers within and near the infected area are located. The investigation should take account of timbers such as noggins, levelling pieces, etc, which may have been incorporated in the masonry. Also, because the dry rot fungus can penetrate masonry, the separation of vulnerable timbers from infected masonry needs to be well considered.

In many cases this investigative work can be a costly and destructive exercise. Additionally, where historic buildings are involved, there is a specific concern about the damage to historic fabric where large areas of timber or plaster are removed for precautionary reasons. Decisions on the scale and scope of investigation and treatment is essentially a professional judgment made in the light of the particular circumstances. Excessive destruction of materials or the unnecessary use of chemicals can be avoided, if the causes and risks associated with dry rot are better understood. The balance between a rightful appreciation of the damage that can be caused by dry rot and that which is often caused by the methods which are sometimes adopted for its eradication should be considered.

A detailed description of remedial treatment is inappropriate here (there are a number of publications produced by the Building Research Establishment, to which further reference can be made). It should, however, be noted that concern over the health of humans, and indeed the wider environment, in relation to the toxicity of timber preservatives has raised concerns about what might be called 'traditional' approaches to the eradication of rot in timbers. Although the use of in situ timber preservatives to remaining timbers may be necessary in certain circumstances, the decision to use them – and where used the amount and type of chemical specified – should be carefully considered. If they are used the extent and intensity of treatment should be commensurate with the particular situation. For example the practice of drilling holes in masonry and saturating the irrigated walls with fungicide has been discredited. This is both because it is ineffective and because it often introduced water (the carrier for the fungicide) into a situation where dampness is the cause of the defect being treated. It is also clear that many treatment specifications still include the use of chemicals for unjustifiable precautionary reasons.

While dry rot can certainly cause severe damage it is a relatively vulnerable organism which can be destroyed by environmental change, particularly in relation to moisture content. It should be remembered that once dry rot is identified the primary treatment, without which all others will fail, is the drying out of the structure and the prevention of further damp conditions. While this may be difficult to achieve in some circumstances, the remedy is in fact no more than to return the building to a fit state.

All photos: John Thatcher

This dry rot outbreak occurred following damage caused by vandals to sanitary fittings and water pipes.

INSECT ATTACK

Insect damage to timber can occur in standing trees, freshly felled logs and other unseasoned timber, as well as in seasoned timber both in storage and in use. Usually those insects that attack unseasoned timber die out as the cut timber dries. It is possible for live insects to be incorporated in a building because they have infested the timber before it has been used. In most cases they cannot reinfest the timber in the condition it is found in buildings and therefore no specific treatment is necessary. Other insects do not feed on wood but bore into timber for shelter. Wood wasps are an example of insects in this latter category.

This section is concerned with insects that infest timber after it is incorporated into a building. The insects which cause problems for timber used in buildings are often referred to as 'woodworm', but they are not worms and, in fact, most can fly; the insects most commonly encountered in timber are beetles.

As with wood-rotting fungi the insects use wood as a food source and as a habitat, and therefore some species can cause damage to building timbers by eating away at it.

Damage to timber caused by wood boring insects.

LIFE CYCLE

The general description of the life cycle is useful for a better understanding of the problem and the remedial treatments.

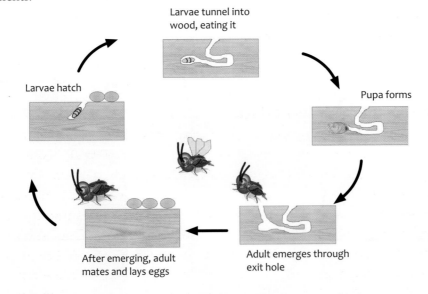

Larvae tunnel into wood, eating it

Larvae hatch

Pupa forms

After emerging, adult mates and lays eggs

Adult emerges through exit hole

Eggs

These are laid by female insects in cracks, splits, rough surfaces or the holes created by previous insect attacks. In the right circumstances the eggs will hatch and emerge as larvae.

Larvae

It is during this stage that the most damage is done, as the larvae burrow into the wood to obtain food and shelter. Unfortunately there is usually no obvious external sign of activity as the damage is taking place within the timber. The length of this period will vary between species, and for each species there will be variance, perhaps between two and ten years, which may be affected by the conditions within the timber.

Larvae of common furniture beetle.

Pupa and adult

During this stage the insect ceases to feed and a process of metamorphosis occurs, at the completion of which the adult insect emerges from the pupal skin and bores its way out of the wood, forming an exit hole. The size and shape of the exit hole is a useful indicator of the type of insect involved, as to a lesser extent is the bore dust (known as frass).

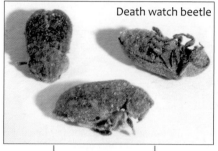

Death watch beetle

Adult death watch beetle – note nearby exit holes.

6 to 9mm

The insects are generally capable of flight but some are more likely to fly than others. The adults mate quite soon after emerging and the female will then look for a suitable place to lay her eggs. Often the most suitable place will be the timber from which she has recently emerged. This means that timber can be continuously reinfected as new insects invade the timber year after year. Soon after laying the eggs the adult will die. The period from emergence to death is generally only a matter of weeks.

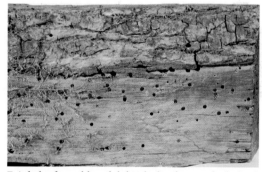

Exit holes formed by adult beetle (in this case both death watch and common furniture) emerging from the timber.

Conditions favourable to insect attack

The likelihood of attack will depend on a variety of factors, including the species of timber, its condition and the temperature and humidity of its environment. It is noticeable for instance that flight holes in the floorboards to suspended timber floors often exist predominantly on the underside; that is facing into the damper void rather than the dryer room. Although the insects generally like damp timber, excessive moisture is by no means a critical factor. However low moisture content in timber can kill larvae by causing them to desiccate (which may be why the introduction of central heating systems into existing property is thought to have an effect on some insect infestations). Ideal requirements will vary between the different insects and between the different life cycles of the same insect. If conditions are good the life cycle will be shorter. This might explain why life cycles are often longer in buildings than they are in controlled laboratory situations.

Sensitivity to environmental conditions as well as the variable food value within timbers (or parts of the same timber) may mean that attacks can be extremely localised.

As the insects are seeking food as well as shelter they will mainly attack sapwood but they may move around any part of the timber while seeking food supplements such as starch or proteins. A much quoted example is the way in which older plywood is readily attacked by the common furniture beetle because it was bonded with animal glue, whereas modern plywoods are rarely attacked.

The table below (from BRE: *Remedial treatment of wood rot and insect attack in buildings*) gives indicators which will help towards recognition.

| TYPE OF BORER | RECOGNITION OF DAMAGE | | | |
	HABITAT	EMERGENCE HOLE SHAPE & SIZE (mm)	TUNNELS	BORE DUST
DAMAGE FOR WHICH REMEDIAL TREATMENT IS USUALLY NEEDED				
Common furniture beetle	Sapwood of softwoods European hardwoods	Circular 1–2	Numerous, close	Cream, and granular, lemon-shaped pellets (x 10 lens)
Ptilinus beetle	Limited number of European hardwoods	Circular 1–2	Numerous, close	Pink or cream talc-like, not easily dislodged from tunnels
House longhorn beetle	Sapwood of softwoods	Oval 6–10 often ragged	Numerous, often coalesce to powdery mass beneath the surface veneer	Cream powder, chips and cylindrical pellets
Lyctus powderpost beetle	Sapwood of coarse-pored hardwoods	Circular 1.5	Numerous, close	Cream, talc-like
Deathwatch beetle	Sapwood and heartwood of decayed hardwoods, occasionally softwoods	Circular 3	Numerous, close, eventually forming a honeycomb appearance	Brown, disc-shaped pellets
DAMAGE FOR WHICH REMEDIAL TREATMENT IS NECESSARY ONLY TO CONTROL ASSOCIATED WOOD ROT				
Wood-boring weevil	Any, if damp and decayed	Ragged 1	Numerous, close, breaking through to surface in places	Brown, fine, lemon-shaped pellets (x 10 lens)
Wharf borer beetle	Any, if damp and decayed	Oval 6	Numerous, close, often coalescing to form cavities	Dark brown, mud-like substance; bundles of coarse wood fibres
Leaf cutter bee/ solitary wasp	Any, if badly decayed	Circular 6	Sparse network	Brown chips, metallic-like fragments, fly wings, barrel-shaped cocoons of leaves
DAMAGE FOR WHICH NO REMEDIAL TREATMENT IS NEEDED				
Pinhole borer beetles	Any in log form	Circular 1–2	Across grain, darkly stained	None
Bark borer beetle	Bark of softwoods in bark, few in sapwood	Circular 1.5–2 some bark and few in wood	Network between bark and few in sapwood	Cream and brown round pellets underlying wood
Wood wasp	Sapwood and heartwood of softwoods	Circular 4–7	Few, widely spaced	Coarse, powdery
Forest longhorn beetles	Any	Oval 6–10 on bark edges only; may be larger in some hardwoods	Few, widely spread; sections on sawn surfaces, oval 6–10 mm	None or, rarely, small piles of coarse fibres
Marine borers	Any	None, but tunnel sections can be exposed by sawing	Circular up to 15mm	None, but tunnels may have white chalky lining

MAIN TYPES OF INSECTS

For the purposes of this book it is not necessary to consider in detail all of the insects which attack timber, but it is useful to consider one or two of the more common species. Some insects will only be able to attack part of the timber's cellular structure and others will be able to attack all parts. For example powder post beetles cannot break down hemicellulose or cellulose and pinhole borers cannot break down cellulose, but others can attack cellulose, hemicellulose and lignin. Examples of this latter, more destructive group, include the following:

Common furniture beetle

This has the Latin name *Anobium Punctatum* and is the most common form of insect to attack timber in this country. Despite its English name it does not attack just furniture (although it is often brought into a building by furniture which has been infested) and it will attack softwoods and hardwoods in many situations, from a damp underfloor area to the roof. The beetle attacks mainly sapwood, heartwood is only usually attacked where it has suffered from fungal decay. If the

Adult common furniture beetle.

sapwood is only a small part of the timber section, as it will tend to be in older buildings (with original timbers) then the amount of damage caused by this insect may be relatively small.

Damage by common furniture beetle can be identified by the round emergence holes which are about 1–2mm in diameter. The adult insect is about 3–5mm long. The female will lay about 30 eggs in a convenient place in the timber. In practice this often means cracks, rough surfaces or even an old exit hole. The larval stage will last between two and five years. The adult beetle that normally emerges between May and August can sometimes be seen in favoured places such as roof spaces.

Deathwatch beetle

This beetle tends to be well known because it is associated with attacks on the type of hardwood, particularly oak or elm, which may be present in more prestigious or historic buildings. Other hardwoods including walnut, chestnut, elder and beech, are also at risk from attack as are softwoods which are in the vicinity of the hardwood.

Adult death watch beetle.

Death watch beetle will usually only attack damp decayed timber. Infestation is often therefore restricted to, for instance, timbers or parts of timbers built into damp walls. Widespread infestation may occur following the type of diffuse dampness associated with condensation. Examples of this may include the badly heated roof areas of medieval halls or churches or the underside of lead roofs. Although the damage caused by this insect may be limited in scope, it is often structurally significant. This is because it attacks heartwood as well as sapwood and because inbuilt timbers are often carrying loads. Its name, incidentally, is supposed to derive from the fact that the beetles sometimes make a knocking sound, with their head, during courtship. It is possible that, in the still of the night, those watching over the dying could hear the noise.

The female will usually lay between 40 and 60 eggs and the larval stage can last 10 years or more. Damage can be identified by the round 3mm holes left by the emerging adult. The adults are 6–9mm long. As with the common furniture beetle, the insect moves towards the outside of the timber to undergo the pupal stage, and this often occurs in July or August, but unlike the common furniture beetle it usually does not emerge from the timber for another nine months or so.

House longhorn beetle

House longhorn beetles attack mainly the sapwood of softwood and are an important insect because of the relatively severe damage which can be caused in a short period of time. They are also interesting in that they are only common in certain geographical areas around London and the Home Counties. The reason that they are only found in this area is thought to be connected with climate. This is acknowledged by the Building Regulations which stipulate that buildings in the prescribed areas should be treated against infestation. The Building Research Establishment asks that suspected outbreaks should be reported to them in order that records of the spread of the attacks can be kept.

The larvae can grow up to about 24mm in length and this, combined with the long larval period, means that the beetle can inflict considerable damage. An attack can be recognised by the relatively large oval-shaped emergence holes which are about 6–10mm in diameter. Also the larvae burrow quite near the surface of the timber, leaving a thin intact skin and this can produce a rippling/blistering effect on the surface. About 140–200 eggs are laid and the larval stage usually lasts between three and six years, with the beetle emerging in the period July to September.

Adult house longhorn beetle.

ERADICATION OF BEETLE ATTACK

Insect attack may die out naturally due to a number of reasons. These might include desiccation of the larvae in timber where the moisture content has been lowered, or attack by parasites, or because the nutrients in the timber have become exhausted.

A detailed description of remedial treatment is inappropriate in this publication although the following brief points are worthy of note: insecticides are usually applied as a liquid or as a paste. In certain circumstances smoke, gas or heat treatments are used. Liquids can be either emulsion or solvent based.

The purpose of anti-insect treatment is to both kill the insects already in the timber and to prevent further attack. The success of a treatment will depend on a number of factors including; the cross-sectional area and porosity of the timber, the type and formulation of the insecticide and the manner of its application. Treatments applied to the surface of the timber will only penetrate to a certain depth and because of this not all of the larvae will be killed immediately. The intention is that those larvae within the area of timber penetrated will be killed immediately – as will the eggs and adults on the surface – and a residual layer of preservative will be left behind to kill newly evolved adults as they tunnel out of the wood. The intention is that this residual layer should also kill any eggs laid at a later date.

The insect should first be identified in order to facilitate judgments about the possible extent of damage to the timber. If the attack is one which is associated with fungal attack, this, and the associated damp problem, needs to be addressed. Attempts should be made to estimate the extent of the damage and whether or not the attack is still active. Establishing either of these factors is often difficult in practice because the damage is essentially taking place in the body of the timber and cannot be accurately assessed without some level of destructive investigation. It is not always easy to determine whether an infestation is active, or has died out naturally, or indeed has been treated in the recent past. However if the flight holes have sharp edges and clean fresh wood can be seen then recent or current activity can be assumed. If structural damage has been caused then obviously repair or replacement of timbers will be necessary.

Appropriate insecticides can be applied in situ if an insect attack is identified as active. As with fungal treatments the focus, extent and intensity of treatment should be appropriate to the particular situation. Pressures on surveyors may mean that timber is treated even when an attack is no longer active. Fixed identification tabs following treatment can be useful in preventing unnecessary double treatments. As with the chemicals used for treatment of fungal attack, care must be taken to ensure the health and safety of occupants and to guard against general environmental damage. Of particular relevance are occupiers and operatives as well as other creatures, such as bats. Bats are a protected species under the Wildlife and Countryside Act 1981. They often nest in roofs and are vulnerable to the toxic chemicals used in pesticides.

With both fungal and insect treatment, the Control of Substances Hazardous to Health (COSHH) Regulations and the Control of Pesticides Regulations – part of the Food and Environmental Protection Act 1985 – will apply.

In recent years some work has been carried out into the use of light traps. The specific targets were deathwatch beetles which were found to be causing damage to historic buildings. The beetles were attracted to a light source and then trapped.

Preservation of new timbers

Because of the problems described above, one might assume that the sensible approach with new timbers that are to be incorporated into a building is to apply preservatives which will prevent attack by fungi or insects. The only statutory requirement for treatment is where the Building Regulations require it in those geographical areas liable to attack by house longhorn beetle. Therefore, not all timbers incorporated in a building will be routinely treated. There are a number of reasons for this, but the main one is cost. Other reasons include the possibility of corrosion to metal fixings if water-borne chemicals are used, and the fact that some chemicals can affect glued timber joints.

The factors which might be taken into account in deciding whether to preserve or not include:

- the required life of the building related to the natural durability of the various timbers within it
- the type and position of the timber component and, therefore, the likelihood of attack. For example tiling battens are more vulnerable than internal door linings
- the environmental implications of chemical treatments.

Although many specifications simply include an instruction to use preserved timber, the subject is more complex than this would imply. There are a variety of preservative chemicals. There are also a variety of processes which can be used to apply the preservative. These different methods offer different levels of effectiveness and must be matched with the type of timber, its position in the building and, therefore, its vulnerability. These factors must then be related to the type and toxicity of preservative.

CHAPTER 13

Condensation

INCIDENCE

Many people associate damp patches on walls with penetrating or rising damp. In fact, most dampness is probably caused by condensation. Unfortunately, for some of those involved in the inspection and maintenance of buildings condensation is a baffling problem. It is compounded by the fact that, in some cases, condensation may be exacerbated by problems of rising or penetrating damp. This chapter explains the phenomenon of condensation and describes how, when and where it can occur. Failure to understand its nature can lead to incorrect diagnosis of damp problems resulting, in many cases, in wasted effort and wasted resources.

Curing or preventing condensation is often much harder than its diagnosis. In practice condensation can be prevented by a careful consideration of a number of factors. These include moisture generation, ventilation, heating and insulation. These factors are discussed in more detail towards the end of the chapter.

Why does condensation occur?

Air contains water vapour, the warmer the air the more vapour it can hold. If moist air comes into contact with a cold surface the air is cooled; if the air is cooled below a particular temperature the water vapour will condense on the cold surface. Therefore, whether or not condensation occurs depends on the amount of water vapour in the air and the temperature of the surfaces in contact with the air.

Everybody is familiar with condensation in one form or another. If you have a cold glass of wine or beer on a summer's day you may have noticed that the outside of the glass is covered in droplets of moisture. This is condensation. When the element in a car's heated rear windscreen is switched on it is to combat the problem of condensation. The warm glass prevents moist air condensing on its surface.

Recognition

Condensation can be confused with penetrating damp and rising damp. Distinguishing the three is not always simple and may require instrument checks to help accurate diagnosis. In some cases all three problems may be occurring at once. Condensation can be particularly difficult to diagnose because, despite evidence of its occurrence, it may not be taking place when a property is inspected. In practice, however, condensation manifests itself in a number of ways.

- The wall has a 'misty' surface.
- Stains or streaks of water running down a wall, particularly in bathrooms, kitchens and below windows.
- Damp patches with no definite edges.
- Dampness behind wall cupboards or inside wardrobes against external walls; areas where air circulation is restricted.
- Patches of mould growth.

In practice, diagnosis cannot just depend on a visual inspection of the dampness; humidity, insulation, ventilation, and heating patterns also need to be taken into account. These are explored in more detail in later sections of this chapter.

These pictures show three examples of condensation. The top left hand picture shows condensation on a kitchen wall; at the time the photograph was taken the kitchen was unheated and had no ventilation other than a small window with a top-opening light. The house was situated on top of a hill and the walls were solid brickwork one brick thick. The top right photograph shows condensation inside a double glazed unit (caused by failure of the seals). The bottom photograph shows condensation on an upstairs ceiling caused by carelessly laid insulation.

Why is it such a significant problem?

Until the 1960s most houses were well ventilated. Houses were built with chimneys in most rooms; it was common to provide vents for small bedrooms without chimneys. When coal fires were burning the room was ventilated by the action of the fire gases rising up the chimney and drawing more air into the room. Even without a fire burning there was some natural convection in the chimney. Windows were single glazed and often quite draughty. Windows were more likely to be left open because of the reduced risks of crime. Single glazed windows acted as 'sacrificial' points. The condensation forming on the windows slowly reduced the humidity of the air. As long as windows and sills were wiped regularly the condensation was just a nuisance; not a risk to the building, or more importantly, to health.

Houses today are very different. A modern double glazed house is almost a sealed unit. Trickle vents in windows (where they exist) are often kept closed so ventilation is limited. The use of balanced flue boilers as opposed to open fires reduces natural ventilation. In addition central heating systems tend to be used intermittently. This means that when the heating is not switched on cold wall surfaces may sometimes coincide with high humidity levels. Even if the heating is fairly constant and walls have high levels of insulation condensation can still occur if there is excess moisture in the air.

Finally, today, there are more appliances in the house generating moisture; the worst culprits are tumble dryers which are not vented to the external air.

Where does it occur?

From the above it should be clear that cold surfaces are most at risk. There are a number of areas where it commonly occurs. Bathrooms and kitchens are the most obvious but there are others, eg:

- solid walls
- windows
- cold water pipes
- the underside of roofing felt.

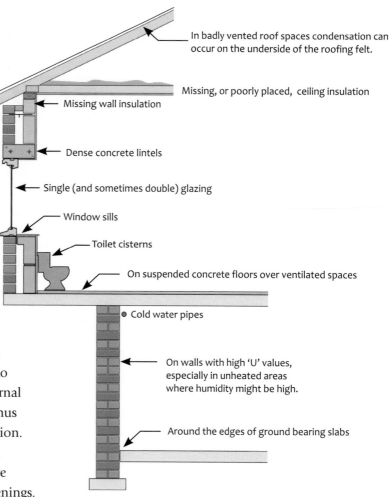

In badly vented roof spaces condensation can occur on the underside of the roofing felt.

Missing, or poorly placed, ceiling insulation

Missing wall insulation

Dense concrete lintels

Single (and sometimes double) glazing

Window sills

Toilet cisterns

On suspended concrete floors over ventilated spaces

Cold water pipes

On walls with high 'U' values, especially in unheated areas where humidity might be high.

Around the edges of ground bearing slabs

Cold bridges are a problem, particularly in well insulated buildings. They occur in localised spots where the nature of the construction allows heat to escape through the structure. Their internal surface temperature can be quite low, thus encouraging patches of local condensation. These damp patches are often confused with penetrating damp, especially where they occur around window or door openings. A few typical situations are shown below.

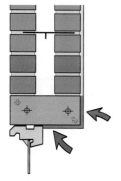

Concrete lintels which cross the cavity will have cold lower and inner surfaces (early 20th century).

Uninsulated box-section lintel (1960s to 1990s).

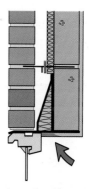

Lower steel member of insulated lintel creates cold bridge.

Improved detail ⟶

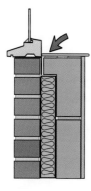

Cold bridges can be created at jambs and sills. Two examples are shown here. The left hand example shows a cavity wall with the dense block inner leaf returned to the brickwork to support a tiled sill. The right hand example shows a similar problem at the jamb. In modern construction insulated cavity closers avoid this problem.

Note that retro filling cavities can increase the risk of cold bridging - see text.

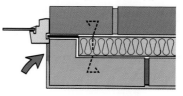

Condensation can also occur on floors.

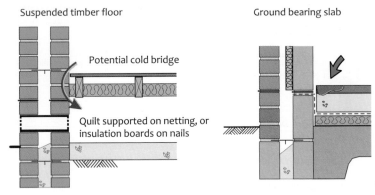

Suspended timber floor

Potential cold bridge

Quilt supported on netting, or
insulation boards on nails

The risk of cold bridging depends, to some extent, on the nature and position
of any wall insulation and the insulating properties of the internal leaf.

Ground bearing slab

Pre-cast ground floor

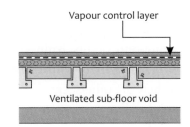

Vapour control layer

Ventilated sub-floor void

Condensation on ground-bearing slabs usually
occurs around the edges, if at all. In suspended
pre-cast floors it's more likely, particularly if
the sub floor void is vented. A vapour control
layer helps prevent it.

Condensation does not always occur where the moisture
is generated. As the amount of water vapour increases so
does the pressure of the air. The increased pressure encourages
the water vapour to travel outwards through the external
wall and to other parts of the property where the vapour
pressure is lower. The latter often results in condensation
in colder rooms well away from the source. A well heated
kitchen, even with high humidity levels, may have walls
well above the dew point temperature (the temperature at
which condensation occurs) but the escaping air may
condense elsewhere.

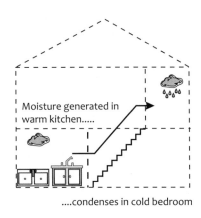

Moisture generated in
warm kitchen.....

....condenses in cold bedroom

Where damp patches mysteriously appear in unheated or partially heated bedrooms the problem
may be condensation. It is often confused with rising or penetrating dampness, particularly when it
occurs, for example, on the bottom half of a wall, around the outer edges of upper floor ceilings (ie just
below the eaves) or around openings. A later section in the chapter considers diagnosis in more detail.

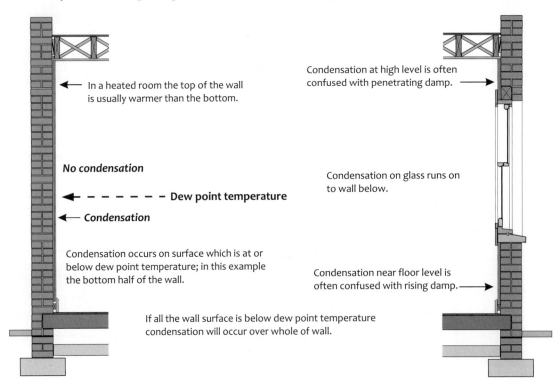

In a heated room the top of the wall
is usually warmer than the bottom.

Condensation at high level is often
confused with penetrating damp.

No condensation

— — — — — — **Dew point temperature**

← **Condensation**

Condensation on glass runs on
to wall below.

Condensation occurs on surface which is at or
below dew point temperature; in this example
the bottom half of the wall.

Condensation near floor level is
often confused with rising damp.

If all the wall surface is below dew point temperature
condensation will occur over whole of wall.

Where does the moisture come from?

Tumbler dryers have already been mentioned but there are, in addition number of other sources of water.

Typical moisture levels produced in a typical three bedroomed house (per day):

	kg moisture		kg moisture
• A family asleep	1.5 – 2.0	• Washing and bathing	1.0 – 1.5
• A family during the day	2.5 – 3.5	• Washing clothes	0.4 – 0.6
• Cooking	2.0 – 3.0	• Drying clothes indoors	3.0 – 5.0

Large dogs produce high amounts of water vapour. Anyone who owns one will have noticed how quickly a car can 'steam up' after a good long walk.

In addition, problems of penetrating and rising damp can add significantly to the amount of water in a house. As the damp evaporates it raises the amount of water vapour in the air.

Fuel burning appliances

Nearly all fuels (coke is an exception) contain hydrogen. As part of the combustion process the hydrogen combines with oxygen to form water. In addition some fuels contain water; coal or coke stored outside can be quite wet. As these burn, the water evaporates. Both of these factors combine to ensure that the exhaust gases from the fuels contain high levels of water vapour. As the water vapour rises up the flue it may condense. In open fires substantial amounts of air are added from the room and, as long as the gases travel up the chimney quickly, condensation will be minimal.

However, in slow burning appliances, the flue gases may rise quite slowly and the lack of additional air from the room will ensure that the exhaust gas retains high levels of moisture and can therefore condense readily on surfaces even if they are quite warm. If there is no flue, as in the case of portable gas or paraffin heaters, high amounts of water are introduced into the air, slowly increasing its humidity.

If moisture condenses in a brick, block or stone flue there is the risk that the condensate can absorb chemicals which are a by-product of combustion. Sulfates and other soluble salts, deposited by the rising exhaust gas, go into solution and travel into the surrounding brickwork. As they slowly evaporate the salts are deposited on the face of the wall (in the form of white crystals) where they can damage plaster and wall decorations. Some of the salts are hygroscopic; in other words they absorb water from the air. These salts will not crystallise as they continually draw water from the air. Damp patches at the top of a chimney breast which seem to grow and shrink according to the weather may be caused by these hygroscopic salts. Damp patches like this are often confused with other problems such as flashing failure or rain water running down the chimney.

Slow burning appliances are often fitted with stainless steel or terra cotta linings to prevent migration of salts into surrounding brickwork. They can sometimes be insulated to prevent condensation occurring in the lining.

Mould growth

Mould growth may form where condensation occurs, particularly if humidity levels remain above 70% for long periods. Condensation is common on cold walls particularly in corners or behind furniture where air flow is restricted. Wardrobes positioned against external walls are often plagued by mould growth. The mould grows from spores which are present in the air all the time. In order to flourish they need to feed on organic matter and water. The former is supplied by the dirt and grease which, even though it cannot often be seen accumulates on porous or rough surfaces such as woodchip paper, emulsioned ceilings and tile grout. Leather goods, particularly when stored in cold, unventilated wardrobes are also affected by mould growth; the mould appears as mildew and occurs because leather goods are often quite dirty.

To the elderly, the very young and asthmatics, mould growth represents a health risk. To others it is just an unsightly inconvenience. It can be removed by dilute bleach although serious cases may require papered walls to be stripped first. Most DIY stores sell a spray which prevents the mould growth from occurring. However this will not cure the underlying problem of condensation on the wall.

Roofs

Condensation is quite common in roof voids. It usually manifests itself as water droplets on the underside of the underlay. This picture (kindly supplied by Surevent) shows a typical problem. In most roofs the problem is caused by inadequate ventilation.

In a traditionally vented roof with insulation at ceiling level ('cold roof') condensation is often a result of insulation blocking the flow of air at the eaves. A similar problem can occur in older roofs which are retro-insulated; it occurs because the original roof had no provision for ventilation.

Mould growth on the edge of the ceiling is often caused by condensation on the soffit of the plasterboard; usually a result of inadequate insulation cover.

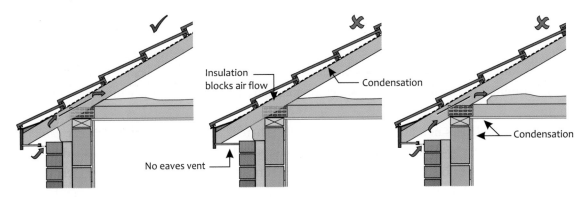

Insulation blocks air flow — Condensation

No eaves vent

Condensation

Eaves-to-eaves or eaves-to-ridge ventilation   Ventilation paths blocked   Gaps in insulation - cold surfaces

In recent years loft conversions have become very common. In a vented roof design, insulation can be provided in between the rafters ('warm roof') as long as a minimum ventilation gap of 50mm can be maintained. In addition, a vapour check (vapour control layer) just behind the new plasterboard will help to reduce vapour transfer. Condensation can occur on the soffit of the underlay if either is missing.

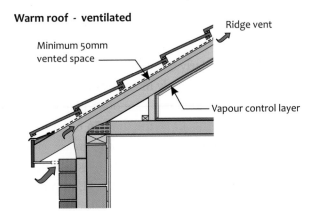

Warm roof - ventilated

Ridge vent

Minimum 50mm vented space

Vapour control layer

To prevent condensation there must be a ventilated space of at least 50mm above the insulation. In addition, an effective vapour control layer should be positioned just above the plasterboard (ie on the warm side of the insulation).

In 'breathing roofs' a vapour permeable underlay allows vapour to escape. These can be designed as 'cold roofs' or 'warm roofs' as shown below. In practice, breathing roofs are sometimes misunderstood; particularly the need for ventilation, vapour control layers and counter-battens. All three are not always necessary; but a vapour permeable underlay on its own may not be enough to prevent condensation.

Cold roof - 'breathing'

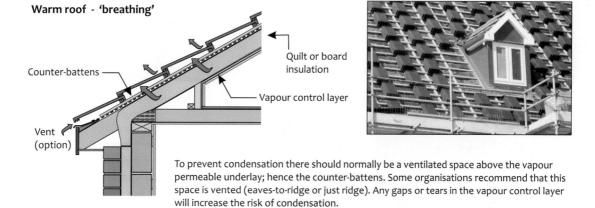

Counter-battens to encourage ventilation

Vapour permeable underlay

Vents at eaves and ridge sometimes recommended

Vapour control layer

There has been much debate about 'breathing roofs' in recent years. If the wrong type of underlay is used or if the vapour control layer is not continuous (taped joints and seals around light fittings and loft hatches) condensation can occur. Counter-battens encourage ventilation; some organisations also recommend eaves and/or ridge vents to help ventilate the void above the underlay.

Warm roof - 'breathing'

Counter-battens

Quilt or board insulation

Vapour control layer

Vent (option)

To prevent condensation there should normally be a ventilated space above the vapour permeable underlay; hence the counter-battens. Some organisations recommend that this space is vented (eaves-to-ridge or just ridge). Any gaps or tears in the vapour control layer will increase the risk of condensation.

Case studies

These three examples show how condensation can occur, all in very different situations. Condensation and damp penetration are easily confused so careful analysis is important; money is often wasted because of incorrect diagnosis.

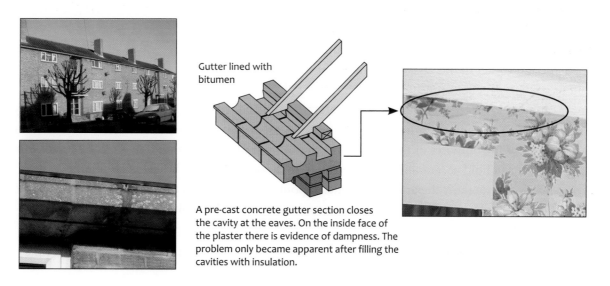

Gutter lined with bitumen

A pre-cast concrete gutter section closes the cavity at the eaves. On the inside face of the plaster there is evidence of dampness. The problem only became apparent after filling the cavities with insulation.

Condensation occurred on these window frames within a few months of the property being occupied. The window surround and mullion are formed in artificial stone (to emulate vernacular buildings nearby). Unfortunately the inner face of the 'stone' is quite cold; hence the mould growth.

End wall

This problem was initially diagnosed as either penetrating dampness through the solid wall or rainwater running down the chimney flue. When the roof space was inspected, however, it was clear that the inner face of the end wall was quite wet and that water was running down the wall causing the damp patches shown above. It was eventually discovered that a bathroom extractor fan which should have discharged through the roof had become disconnected and was actually discharging into it! The moist air was condensing on the cold end wall.

ANALYSIS

Measuring the amount of water vapour in the air

As established earlier in this chapter, the amount of water (in the form of water vapour) that air can hold or contain depends on its temperature. If, for a given temperature, the amount of water vapour the air can hold is exceeded the air becomes saturated and condensation will occur. Relative humidity is a term used to quantify the amount of vapour in the air. It can be defined (not a strict definition but one suitable for the purposes of this chapter) as the amount of water vapour that air holds relative to the amount it can hold at saturation point. Saturation occurs at 100% RH. Relative humidity in itself cannot be used to quantify the amount of water vapour in the air unless the temperature is taken into account. For example, at different air temperatures, say 10°C and 20°C an RH of 50% will represent different amounts of water vapour.

Air temperature	RH	Amount of moisture in the air
10°C	50%	4.0g/100g dry air
20°C	50%	7.4g/1000g dry air

It should be clear that air at the higher temperature (20°C) is capable of holding higher quantities of moisture or water vapour.

Relative humidity affects comfort. Very dry or very humid conditions can both provide an uncomfortable environment. An RH of 45–65% is typical of a well heated and ventilated lounge, with a temperature of 18–20°C. Levels below 45% will feel very dry; levels over 70% are likely to feel 'close' and will increase the risk of condensation and mould growth.

Establishing if condensation is taking place

A psychometric chart enables the incidence of condensation to be predicted (psychometric – from the Greek 'a measure of coldness'; in fact the Greek name for a thermometer).

The horizontal lines show moisture content, the vertical lines show air temperature and the curved lines show relative humidity. In practice measuring the moisture content of the air during a building inspection is difficult but it is fairly easy to establish the RH and the air temperature. If these two variables can be measured, establishing the moisture content of the air, and therefore the dew point, is easy.

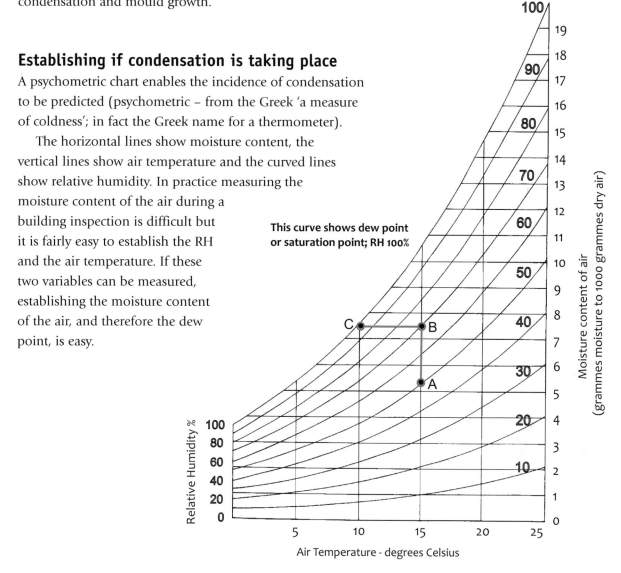

This curve shows dew point or saturation point; RH 100%

Relative Humidity %

Air Temperature - degrees Celsius

Moisture content of air (grammes moisture to 1000 grammes dry air)

Example

If the air inside a house is 15°C and it contains 5.3g of water per 1000g of dry air its RH is 50% (Point A). If additional moisture or water vapour is added, say another 2.2g per 100g of dry air (by activities in the house such as washing or cooking) its RH will increase to 70% (Point B). If this air comes into contact with a cold surface such as glazing or a solid uninsulated wall the air will be cooled. As it cools the RH increases. If the RH Reaches 100% (dew point) saturation will occur. From the chart it should be clear that this temperature (Point C) is 10°C. Therefore, any surface at 10°C or less will cause condensation. Condensation in the form of a fine mist, or droplets of water, will form on the wall's surface. Although warm air can hold more water than cold air the dew point is directly related to the actual amount of water in the air. If the air contains 7.5g of water per 1000g of dry air its dew point will always be 10°C whatever the internal air temperature.

This last point is commonly misunderstood. It is often assumed that heating up the air, by possibly turning up the heating system, prevents condensation because the warm air can hold more moisture.

Turning up the heating will certainly reduce the RH, but however high the internal air temperature, condensation will still occur if the wet air comes into contact with a surface at 10°C or less. In fact, the purpose of turning up the heating system is to raise the temperature of the walls above dew point.

The psychometric chart identifies the point at which condensation will occur given a specific RH and internal surface temperature. But how do you measure these two items?

Inner surface temperatures

The temperature of the internal surface of glazing or walling will depend on three factors, the thermal transmittance of the material ('U' value – a measure of the rate at which heat is transmitted through a wall), the outside air temperature and the inside air temperature. Well insulated walls (with low 'U' values) will have high internal surface temperatures. In a house the coldest surface will probably be single glazing. In practice, surveyors nowadays use electronic instruments to measure internal surface temperature. A modern instrument with a digital readout will give an accurate surface temperature in a few seconds. Before these instruments were available the internal surface temperature was established using the chart shown below. The chart assumes that 'U' values are known.

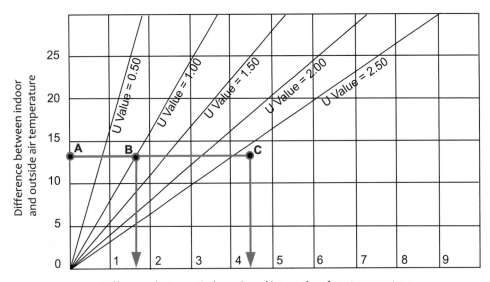

Difference between indoor air and internal surface temperature

Although this chart is virtually redundant these days it does show how poor insulators result in lower internal surface temperatures. Consider a situation where the temperature externally is 7°C and the temperature internally is 20°C. The temperature difference is therefore 13°C. This is shown as point A on the chart above. In addition imagine we are considering two types of building; a 1920s house with a solid 1 brick wall ('U' value of 2.60) and a 1980s cavity wall house ('U' value about 1.00).

Point B is the 'U' value of the 1980s house. Point C is the 'U' value of the 1920s house. If the red lines are extended downwards the difference between indoor air and the internal surface temperature can be established. These are 1.6 for the 1980s house and 4.3 for the 1920s house. Subtract these values from the internal air temperature to find the internal surface temperature.

Thus the internal surface temperature of the old wall is 15.70°C and the new wall is 18.40°C. As you would expect, the 1980s house, with its better insulation has an internal surface temperature closer to the internal air temperature.

Now if we refer to the psychometric chart shown previously we can assess the risk of condensation. If the RH is 80% and the internal air temperature is 20°C the dew point is about 16–17°C. In other words any surface at 17°C or less will cause condensation. In our examples the cavity wall is above dew point (no condensation) and the solid wall is below dew point (condensation will occur).

Measuring relative humidity

Relative humidity can, nowadays, be measured quite simply. Thermohygrometers (sometimes electronic) measure air temperature and relative humidity. Dew point can then be read by reference to the psychometric chart or by reference to a simple table based on the chart.

Air temp	RH 20%	RH 30%	RH 40%	RH 50%	RH 60%	RH 70%	RH 80%
10				0	2	5	7
11				1	3	6	8
12				2	4	7	9
13			0	3	5	8	10
14			1	4	6	9	11
15			2	4	7	10	12
16			2	6	8	11	13
17			3	6	9	11	13
18		0	4	7	10	12	14
19		1	5	8	11	13	15
20		2	6	9	12	14	16
21		3	7	10	13	15	17
22		4	8	11	14	16	18
23		4	9	12	15	17	19
24	0	5	10	13	16	18	20
25	0	6	10	14	17	19	21

If the internal air temperature is 20°C and the RH is 80%, condensation will occur if the moist air comes into contact with surfaces which are 16°C or below. Thus 16°C is the dew point.

The blank spaces indicate dew points below freezing point. Internal surface temperatures below 0°C are almost impossible in a house.

Interstitial condensation

In most cases condensation will occur on the surface of a material. Most walls, however, are porous and will therefore allow the water vapour to slowly pass through the wall. If the internal surface temperature is above dew point but the centre of the wall is below dew point there is a risk that interstitial condensation will occur. The water vapour therefore condenses inside the wall rather than on its surface. Impervious wall finishes such as gloss paint will limit the passage of water vapour into the wall but emulsion paints and porous wallpapers will be less effective.

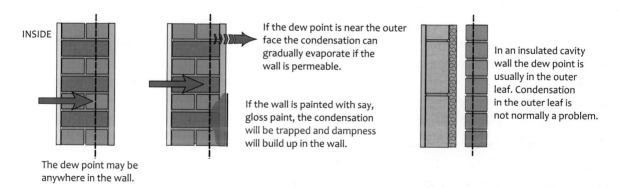

INSIDE

The dew point may be anywhere in the wall.

If the dew point is near the outer face the condensation can gradually evaporate if the wall is permeable.

If the wall is painted with say, gloss paint, the condensation will be trapped and dampness will build up in the wall.

In an insulated cavity wall the dew point is usually in the outer leaf. Condensation in the outer leaf is not normally a problem.

Interstitial condensation is a common problem with one brick thick solid walls. It is not normally a problem in cavity walls.

The temperature of the wall obviously falls towards the outside. This temperature gradient depends on the nature of the walling material and the presence, and position, of insulation. If a one brick wall is insulated by dry lining the temperature drops quickly and the wall itself will be quite cold. Where the wall is insulated externally the wall will remain at a high temperature. These temperature gradients can be calculated accurately if the characteristics of the various materials making up the wall are known. However, this level of scientific analysis is beyond the scope of this book.

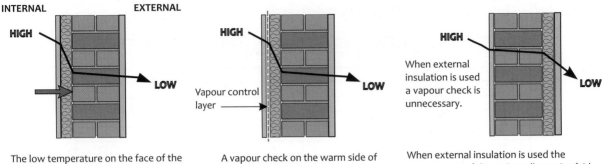

INTERNAL   EXTERNAL

HIGH   LOW

The low temperature on the face of the brick is likely to be below the dew point. If so, interstitial condensation will occur.

HIGH   LOW

Vapour control layer

A vapour check on the warm side of the insulation prevents the moist air reaching the cold surface.

HIGH   LOW

When external insulation is used a vapour check is unnecessary.

When external insulation is used the temperature of the main wall remains fairly close to internal surface temperature.

As the temperature drops through the thickness of the wall so does the dew point. This is because the wall's degree of porosity will provide some resistance to the passage of water vapour through the structure. The dew point will drop because air containing lower amounts of water vapour has a correspondingly lower dew point. Interstitial condensation will only occur if the dew point temperature, in other words the temperature at which saturation occurs, is higher than the surface temperature.

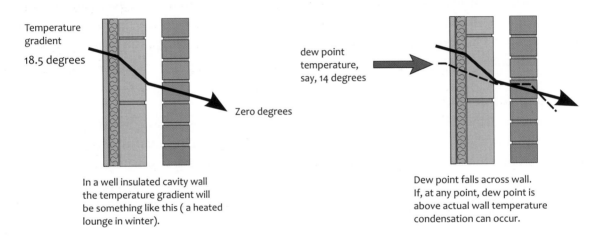

Temperature gradient

18.5 degrees

Zero degrees

In a well insulated cavity wall the temperature gradient will be something like this (a heated lounge in winter).

dew point temperature, say, 14 degrees

Dew point falls across wall. If, at any point, dew point is above actual wall temperature condensation can occur.

Interstitial condensation may cause problems of dampness but it should not affect the durability of brick or block walling materials. However, if interstitial condensation occurs in a material such as timber, action will be needed to prevent the risk of rot.

In timber frame housing the risk of interstitial condensation is minimised by placing a vapour control layer on the warm side of the insulation. The layer (usually polythene) should have taped joints and any holes for socket boxes etc need to be made with care.

Some occupiers of concrete pre-cast houses first noticed damp problems when their houses were insulated in the 1980s. Condensation occurred on the cold inner face of the external concrete panels and ran down to ground level. In extreme cases the build-up of water eventually soaked the base of the fibreboard/plasterboard linings; this is easily mistaken for rising damp.

'Reverse' or 'summer' condensation

Very occasionally a phenomenon known as 'reverse' or 'summer' condensation occurs. It only affects solid walls where vapour checks have been installed. It manifests itself as wet patches on the floor at the base of the wall and is caused by condensate running down the back of the vapour check. If the external air temperature is high and the air contains high levels of moisture the resultant high pressure will tend to move into areas of lower pressure. This can result in warm moist air reaching the back of the vapour check which may have a surface temperature below dew point. Fortunately, our climate is such that this can only be a temporary nuisance.

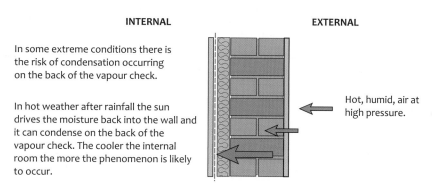

INTERNAL

In some extreme conditions there is the risk of condensation occurring on the back of the vapour check.

In hot weather after rainfall the sun drives the moisture back into the wall and it can condense on the back of the vapour check. The cooler the internal room the more the phenomenon is likely to occur.

EXTERNAL

Hot, humid, air at high pressure.

Effect of wall structure and heating pattern

In a well heated dwelling, with average levels of humidity, condensation will only occur locally after long hot baths, cooking or drying clothes. As you would expect, kitchens and bathrooms are most at risk. However, when houses are heated intermittently condensation may also appear on living room and bedroom walls; even houses with central heating are at risk. Few people keep their central heating on permanently and a more typical arrangement is to have heating for a few hours in the morning and a few hours in the evening. This pattern of heating can cause condensation at specific times of the day or night. For example, a house with one brick walls has a fairly high heat capacity – in other words heat is stored in the walls which cool slowly when the heat is turned off. If the heating has been turned off over night, condensation may occur in the morning as the house comes to life and moisture generating activities cause condensation on the cold walls.

Modern lightweight construction, with say an insulated dry lining, acts in the opposite way. When the heating is turned on the wall will rise in temperature very quickly so condensation in the morning should not be a problem. The problem with these walls occurs at night. When the heating is turned off the wall cools rapidly. A family asleep produces large amounts of water vapour. Thus condensation occurs throughout the night. This problem of intermittent condensation makes its diagnosis difficult and is another reason why a broad understanding of condensation is required to help interpret instrument readings. Protimeter manufacture a condensation tester which is designed to operate over a 24 hour cycle. The instrument will show if condensation has occurred in any time over this cycle.

HOW CAN IT BE AVOIDED?

Dealing with condensation effectively can be a problem particularly where people have limited incomes and cannot therefore afford continuous heating. There are four areas to consider:
- limiting the amount of moisture generated
- insulation
- ventilating the property
- heating.

Moisture generation
As we have already seen moisture generation in a typical family house can be considerable. In order to limit the amount of water generated there are a number of simple precautions:
- keep lids on pans
- dry clothes outside whenever possible
- ensure tumble dryers are ventilated externally
- do not use paraffin or bottled gas heaters
- put cold water in the bath before hot water
- keep the water vapour in the room where it is generated to prevent it from travelling to cooler areas (close bathroom doors for example after taking a shower)
- ensure that trickle ventilators are kept open, particularly at night
- open windows when cooking or taking a bath.

Minor life style changes, as set out above, can significantly reduce the incidence of condensation. Because there are no cost implications these are the areas to consider first.

Ventilation
Ventilation plays an important part in combating condensation. However ventilation alone will not prevent condensation from occurring and may, in fact, cause lower air temperatures. The BRE has established that ventilation rates of over one air change per hour have little effect on relative humidity but a marked effect on air temperature. Limited ventilation will help to keep relative humidities below 70% (see Chapter 14: Damp). This includes ensuring that trickle ventilators in windows are left open and that extractor fans are used in kitchens and bathrooms. Where fireplaces are removed a hit and miss ventilator in the chimney breast will provide additional ventilation (although care should be taken to ensure that defective or capped flues cannot transfer water vapour into the roof space). Tumble driers should be vented to the external air and, wherever possible, clothes should not be left to dry in rooms without adequate ventilation.

Insulation and heating
These two areas cannot be considered in isolation. Wall insulation on its own will have little success in combating condensation unless heating is provided. However thick the wall insulation it can only be effective if heat is provided. If a house is poorly heated, well insulated walls can still be below the dew point. If the internal air temperature is 10°C, a well insulated wall will have an internal surface temperature of about 8°C. If the air contains 7g of moisture per 1000g of dry air (a fairly modest amount), the dew point is about 9°C; if the dew point is above the internal surface temperature condensation will occur. Where insulation is retro-fitted consideration must be given to problems of cold bridging and interstitial condensation. As already explained external insulation is more effective in maintaining

consistently high internal temperatures. It will also reduce the risk of cold bridging. However, external insulation is generally more expensive than internal insulation and there are a number of situations where it is not a suitable option. Where the external facade is quite decorative or where there are planning restrictions internal insulation may be the only option.

Heating, on its own (assuming correct levels of ventilation and controlled moisture generation) will usually prevent condensation from occurring. However, few people can afford to run their heating systems continuously. By providing some form of insulation occupiers will be encouraged to use their heating system safe in the knowledge that fuel bills will not be extortionate. There are also obvious environmental savings.

The role of dehumidifiers

Dehumidifiers can be effective in combating condensation but to recognise their limitations it is important to understand how they work. In a dehumidifier air is drawn over a cold element where it condenses. The collected moisture is either collected in a container or discharged through a drain. Dehumidifiers are most effective when the air contains high amounts of water vapour, in other words well heated houses with high amounts of water vapour in the air. Where air is at a low temperature the amount of vapour in the air will be less and the dehumidifier will only be able to extract low amounts of moisture. Most problems of condensation are, in fact, caused by the latter scenario. Where people cannot afford to heat their houses properly dehumidifiers are therefore of little use. In addition they can be noisy, are fairly bulky, and of course, require continuous power.

CHECK LIST

A simple check list is shown below to help identify condensation or confirm a diagnosis.

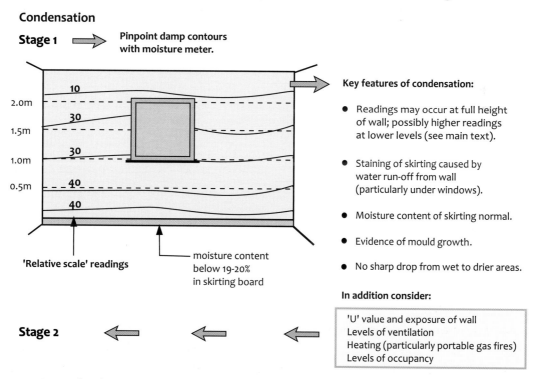

Condensation

Stage 1 → Pinpoint damp contours with moisture meter.

10
30
30
40
40

2.0m
1.5m
1.0m
0.5m

'Relative scale' readings

moisture content below 19-20% in skirting board

Key features of condensation:

- Readings may occur at full height of wall; possibly higher readings at lower levels (see main text).

- Staining of skirting caused by water run-off from wall (particularly under windows).

- Moisture content of skirting normal.

- Evidence of mould growth.

- No sharp drop from wet to drier areas.

In addition consider:

'U' value and exposure of wall
Levels of ventilation
Heating (particularly portable gas fires)
Levels of occupancy

Stage 2 ⇐ ⇐ ⇐

- Deep wall probes indicate low readings in centre of wall, dampness unlikely to be rising or penetrating damp.

- Salt analysis is negative (no chlorides or nitrates) - suggests condensation.

- Measure internal wall surface temperature using surface thermometer and establish dew point temp by measuring air temperature and RH. Surface temperature below dew point temperature = condensation.

- Consider use of Protimeter Damp Check (records incidence of condensation over 24 hours or so).

CHAPTER 14

Damp

INTRODUCTION

This chapter explains the main causes of penetrating and rising damp. It is mostly concerned with external walls (floors and roofs have been dealt with in other chapters). The chapter defines dampness, explains how dampness can occur, how it can be identified and how it can be measured. Incorrect diagnosis is a common problem caused, in many cases, by insufficient thought and over reliance on inconclusive instrument checks. Incorrect diagnosis at best leads to wasted effort and wasted resources, at worst, it can lead to 'repair' works which can actually exacerbate the initial problem. There are two diagnostic check lists (penetrating and rising damp) included in the chapter which should help the reader approach a suspected problem of dampness in a systematic way. A similar check list for condensation can be found at the end of the previous chapter.

The problem of dampness

The presence of excess water in buildings probably gives rise to more defects than any other cause. Dampness can:

- be a health hazard
- reduce the strength of building materials such as chipboard and plasterboard
- cause movement in building elements, eg block floors
- lead to outbreaks of dry and wet rot *(see chapter on Timber Pests)*
- cause chemical reactions in building components *(see chapter on Walls)*
- reduce the effectiveness of insulation
- damage decorations.

Unfortunately problems of dampness can be quite complex; a sensible starting point is to consider a definition of dampness and to examine how water is held in materials.

DAMPNESS AND MOISTURE CONTENT

The word 'damp' is potentially confusing because most building materials (with the exception of plastics and metals) are porous and will always contain some moisture. Assuming there are no problems of condensation, spillage, or damp penetration through the fabric, the moisture they contain will vary according to the relative humidity of their surroundings. This is why it is so important to ventilate timber ground floors. Inadequate ventilation, particularly if the ground below the void is wet, will allow the relative humidity of the air below the floor to rise. This occurs because the air steadily absorbs water.

At the same time the moisture content of the floor structure will slowly settle into equilibrium with the damp air, and, as it does so, its moisture content will also rise. If the air below the floor has a relative humidity of 85% or more, *(see Chapter 13, page 252 for a definition of relative humidity)*, the timber's moisture content will rise to a point where biological decay is likely in most untreated softwoods; this is about 20%.

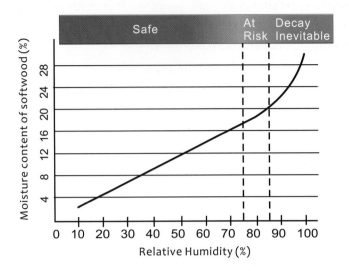

The relationship between relative humidity and moisture content of softwood is shown in the diagram below.

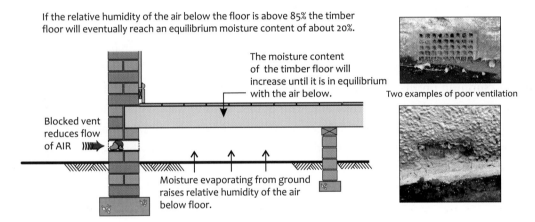

If the relative humidity of the air below the floor is above 85% the timber floor will eventually reach an equilibrium moisture content of about 20%.

The moisture content of the timber floor will increase until it is in equilibrium with the air below.

Blocked vent reduces flow of AIR

Moisture evaporating from ground raises relative humidity of the air below floor.

Two examples of poor ventilation

Ventilation of the under floor void should ensure that the relative humidity of the air remains below 75%. Timber in equilibrium with a relative humidity of 75% has a moisture content of about 17–18%.

Moisture content of timber

Timber with a moisture content of less than 17–18% is safe from decay. At 18–20% the risk of decay is more likely and, above 20%, it is inevitable (assuming the dampness is chronic and the presence of spores – *see Chapter 12: Timber Pests*). The moisture content can be calculated as shown below:

$$\frac{\text{Wet weight of material – dry weight of the material}}{\text{dry weight of the material}} \times 100$$

So, for example, if a sample of timber weighs 150g and if the same sample weighs 130g when dried its moisture content will be: $\frac{150 - 130}{130} \times 100 = 15\%$

The method is often referred to as oven drying. The sample of timber is weighed, dried in the oven, and then weighed again. However, this method is of little practical use and site testing usually requires the use of a moisture meter.

It should also be remembered that moisture content will vary with the seasons; roof timbers in a well ventilated roof, for example, can vary from 15% in a wet January to 5% in a dry July.

Moisture meters

When materials absorb moisture they can conduct electricity. By measuring a material's ability to conduct electricity a measure of the amount of water in it can be assessed. Thus it should be remembered that most moisture meters measure conductivity not moisture content. It should also be remembered that the ability of a material to conduct electricity is not in direct proportion to the amount of water it holds. Very low levels of moisture cannot be detected. This is because low levels of water cannot dissolve the soluble chemicals required to form the solution which actually conducts the electricity.

If there is sufficient 'free' water, soluble salts will dissolve. When soluble salts are dissolved in water each molecule splits into two parts, one with a negative charge and one with a positive charge. These charged particles (known as ions) are attracted to the negative and positive electrodes of a conductance meter (the most common form of moisture meter – see below) where they give up their charges. This results in a flow of electricity, the level of which is measured by the moisture meter.

Conductance meters

These instruments, which are by far the most common form of meter, usually have two metal probes (the electrodes) which are firmly pressed into the material being tested. The electrical resistance between the two probes can then be measured. The meters (depending on type) can provide analogue or digital readings. The Protimeter 'Mini' meter, for example, has coloured LED lights (green, yellow and red) representing 'safe', 'borderline' and 'decay inevitable' situations. More sophisticated meters provide specific percentage measurements and can be used with attachments to provide relative humidity and surface temperature readings to help identify condensation.

Protimeter Mini, Survey Master & Timber Master

Most moisture meters are specifically calibrated for timber and cannot give accurate readings (ie percentage moisture content) for other materials. Even with timber there are differences between timber species, and between hardwoods and softwoods. However, the differences are not particularly significant and measurements can be adjusted if an accurate moisture content is required for a known species. There are, however, one or two pitfalls which should be recognised by anyone carrying out testing to assess moisture content. For example, readings from an electric moisture meter can be affected (readings will be too high) if the timber has been treated with water based preservatives. The temperature of the wood will also affect readings.

A definition of damp

In a well ventilated room with a fairly constant relative humidity of about 30% a softwood timber skirting board may contain as little as 6% moisture *(see previous chart)*. In a kitchen with a constant relative humidity of about 70% the moisture content of the skirting will be approximately 16%. In both cases the timber contains moisture but, because its moisture content fluctuates according to the relative humidity, and because a relative humidity of 70% will not encourage decay in timber, it can be referred to as 'air dry'. However, if the relative humidity reaches 85% and stays at, or above, this level for a length of time conditions are created which can cause decay in the timber, damage to decorations, the development of moulds and the appearance of mites.

Thus, when a timber skirting is in equilibrium with this high relative humidity (over 85%) the material can be regarded as 'damp'. Its moisture content will be about 19% or even more. Thus dampness can occur through persistent levels of high humidity where the timber in contact with the moist air reaches a point of equilibrium. Similarly timber can be regarded as 'air dry' (even if it contains some moisture) if it is in equilibrium with a 'normal' atmosphere which is typically 30–70% relative humidity. At these humidities its moisture content will be 7–18%.

Although dampness can be caused by persistent levels of high humidity it is more likely to be caused by penetrating damp, and rising damp. The main causes of these are discussed later in the chapter.

Both these problems are caused by long term, persistent, dampness.

Materials other than timber

Of course, the external wall of a house contains a number of elements in addition to timber skirtings. These elements vary considerably in their acceptable percentage moisture content. Timber, as demonstrated above, can vary from 4% to 18%, depending on the relative humidity, and still be regarded as 'air dry'. Brickwork (depending on specific type), however, is wet at 5% and saturated at a moisture content of 10%; a more acceptable level would be between 1.5% and 2.5%. Some plasters are dry at 1% and saturated at 5%. Like timber, all porous materials will settle into equilibrium with the air surrounding them. The higher the relative humidity the greater their moisture content. If a moisture meter is used to measure dampness in these materials it must be remembered that the moisture meter is calibrated to measure moisture content in timber. A reading on any other surface will only give a comparative reading, not an accurate indication of moisture content.

Measuring the moisture content of concrete, plaster and brickwork etc, can be carried out using the oven drying method mentioned earlier. Another option is to use carbide testing. Whichever technique is used, it must be recognised that establishing moisture content provides little help unless the particular characteristics of the material are known. In addition, it should be recognised that spot readings on their own, may show high levels of moisture, but will give little indication as to their cause. One of the first stages in any diagnosis, therefore, should be to establish contours, or patterns, of dampness across the face of a wall. The section on diagnosis explains this in more detail.

Carbide testing

For masonry products, bricks, blocks, mortars etc, carbide testing can be used to assess moisture content. Its use should be regarded with caution because, as already mentioned, these materials will differ in the amount of moisture they can contain and still be regarded as dry. By drilling into a wall with a large slow bit (to minimise heating and, therefore, drying) a sample of suspect material can be collected. Part of the sample is weighed and tipped into a cylinder. A specific amount of calcium carbide is added and the container sealed. Vigorous shaking of the cylinder will then mix the two materials. Moisture in the sample reacts with the calcium carbide to form acetylene gas. As the gas forms the pressure in the sealed cylinder increases. A pressure gauge within the cylinder is calibrated to show the moisture content of the sample.

Although these instruments will fairly accurately measure moisture content they do not help diagnosis of dampness unless several readings are taken across a wall. This is obviously a fairly destructive exercise.

The oven drying method of calculating moisture content and the carbide tests show that if two materials of differing densities contain the same amount of water they will have different moisture contents. Even similar materials can be different, eg weak and strong renders. Therefore, to use this method to interpret whether a material is wet or not would require a detailed knowledge of a whole range of building materials. Gypsum plaster with a moisture content of 2% would in fact be very wet, lime mortar at 4% would be fairly dry. In practice this is not a practical option and most surveyors tend to rely on conductance meters.

Interpreting a conductance meter

Conductance meters can be used to help determine levels of dampness in materials other than timber although, as indicated above, they should be used with caution. A simple laboratory based experiment illustrates this.

A sample of facing brick and a sample of cement/sand plaster were oven dried for 36 hours and then weighed. They were then soaked in water for 48 hours. Surplus water was removed and they were weighed again. This allowed the moisture content to be calculated in their saturated state. Over the next few days the samples were allowed to dry in a well ventilated room of average humidity. At regular periods, while they were drying, the samples were wrapped in polythene for several hours to ensure water was evenly distributed within each sample. After unwrapping, the samples were weighed again, and measured with a moisture meter. The graphs below show the relationship between their actual moisture content and the relative readings on the moisture meter.

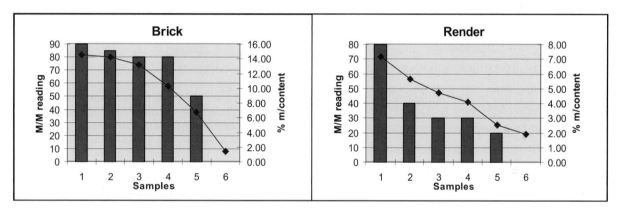

It can be seen from the above graphs that the relative readings (the vertical bars) roughly follow the curves of the actual moisture content (the blue line). In ambient conditions (18°C with an RH of 50%) both samples had a moisture content slightly below 2%, and in these conditions the moisture meter showed them both to be dry. Similarly, when both materials were saturated (7% render, 14% brick) the meter readings were both very high. The meter was also fairly accurate in providing a comparative assessment of dampness. However, it should be clear from the above results that the relative readings provided by the meter should not be confused with actual moisture content.

Conductance meters are therefore ideal for establishing contours or patterns of dampness and for establishing whether or not materials are dry. Establishing contours of dampness should therefore be an operation which should be one of the first stages in diagnosing dampness.

Measuring dampness within a wall

Carbide testers, by their very nature, measure the dampness within a wall rather than at its surface. Conductance meters, however, measure dampness at, or just below the surface. Surface dampness can be caused by condensation. However, a range of deeper probes can be used if the internal part of a wall needs to be measured. Deeper testing is often necessary if inaccurate surface readings are to be avoided. The deep probes have insulated sides to make sure that readings nearer the surface are ignored.

Salts

Another problem which affects the accuracy of conductance meters is related to the salt content of the material being tested. Salts, such as nitrates and chlorides, are commonly deposited on the inner surface of walls as dissolved groundwater rises up a wall and evaporates. Other salts deposited on the surface or near the surface of the plaster can come from chemicals naturally present in the walling materials which are dissolved by dampness and deposited as salts on, or near, the surface of the wall. These salts readily conduct electricity and, when tested with a conductance meter, can indicate levels of moisture far higher than the actual one. In addition, nitrates and chlorides are hygroscopic; in other words they absorb moisture. Thus a high moisture reading may not indicate problems of ongoing rising or penetrating damp but may be measuring water from the air absorbed by the hygroscopic materials left by previous dampness. Fortunately, there are instruments which can detect these salts either through site testing or laboratory analysis. The nature and detection of these salts is covered in more detail later in the chapter.

SOURCES OF DAMP

There are four main sources of dampness in housing; rising damp, penetrating damp, water from leaks (pipes bedded in screeds etc) and condensation. In new buildings an additional problem is trapped construction water. This next section explains the most common ways in which penetrating damp can occur.

Structural movement and consequential cracking can form obvious paths for water to penetrate a wall. Similarly, frost attack, sulfate attack and surface erosion caused by atmospheric impurities can reduce a wall's resistance to damp penetration. These phenomena, and their potential consequences in terms of chemical attack and wall deterioration, have been considered in earlier chapters.

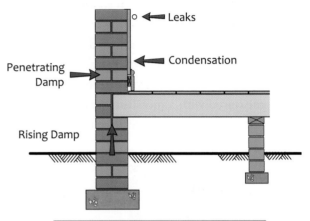

Condensation, particularly interstitial condensation, is often mis-diagnosed as penetrating or rising dampness.

PENETRATING DAMP

Penetrating dampness occurs in a variety of ways; the main 'mechanisms' are shown on the right.

Mechanisms for water ingress include:
- gravity
- capillary action
- surface tension
- kinetic energy (splashing)
- wind force
- differential air pressure (inside and out)

Water is also absorbed, ie taken directly from the air. The amount of water absorbed depends on the ambient humidity and the nature of the material.

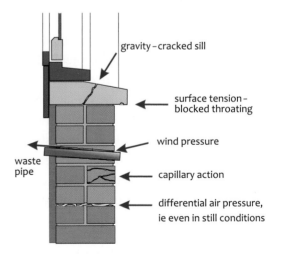

Solid external walls

Until the 1920s the majority of buildings in the UK were constructed from solid, ie non-cavity masonry. The materials used, mainly brick and stone, varied widely in their resistance to moisture absorption. Some materials, such as engineering brickwork, salt glazed brickwork and granite are virtually moisture resistant. Others, such as limestone, sandstone and most clay facing bricks are more permeable.

Contrary to what one might expect, it is often walls constructed from the denser materials that are most at risk from damp penetration. The diagram below shows the effect of rainwater on two solid walls. The moderately porous wall will absorb a proportion of the rainwater running down its face. As long as the wall is of adequate thickness (usually one brick) and contains no significant cracks it will resist damp penetration in all but the worst of exposures. In wet conditions the wall will absorb substantial amounts of rainwater but when the weather changes the absorbed water will slowly evaporate from the wall. Minor imperfections or cracks in the wall will not necessarily lead to damp penetration due to the limited amount of water running down the wall's face.

Walls built from denser materials such as granite or engineering brick lack the ability to absorb any rainwater running down the wall's face. Minor cracks in the walling material or in the mortar itself are therefore more critical and can allow moisture ingress. In addition, the dense, smooth nature of some of these materials precludes a good bond with the mortar. The interface between walling material and mortar can therefore provide a capillary route for damp penetration. In some properties the effect of this can be seen when observing the internal plasterwork; 'lines' of dampness in the plaster correspond with the position of the mortar joints.

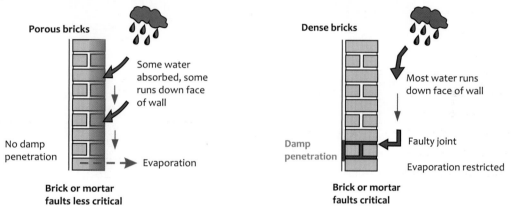

CAUSES OF DAMPNESS

General defects

The diagram below shows some of the common routes for damp penetration in houses with solid and, in some cases, cavity walls. The photos overleaf show a few more examples.

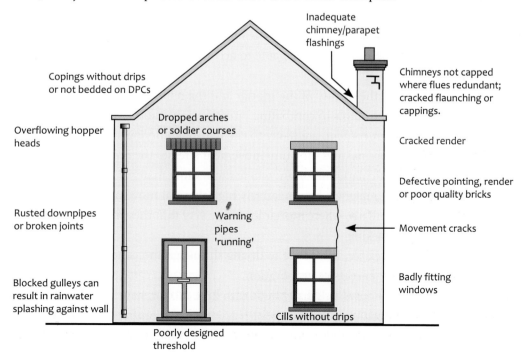

Cracks in masonry

Mortar flashings

Slipped tiles

Patched downpipes

Overflowing warning pipes

Gutters full of vegetation

Sills without drips

Render over DPC (and cracked)

Patched lead work; inadequate steps

Decayed pointing

Leaking gutter union

Mortar joints

The purpose of the mortar joint is to:

- provide an even bed to distribute the loads of the wall
- to bond the bricks or stones together
- to seal the joint against rain penetration.

An effective seal depends on a good bond between the mortar and the bricks or stones. In addition, to be effective over a long period, the mortar must be able to accommodate thermal and moisture movement without cracking.

It is generally accepted that the quality of the mortar and the nature of the pointing are of more significance than the bricks themselves in combating moisture penetration. Strong mortars, in other words those with a high cement content, are likely to crack through drying shrinkage and they are less capable of withstanding the thermal and moisture movement stresses the mortar will have to cope with during its life. This is particularly important where the walling material is concrete blockwork or artificial cast stone. Both these materials have relatively high thermal movement which places considerable stresses on the mortar. Similarly, where mortar joints are very thin they may not have the capacity to tolerate movement without cracking.

Weaker mortars are more flexible, less prone to drying shrinkage and generally more durable. They 'breathe' more readily and will encourage evaporation.

Unfortunately, many older houses, originally built with lime mortar, have been repointed with strong cement mortars. This is often counter productive; for reasons already outlined they may encourage penetrating damp. At the same time their dense nature restricts evaporation. Minor problems of penetrating or rising damp become more significant as the reduced rate of evaporation encourages water to dry out on the inner face of the wall. At the same time rising dampness may move further up the wall as its ability to evaporate is impeded. Changing the 'equilibrium' of a building's external fabric is a common cause of damp problems.

Where old solid walls are damp many builders will suggest external rendering as a solution. However, it should be recognised that renders, like mortars, are more effective if they are made from relatively weak, porous materials *(see Chapter 9: External Rendering)*. Strong renders are more likely to crack (drying shrinkage) and will impede evaporation. This can actually increase the likelihood of penetrating damp.

The nature of the pointing finish will also affect a wall's ability to withstand rain penetration. Generally, experience shows that tooled joints are more effective. Tooling the joint helps seal the mortar against the brick, and compresses the mortar slightly, thus increasing its impermeability. Some joints, such as the recessed joint, are very vulnerable. Whichever joint is used the mortar must be laid as a continuous bed and perpend joints should be well filled.

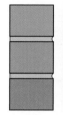  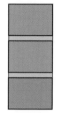 

Bucket handle - a good bed and the face of the mortar is compressed by tooling.

Flush pointing - a good bed but the face mortar is not compressed because the joint is not tooled.

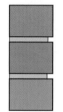  

Recessed - only a partial bed and poor at shedding water. Brick edges vulnerable to frost attack; algae growth common.

Struck joints - good at shedding water but very skilled work. The re-pointing on the right is vulnerable to physical damage and frost attack.

Tooled joints, full mortar beds and well filled perp joints help resist damp penetration.

String courses, parapets and subsills

These are all vulnerable areas. If copings are not weathered (sloped) they will be slow to shed water. Cracked, pervious or loose pointing in between the copings will allow vertical penetration if the copings are not bedded on a DPC. Copings are obviously in a very exposed situation and, over the years, erosion and chemical attack can cause deterioration in the stone which reduces the profile of the drips. The drips prevent water running back under the coping and into the wall. In some rehabilitation projects paving slabs are used as a substitute for proper copings. These are often bedded level and, by their very nature, do not contain drips.

In many parts of the country string courses are cut from fairly soft stone. These are often fairly porous materials which readily absorb water. If the top is not weathered or not protected by a lead or copper flashing, problems of damp penetration can occur. String courses are often subject to frost attack; as the ice forms, parts of the stonework break away. Not only does this further reduce the ability of the string course to resist water penetration, it also places pedestrians below in very real jeopardy.

Defective stone or concrete subsills can also lead to problems of damp penetration. Problems include insufficient fall, gaps at junction with window, deterioration of drips (often blocked through painting) and lack of adequate projection. The latter often occurs where buildings are rendered at some point after their original construction.

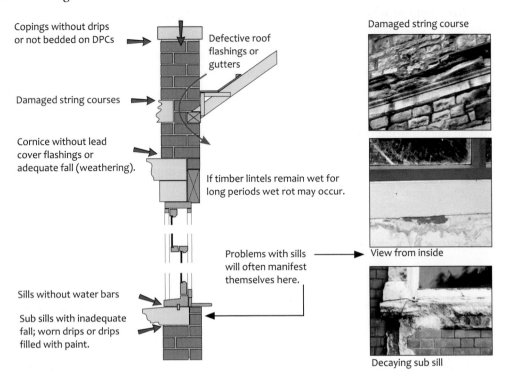

Copings without drips or not bedded on DPCs

Defective roof flashings or gutters

Damaged string courses

Cornice without lead cover flashings or adequate fall (weathering).

If timber lintels remain wet for long periods wet rot may occur.

Problems with sills will often manifest themselves here.

Sills without water bars

Sub sills with inadequate fall; worn drips or drips filled with paint.

Damaged string course

View from inside

Decaying sub sill

Chimneys

Chimneys are very vulnerable to damp penetration, particularly if they are no longer in use. Correct diagnosis of the source of the moisture can be difficult as chimneys are of relatively complex construction.

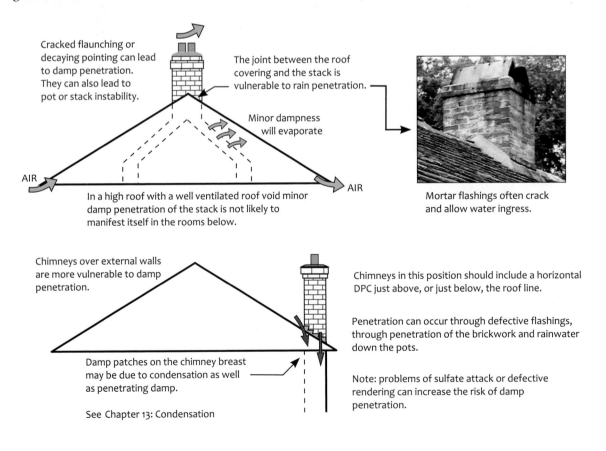

Cracked flaunching or decaying pointing can lead to damp penetration. They can also lead to pot or stack instability.

The joint between the roof covering and the stack is vulnerable to rain penetration.

Minor dampness will evaporate

AIR

AIR

In a high roof with a well ventilated roof void minor damp penetration of the stack is not likely to manifest itself in the rooms below.

Mortar flashings often crack and allow water ingress.

Chimneys over external walls are more vulnerable to damp penetration.

Chimneys in this position should include a horizontal DPC just above, or just below, the roof line.

Penetration can occur through defective flashings, through penetration of the brickwork and rainwater down the pots.

Damp patches on the chimney breast may be due to condensation as well as penetrating damp.

See Chapter 13: Condensation

Note: problems of sulfate attack or defective rendering can increase the risk of damp penetration.

Damp penetration can occur from:
- rainwater running down through the pots – a particular problem on low chimneys or low pitched roofs
- failure of the flashings where the chimney runs through the roof
- lack of DPCs in stack itself
- sulfate attack in the brickwork
- inadequate detailing around the top of the stack.

On buildings with fairly steeply pitched roofs problems of damp penetration may not manifest themselves in the rooms below. More at risk are buildings with shallow roofs particularly where the chimneys are situated above external walls at eaves level.

Basements

In modern construction basements are fairly rare although, until the 1930s, they were quite common. The examples below show typical late Victorian construction – all designed for areas where the ground was well drained and where the water table was well below the foundations.

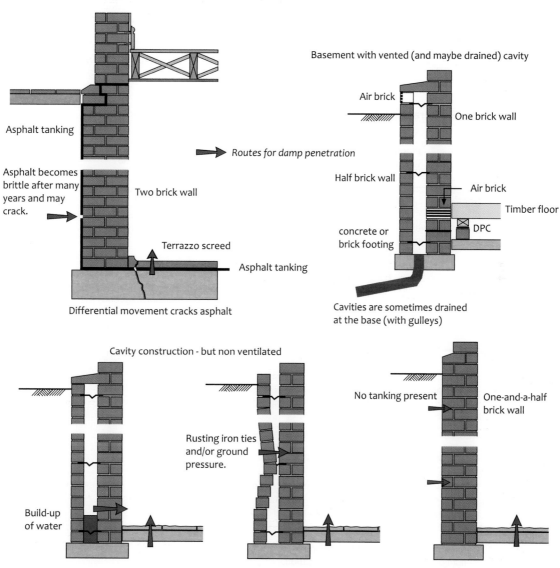

Asphalt tanking

Asphalt becomes brittle after many years and may crack.

Two brick wall

Terrazzo screed

Asphalt tanking

Routes for damp penetration

Differential movement cracks asphalt

Basement with vented (and maybe drained) cavity

Air brick

One brick wall

Half brick wall

Air brick

Timber floor

DPC

concrete or brick footing

Cavities are sometimes drained at the base (with gulleys)

Cavity construction - but non ventilated

Rusting iron ties and/or ground pressure.

Build-up of water

No tanking present

One-and-a-half brick wall

Most basement construction methods depend on well drained ground and a low water table. In this situation the building techniques protect against capillary action. Protecting existing basements against direct water pressure requires very expensive tanking works.

Other problems

There are a number of other problems which can cause penetrating damp. In many cases these are fairly easy to diagnose because the dampness will be localised and this will help to pinpoint the external defect. Problems include:

- leaking gutters and downpipes – cast iron downpipes often rust and develop leaks at the back, ie against the wall (this part of the downpipe is the hardest to paint). Gutters can fail through lack of correct support which allows backfalls, damaged joints and lack of regular clearing
- overflowing warning pipes which do not throw water clear of the wall – on some buildings lead overflows have even been cut off for their scrap value!
- poorly fitting windows
- door frames with poorly designed thresholds
- cracked render.

Aid to diagnosis

The chart below, which should be read in conjunction with the charts on pages 259 and 281, explain the basic steps in trying to identify and pinpoint problems of penetrating damp. Mistakes in diagnosis are common, and where the problem is not obvious, remedial action should not be considered until all the steps have been carried out. Specialist advice is often necessary although it should be remembered that many of the so called 'Damp Specialists' have a vested interest in promoting their own products.

Diagnosing Penetrating Dampness

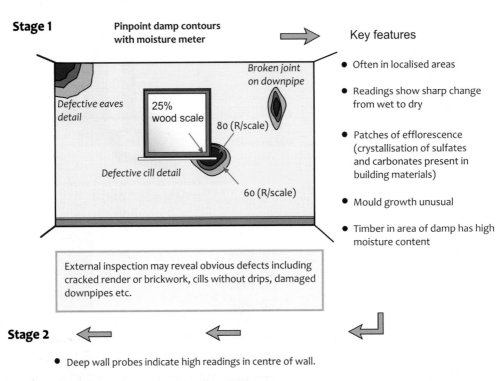

Stage 1   **Pinpoint damp contours with moisture meter**

Key features

Broken joint on downpipe

Defective eaves detail

25% wood scale

80 (R/scale)

Defective cill detail

60 (R/scale)

- Often in localised areas
- Readings show sharp change from wet to dry
- Patches of efflorescence (crystallisation of sulfates and carbonates present in building materials)
- Mould growth unusual
- Timber in area of damp has high moisture content

External inspection may reveal obvious defects including cracked render or brickwork, cills without drips, damaged downpipes etc.

Stage 2

- Deep wall probes indicate high readings in centre of wall.
- Salt analysis shows zero level of nitrates and chlorides (eliminating rising damp).
- Measure wall temperature, air temperature and establish humidity levels to eliminate condensation.

Cavity walls

The introduction of cavity walls in the 1920s was intended, in part, to improve the resistance of a wall to damp penetration. Unfortunately, failure of cavity walls, in terms of damp resistance, is a not uncommon problem. The defects, in the main, relate to poor design and poor workmanship rather than inadequate materials. A cavity wall may have a saturated external leaf after prolonged rainfall. Water should, in theory, run down the inner face of the brick and disperse back into the ground or substructure.

Cavity walls 1920s to 1960s.

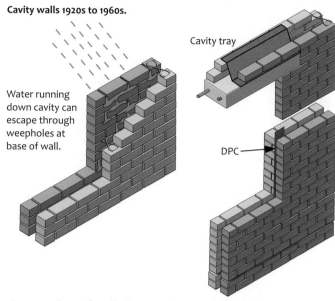

Water running down cavity can escape through weepholes at base of wall.

Cavity tray

DPC

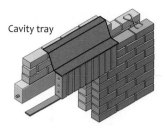

Cavity tray

Vertical DPCs at the sides of openings prevent water crossing to the inner leaf. Cavity trays over windows and doors direct any water outwards. Both were sometimes omitted in early cavity walls.

View inside cavity

Inner leaf   Closer

In exposed areas (usually the west of the country) heavy rain can result in water running down the inner face of the outer leaf. In sheltered areas the outer leaf can be come damp. Cavities must be kept clear of any debris in order to keep the inner leaf dry.

Unfortunately, there are a number of ways in which water can bridge the cavity and cause problems of dampness on the inner leaf of the wall. Careless bricklaying is probably the most common cause. Mortar droppings in the cavity can cause a build-up on wall ties or on floor joists which project beyond the internal leaf. It only needs a relatively small amount to bridge the cavity. Wall ties, if laid upside down or laid sloping towards the inner leaf, can also allow water to cross the cavity.

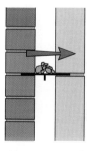

Mortar on ties bridges cavity

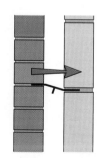

Uneven courses result in ties sloping towards internal leaf.

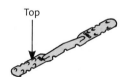

Top

Some ties, if laid upside down, can actually channel water across the cavity.

These joists should not project beyond the inner leaf - the ledge forms a bridge across the cavity.

These photographs were taken using an endoscope. They all shows wall ties covered in mortar.

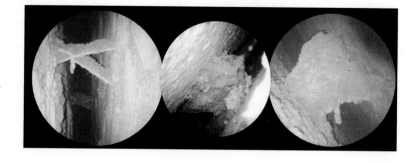

Cavity insulation

Most forms of cavity insulation (complete cavity fill) are treated to make them water repellent. Unfortunately, if the manufacturers' recommendations regarding the sequence of construction are not followed, situations can occur which permit damp penetration. Where cavity batts are used there is the risk that mortar can collect on the top of each batt and create a bridge. This is a particular problem where batts are inserted down into the cavity rather than being positioned against the wall as it is being built. Evidence suggests that new houses with filled cavities suffer more failures than older houses with 'retro' filled cavities – this suggests that construction errors might be the cause.

When cavities are filled the insulation which fills the cavity will restrict evaporation. If an outer leaf becomes saturated through prolonged rainfall, it will dry more slowly as evaporation into the cavity is restricted. At the same time a wall which remains wet is at greater risk from frost attack. In addition, any mortar droppings in the cavity will prevent complete cavity fill and create a cold bridge. This can result in localised spots of condensation on the inner face of the wall.

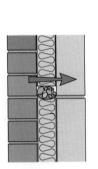

Water can pond in depression caused by mortar or bridge cavity by capillary action.

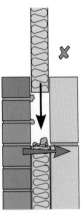

Batts should never be pushed down into cavities. They will dislodge mortar forming a potential bridge for damp.

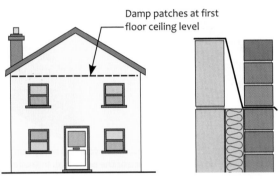

Damp patches at first floor ceiling level

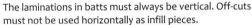

If batts stop at ceiling level a cavity tray must be provided. Otherwise water can run across top of batts.

The laminations in batts must always be vertical. Off-cuts must not be used horizontally as infill pieces.

At gable ends batts should be taken right up to the verge unless a cavity tray is provided.

Note that current Building Regulations do not normally permit the use of cavity fill in exposed situations unless some form of impervious cladding such as weather-boarding protects the masonry. This obviously reflects concerns about long term resistance to damp penetration.

Where cavity boards are used (cavity boards do not fill the cavity), the special ties, which include large washers to hold the insulation in place, must not be omitted – the boards can fall against the outside leaf and form a path for water penetration. They can also form ledges which may collect mortar droppings.

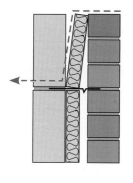

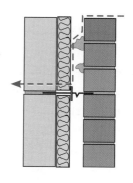

Cavity boards must be held securely against the inside leaf. If they touch the outer leaf a path may be created for water penetration.

Cavities must be kept clean - if mortar touches an insulation board water can run down its face, into the blockwork.

Parapets

The problems of parapets in solid walls have been mentioned earlier in the chapter. Cavity walls are equally at risk. Damp stains below the parapet at ceiling level may be caused by failure of the roof covering, be it pitched or flat. They can also be caused by water penetrating the coping stones. Coping stones without drips or which are not bedded on a DPC, supported over the cavity, can permit water to run down the outer face of the inner leaf. Similarly, if a cavity tray is omitted at roof level water can penetrate the inner leaf and run down into the main building.

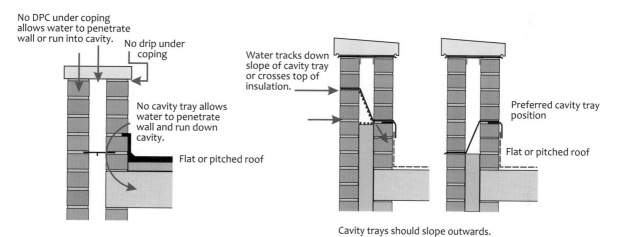

No DPC under coping allows water to penetrate wall or run into cavity.

No drip under coping

No cavity tray allows water to penetrate wall and run down cavity.

Flat or pitched roof

Water tracks down slope of cavity tray or crosses top of insulation.

Preferred cavity tray position

Flat or pitched roof

Cavity trays should slope outwards.

Cavity trays

Problems of missing cavity trays or mortar build-up on trays have already been mentioned. There are two common situations, apart from over windows and in parapets, where correct installation of cavity trays is important. These both occur where structures abut a cavity wall. Problems may occur through incorrect installation but are more likely to be caused by subsequent building alterations or additions.

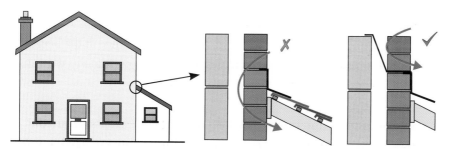

If there is no cavity tray (in a cavity wall) water can penetrate the new addition. When an extension is added a tray is sometimes omitted because it's difficult to fit.

Cavity trays and flashings are also required where two roofs join at a setback.

Windows and doors

In cavity walls the cavity is normally closed to form openings for windows and doors. At the jambs it is normal practice to return the blockwork to the brickwork and insert a vertical DPC between the two materials to prevent damp penetration. If built correctly problems of damp penetration will not occur. However, the construction around windows and doors can be quite complex. Mistakes in construction and detailing occur quite commonly. There are three vulnerable areas: the head, the jambs and the sill.

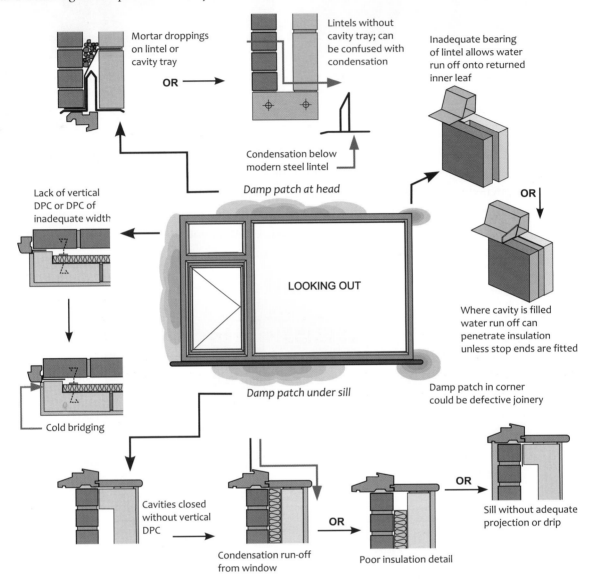

Mortar droppings on lintel or cavity tray

OR

Lintels without cavity tray; can be confused with condensation

Inadequate bearing of lintel allows water run off onto returned inner leaf

Condensation below modern steel lintel

Damp patch at head

Lack of vertical DPC or DPC of inadequate width

OR

LOOKING OUT

Where cavity is filled water run off can penetrate insulation unless stop ends are fitted

Cold bridging

Damp patch under sill

Damp patch in corner could be defective joinery

Cavities closed without vertical DPC

Condensation run-off from window

OR

Poor insulation detail

OR

Sill without adequate projection or drip

RISING DAMP

This section explains the phenomenon of rising damp and describes its most common causes. It also identifies the various steps which may be required in order to confirm (or otherwise) a diagnosis. Rising damp is often confused with penetrating damp and condensation, and, as already explained at the beginning of the chapter, incorrect diagnosis often leads to remedies which are wholly inappropriate to the cause.

How rising damp occurs

In liquids, surface tension is a result of cohesive intermolecular forces. These cohesive forces are the reason why rain falls as small droplets. The molecules at the surface of the liquid are attracted inwards; this attraction results in the liquid trying to minimise its surface area and forming the shape of a sphere. If the liquid is in contact with another material there will also be some adhesion between the molecules in the liquid and the material itself. So, there are two forces; the cohesion (surface tension) within the liquid, and its natural inclination to adhere to another material (wetting). If the solid material is a surface such as plaster, brick or block (known as wettable surfaces) the adhesion forces will be greater than the cohesion forces and the water will tend to spread outwards and form a thin surface layer. However, if the adhesion forces are less significant than the cohesive forces, as occurs when water lies on a waxed surface, the liquid will remain in droplets. This explains, for example, why a newly waxed car is covered in tiny droplets after rainfall.

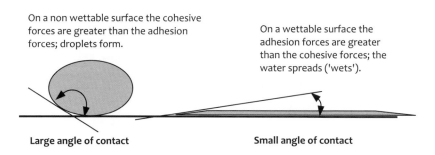

On a non wettable surface the cohesive forces are greater than the adhesion forces; droplets form.

On a wettable surface the adhesion forces are greater than the cohesive forces; the water spreads ('wets').

Large angle of contact

Small angle of contact

This effect is seen after waxing a car or varnishing garden furniture; the surface becomes non-wettable and, after rainfall, is covered with small droplets of water.

Most building materials are porous; in other words they contain tiny spaces or pores. The pores can occur naturally and may also be created by the shrinkage of materials such as render and mortar. Water's natural inclination to adhere to wettable surfaces will cause it to spread into these tiny spaces and, the smaller the diameter of the pore, the greater the attraction (capillary action). This occurs because in narrow pores the adhesive forces (wetting) at the interface of the pore sides and the water are greater than the cohesive forces (surface tension) As the pore gets wider, the changing ratio between the cohesive forces and the adhesive forces results in the water only rising to lower levels.

If the pores are coated with a non-wettable solution (the basis of chemical DPC injection) the adhesive forces are reduced and the water will settle below its natural level.

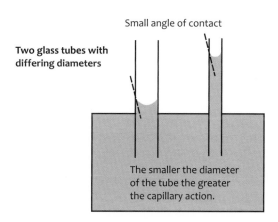
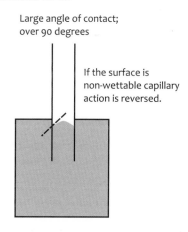

Small angle of contact

Large angle of contact; over 90 degrees

Two glass tubes with differing diameters

If the surface is non-wettable capillary action is reversed.

The smaller the diameter of the tube the greater the capillary action.

Rising dampness in a building is caused, therefore, by water in the subsoil rising up the pores or capillaries of the materials in the wall. Since the early 1900s most houses have been constructed with damp proof courses. When functioning correctly they provide an effective barrier to rising damp. However, DPCs can fail for a number of reasons. These include deterioration of the original material or physical damage caused by building movement. Early DPCs included tar and sand, lead, copper, bitumen felt, lead cored felt, asphalt, slate and engineering bricks. Some of these materials are quite brittle and even minor building movement can cause damage. However, in many cases, rising damp is caused by bridging of the DPC or changes in ground water levels, rather than failure of DPC material. The diagrams overleaf show a number of ways in which DPC bridging can occur.

There are a number of ways in which ground water levels can change. These include natural or man-induced changes to the water table, leaking drains, blocked land drainage systems, leaking water mains and springs. If rising dampness occurs fairly suddenly in a wall that has hitherto been dry, changes in water levels are more likely to be the cause than any breakdown of an existing DPC. Changes in water levels may ultimately cause the expansion of certain sub-soils with obvious implications for the building structure.

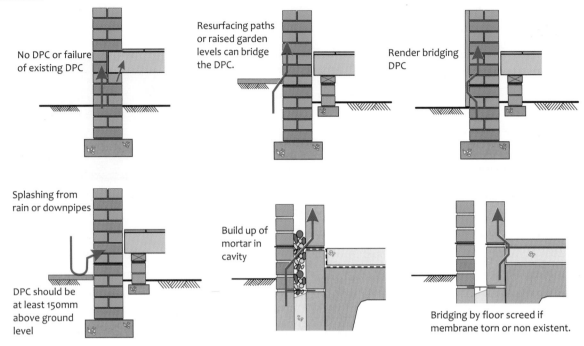

No DPC or failure of existing DPC

Resurfacing paths or raised garden levels can bridge the DPC.

Render bridging DPC

Splashing from rain or downpipes

DPC should be at least 150mm above ground level

Build up of mortar in cavity

Bridging by floor screed if membrane torn or non existent.

The height of rising damp

The height to which damp rises depends on a number of factors.

The porous nature of the material Research has shown that pores of about 0.01 mm provide the main routes for rising damp and that, where these are present, damp can rise about 1.2–1.5 metres. Occasionally, it may be higher. Theoretically, finer pores should allow damp to rise to even greater heights but, in practice, finer pores can actually resist the flow of water.

The amount of water present in the ground The height to which damp rises will, to some extent, be influenced by the amount of water in the ground. This, in turn, will be affected by the moisture content of the soil and the level of the water table; both of these will depend on the season and the amount of recent rainfall. During wet, humid conditions, for example, evaporation from the wall will be reduced and this will encourage the damp to rise further up the wall.

Limits on evaporation The impervious nature of wall finishes such as internal plastering, external rendering or painted finish will limit evaporation from a wall. A reduction in evaporation will lead to an increased height of rising damp. Thick walls can also limit evaporation because the surface area is comparatively low in relation to the volume of the wall. Conversely, thin walls, where the surface area is comparatively high in relation to volume, are less likely to have high levels of rising damp.

The level of heating inside the building A well heated dry environment will encourage evaporation and thus limit the height of rising damp.

Chemicals in the ground and walls The ions of dissolved mineral salts affect patterns of rising damp in a number of ways. Sodium, calcium and potassium salts can be present in the ground or in the masonry. In solution they are drawn up into the wall through capillary action. Where evaporation occurs these salts may crystallise (and cause efflorescence on the surface) and block the capillaries through which the water is evaporating. This can reduce evaporation and drive the water further up the wall. Perhaps the worst culprit is calcium sulfate. This chemical is not readily soluble and is most likely to crystallise.

Chemical analysis

Moisture meters are helpful in providing a first step towards the diagnosis of rising damp. However, as described in the first part of this chapter, it should be remembered that they have limitations when used on materials other than wood and are easily confused by the presence of salts. The various stages in achieving a diagnosis are included in a chart later in this chapter. Definitive evidence can be provided by chemical analysis.

Sub-soil naturally contains nitrates, from decaying plant matter and fertilisers, and chlorides. These chemicals are very soluble and are drawn up into the wall with the water. When the water evaporates from the wall these chemicals are left behind. They form heavy concentrations on the plaster surface and wallpaper. Nitrates and chlorides are both hygroscopic – in other words they absorb moisture from the air. As they absorb water they continually re-dissolve and this prevents any crystallisation. Research has shown that these hygroscopic salts are most abundant at the peak of rising damp because this is where most evaporation takes place. In nitrates and chlorides hygroscopic action commences at very low relative humidities; 55% or less.

Dampness lower down the wall is usually the result of capillary action and analysis has shown that the amount of hygroscopic salts is usually quite low at the base of the wall. Rising dampness is often accompanied by efflorescence. This is caused by a number of sulfates and carbonates which are always present in building materials. These materials crystallise quite readily because they are not that soluble. They are not hygroscopic and merely indicate that moisture is evaporating from a structure. They can sometimes block pores in brickwork thus preventing evaporation. The effect of this may be to drive damp higher up a wall.

A scraping of wallpaper and finish-coat plaster can be sent for chemical analysis. The degree of contamination by chlorides and nitrates will give an indication of how long the damp has been rising. Condensation or penetrating damp should show zero levels of nitrates and chlorides.

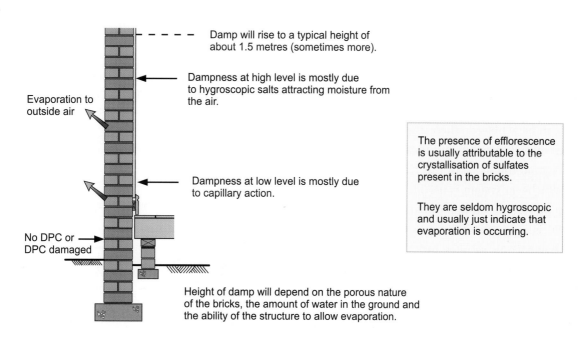

Damp will rise to a typical height of about 1.5 metres (sometimes more).

Dampness at high level is mostly due to hygroscopic salts attracting moisture from the air.

Evaporation to outside air

Dampness at low level is mostly due to capillary action.

No DPC or DPC damaged

The presence of efflorescence is usually attributable to the crystallisation of sulfates present in the bricks.

They are seldom hygroscopic and usually just indicate that evaporation is occurring.

Height of damp will depend on the porous nature of the bricks, the amount of water in the ground and the ability of the structure to allow evaporation.

DIAGNOSIS

The diagnosis of rising damp needs careful and systematic thought because it can easily be confused with penetrating dampness and condensation. The BRE have suggested that only 10% of the dampness problems it investigates are attributable to rising damp. Unfortunately, there are a number of companies specialising in DPC replacement who obviously have a commercial interest in finding problems of rising damp. Their diagnosis needs to be treated with caution. Although there are several reputable companies working in this field it may be wise to seek independent advice. Further 'encouragement' to find problems of rising damp is provided by banks and building societies who often request a damp report as a condition of a mortgage advance.

The diagnostic stages and key features of rising damp are shown in the diagram (top of opposite page).

Typical patterns of dampness

The contours of dampness shown in the diagram, ie high readings (on the relative scale) at the base of the wall and at 0.50 metres, slightly lower readings at 1 metre, and virtually zero levels at 1.50 metres, are typical of rising damp. There are, however, other common patterns which can confuse an inexperienced surveyor. The bottom diagram, opposite page, shows some typical patterns and suggests likely causes.

Problems with replacement DPCs

Replacement DPCs or new DPCs (where none existed before) can be of various types. In the past, electro-osmotic systems and atmospheric tubes have been popular. Electro-osmotic systems can normally be identified by an horizontal continuous copper strip or thick wire running along the base of the wall (they may be hidden by render). Atmospheric tubes are normally made from ceramic or clay and are about 50mm in diameter. They are usually fixed at centres between 200mm and 400mm and penetrate approximately 2/3 of the wall. In many cases a plastic vent will cover the end of the tube.

Nowadays the two options recommended by the BRE are physical DPCs and chemical DPCs. The former are more expensive and can only be laid in horizontal courses, ie in brickwork or coursed stonework. They are not suitable, therefore, for rubble walls. In practice chemical DPCs are the cheapest and most common form of remedial treatment. There are a number of patented methods but most work on the same principle, ie they line the pores with a non-wettable surface to reduce capillary action. Their effectiveness depends on their successful penetration of the wall and a number of concerns have been raised about how successfully this can be achieved. There are other specialist repairs but these are beyond the scope of this text.

Whichever system is used it must be recognised that providing a DPC is only part of the solution. The hygroscopic nitrates and chlorides present in the wall structure and plaster need to be dealt with as well. Most of the hygroscopic salts will be present

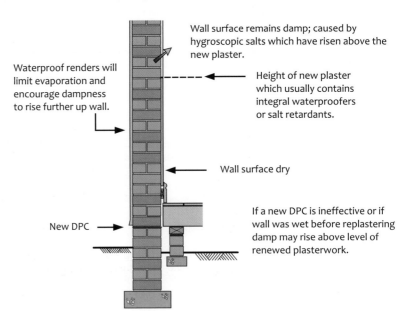

Wall surface remains damp; caused by hygroscopic salts which have risen above the new plaster.

Waterproof renders will limit evaporation and encourage dampness to rise further up wall.

Height of new plaster which usually contains integral waterproofers or salt retardants.

Wall surface dry

New DPC

If a new DPC is ineffective or if wall was wet before replastering damp may rise above level of renewed plasterwork.

Rising dampness

Stage 1 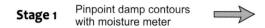 Pinpoint damp contours with moisture meter

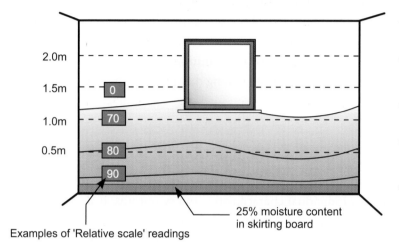

2.0m

1.5m   0

1.0m   70

0.5m   80

90

25% moisture content in skirting board

Examples of 'Relative scale' readings

Key features

- Limit is usually 1 - 1.5m above ground level

- Readings will drop quickly above peak of dampness

- Stains and/or tide mark on wall paper/plaster

- High % moisture content of timber skirtings

- Areas of dampness appear to get wetter in humid conditions (hygroscopic salts)

External inspection may indicate missing or bridged DPC

Stage 2 ← ←

- Deep wall probes indicate high readings in centre of wall; eliminates condensation.

- Salt analysis shows high levels of nitrates and chlorides

- Establish internal surface temperature and relative humidity to determine whether condensation is a possibility

Tide marks are common in rising damp but do not confuse them with leaks/spills.

	A	B	C	D
1.75	20	10	00	100
1.50	30	80	10	100
1.25	40	70	60	100
1.00	50	00	50	100
0.75	60	00	40	100
0.50	70	00	20	100
0.25	80	00	20	100
0.00				

A   Typical of condensation. The lower part of the wall is often colder; hence the higher readings. Condensation running down the wall also accounts for higher readings.

B   Typical of a failed replacement DPC. Replastering works ensure that the base of the wall remains dry. The dampness manifests itself above the limit of the replaced plasterwork. There may also be contamination by salts.

C   Typical of long-term, but controlled rising damp. The low readings at the base indicate minor problems of rising damp. The high readings further up the wall suggest long-term contamination by salts.

D   Consistent maximum readings suggest that wall is lined with metal foil. This is probably a past remedial measure against dampness.

in the plaster and this should be renewed (usually to a height of 300mm above the level of rising damp). The new plaster must prevent any remaining salts in the wall from migrating to the plaster surface. This usually requires the use of cement based plasters as most gypsum plasters cannot prevent the passage of hygroscopic salts and quickly break down in wet conditions. Thus, apparent failure of a new, replacement DPC may, in fact, be caused by failure to renew the plasterwork or incorrect choice of plastering materials.

Whichever plaster system is used, it should be recognised that limiting evaporation from the wall, eg by using an external render system, may encourage any remaining dampness in the wall, or dampness caused by an ineffective DPC to rise even higher.

CONCLUSION

Rising damp, penetrating damp and condensation are easily confused. Correct diagnosis requires a systematic approach which recognises their individual characteristics and symptoms. It should also be recognised that there may be more than one source of dampness. In an old solid wall, penetrating dampness and rising damp may both occur at the same time. In addition, a wet wall conducts heat more readily than a dry wall; if the wall surface temperature drops below the dew point, condensation can occur.

Although carbide testing and moisture meters will help to determine the relative intensity and contours of the dampness, they will not, on their own, provide conclusive information as to its causes. A detailed building inspection to highlight any obvious building defects, together with chemical analysis and an assessment of humidity levels, wall surface temperatures and patterns of occupation are all of vital importance to a correct diagnosis. In practice, mistakes are common and can result in wasted effort, wasted resources and, in some cases, incorrect remedial work which exacerbates rather than resolves the original problem.

CHAPTER 15

System Building

SECTION 1

BACKGROUND

INTRODUCTION

Traditionally, most of the construction process took place at the building site (although some elements were formed elsewhere, eg the hardwood members of timber-framed buildings were shaped and jointed in yards and taken to the site for erection). However, the first half of the 20th century saw industrialised processes introduced to the UK construction industry. At their simplest,

Possibly the earliest example of a 'modern' system building in England, this concrete house on the Isle of Wight was erected in the 1850s.

these involved the use of factory produced components within the traditional building process, eg roof trusses. More wide ranging methods of non-traditional or system building were also introduced. These ranged from the use of in situ factory techniques, such as shuttered and poured concrete walling, to the site assembly of pre-fabricated components, eg timber, steel or concrete frames, and cladding. It is worth noting that most system building involves elements of both traditional and industrialised construction methods. In particular, the substructure is usually formed in situ, rather than assembled from delivered components.

There were a number of apparent beneficial factors that led to the introduction of system building.

Efficiency – factory processes would make the best use of available labour and materials to produce as much of a building as possible.

Speed – factory prefabrication combined with simple site assembly would result in considerable savings in time.

Specialisation – standardisation of components together with mass manufacture would give economies of scale.

Better quality – use of the workshop production line rather than in situ construction techniques would allow higher levels of quality to be achieved.

Economy in cost – the combination of the above factors would lead to cheaper housing.

Unfortunately, there were also a number of problems that needed to be avoided if the system-built housing stock was to perform adequately over its intended life span. The following chapter examines some of the more important of these problems and their effect upon the buildings involved.

BRIEF HISTORY

The aftermath of the First World War saw an urgent need for thousands of houses. The immediate reasons for this were a lack of any new construction and maintenance during the war years combined with the need to house large numbers of demobilised soldiers and their families. Post-war shortages of materials, such as bricks, timber and labour, in particular skilled labour (as the result of war casualties and the lack of training), meant that new methods of construction were encouraged by the Government. System-built housing was undertaken by a number of local authorities throughout the UK, although the Government did not offer any special financial assistance towards its construction.

Ultimately, some 50,000 system-built domestic units were constructed during the inter-war period, about 1% of the total number of houses built between 1919 and 1939. Large numbers of different systems were approved, including timber frame, steel frame, concrete frame/slab and in situ wall slab. Most have subsequently been demolished.

The Second World War resulted in similar shortages of building materials and skilled labour. The situation was exacerbated by the loss of some 200,000 houses due to enemy bombing, as well as damage to about 25% of the entire building stock. An increasing population (the post-war 'baby boom') caused further pressures.

Government response was much more positive, there being generous funding of both traditional and non-traditional (system) building by successive Labour and Conservative Governments between 1945 and 1955. During this period about 20% of the new housing in England and Wales and 50% in Scotland was system built – approximately 500,000 units in total. Initially, there were two approaches – a permanent housing programme and a temporary housing programme (below). The former produced houses of two or more storeys, many thousands of which are still in use. Relatively few remain of the 150,000 plus temporary houses (or 'pre-fabs'). A large number of different systems were produced including many with frames and/or claddings of steel or pre-cast reinforced concrete (PRC), and others formed with cast in situ concrete walls. The numbers of each system ranged from a mere handful to many thousands. Some local authorities invested very heavily in system building to meet their housing needs, eg Birmingham and Leeds Metropolitan Councils, Bristol City Council.

Bristol Record Office

Between 1955 and the late 1970s, the techniques learned in low-rise system-built housing were developed and used in the construction of medium-rise and high-rise flats. During this period, it is believed that between ¾ million and one million 'industrialised' (system) dwellings were constructed, of which some 500,000 were low/medium-rise and 140,000 high-rise (available statistics are, unfortunately, incomplete). However, the 1970s saw a reaction against system building in general. This was mainly as a result of problems of maintenance and repair caused by one or more of the following:

- poor design
- inadequately controlled prefabrication processes
- poor construction as a result of the use of unskilled labour and/or poor site management.

These problems led to considerable failure of structural and cladding components. Many houses and flats also suffered from condensation as well as poor thermal and noise insulation.

The early 1980s produced another problem. The Conservative Government of that period had introduced 'Right to Buy' legislation which led to the purchase of many thousands of local authority houses by sitting tenants. This included a great number of system-built houses. At the same time, but too late to prevent such properties being purchased by the tenants, it became apparent that those houses constructed with pre-cast concrete components were often affected by carbonation or chloride attack *(see following section for details)*. Building societies refused to mortgage or remortgage these houses and they were, effectively, blighted. The Government responded to the resultant outcry by introducing the Housing Defects Act of 1984, which provided repair grants for those houses that were designed before 1960 (ie not high-rise homes, which, mostly, tend to be of later design), were of certain designated classes (all of pre-cast concrete construction), and had recognised qualifying defects. These grants were controversial as they were only available to private owners of such properties and not to local authorities. In addition, grants were available even if the houses were not themselves defective but were of a potentially defective type.

Limited numbers of steel or concrete framed/clad houses have been constructed since 1970, but the following decade saw a rise in the popularity of timber-framed houses. Unfortunately, many of those built in the 10–15 years leading up to 1984 were found to have inherent defects as a result of poor design/construction. This led to a short period of decline but, since the mid-1980s, improved marketing together with better design, construction techniques and site control have resulted in timber-framed housing regaining its popularity.

SECTION 2

DEFECTS IN CONCRETE HOUSES AND STEEL-FRAMED HOUSES

INTRODUCTION

It would be impossible in a textbook of this nature to detail the problems experienced by the complete range of concrete and steel houses currently in use. The following sections, therefore, initially describe some important defects that are commonly found across a number of different types of system-built houses and then briefly examine, in more detail, a representative sample of systems. It should be noted that the individual housing systems frequently had variations in design and/or construction that could result in minor changes of detail and the use of different materials.

DEFECTS IN HOUSES CONSTRUCTED BETWEEN 1945 AND 1955

This section concentrates on those houses constructed with components of either pre-cast reinforced concrete (PRC), cast in situ concrete or steel. Many thousands of these were erected in this period. A great number have subsequently been demolished for the reasons briefly referred to above. The remaining houses, which number in their thousands, are still in use. Some are in good condition, although many suffer from problems similar to those already demolished. Their continuing use is normally due to the owner (usually a local authority, sometimes a private person) being unable to replace them because of a lack of finance.

Carbonation of concrete

This is a problem that affects reinforced concrete and has been discovered across a broad range of PRC houses. Carbonation is a natural process that takes place in all concrete but, where insufficient allowance has been made for its effect, it can have disastrous consequences on reinforced concrete, as it can lead to corrosion of the steel and cracking/spalling of the concrete.

Fresh concrete contains calcium hydroxide which is highly alkaline and protects the steel reinforcement, preventing it oxidising. The concrete will, over time, react with carbon dioxide in the air to slowly form calcium carbonate, which is insufficiently alkaline to protect the steel. This process will slowly penetrate the concrete to a depth of 50mm or more. Where the outer layer is also affected by moisture, this can accelerate the process. The effect is to lead to the rusting of any steel within it.

The rate of carbonation will depend on a number of factors:

- quality and density of the concrete – improperly compacted lightweight concretes are particularly prone to the problem
- exposure of the building – the process requires moisture
- relative humidity of the atmosphere – carbonation is encouraged where it is between 25 and 75%, in particular, the higher range of 50–75% (this often results in internal components being affected more quickly than wetter external members).

The design of any reinforced concrete component should ensure that the steel is placed at a sufficient depth to prevent carbonation reaching it during the anticipated life of the building. Similarly, the manufacturing process must strictly follow this criterion.

Unfortunately, many prefabricated and cast in situ reinforced concrete components of system houses (and medium and high-rise blocks of flats) were poorly manufactured, with insufficient depth of cover to the steel, as well as poor quality concrete. This inevitably leads to carbonation problems over a period of time. Inadequate thermal insulation was also a common problem with these buildings and, when combined with the former defects, tends to accelerate the carbonation process. Furthermore, the presence of chlorides in the concrete (*refer to the next sub-section*) increases the depth and rate of carbonation.

In appearance, the affected component will tend to show longitudinal cracking along the line of any steel reinforcement. Its initial appearance will be in the form of hairline cracking, which can occur as early as a few months after construction. Over time, expansion of the rusting steel results in the component cracking along its length, as well as spalling of the surface concrete.

The presence of carbonation can be determined by in situ testing of the concrete. A chemical indicator (manganese hydroxide or phenolphthalein solution) is applied to the surface of the suspected area and will show both the extent and depth of any attack. Determination of the effect upon the steel is then necessary. It is normal to assume that the current rate of carbonation will continue and that, once it reaches the steel, corrosion and longitudinal cracking will commence immediately.

Two examples of carbonation, one in a concrete lintel and one in the loadbearing column of a Cornish Unit.

System buildings that have been shown to have instances of carbonation problems include:

- Airey
- Boot pier and panel cavity (in England, where breeze aggregate was used)
- Cornish Units (in South West England, where high levels of chloride additive were used)
- Easiform cavity wall
- Reema hollow panel
- Orlit (for similar reasons to Cornish Units)
- Parkinson framed
- Smith system
- Unity
- Woolaway.

Cornish units are easily recognised. They have mansard roofs and were made in 2, 3 and 4 storey versions.

Chloride attack of concrete

Calcium chloride was commonly added to concrete up to the late 1970s to accelerate its curing time, especially during cold weather, and thus speed the construction process. Unfortunately, it can break down the alkaline content of the concrete, especially where it has been introduced as an on-site additive (rather than during cement manufacture). Quality control of on-site additives is always difficult. It often resulted in uneven distribution of the chemical throughout the concrete which tended to exacerbate the problem.

The loss of alkalinity within the surface of the concrete removes its protective capability to stop the encased steel from oxidising. Where carbonation is present, chloride attack can increase the rate of oxidisation of any steel reinforcement. However, chloride attack can lead to the steel suffering corrosion even if carbonation is not present.

The appearance of chloride attack differs from carbonation in that it tends to induce large cracking or bulging within the concrete of a more localised nature. The steel can also suffer sudden failure, especially in the presence of both chloride attack and carbonation, as it can become relatively brittle due to extreme corrosion. Low levels of chloride ions (below 0.4% by weight of cement) are not considered to be of concern, unless carbonation is present. Between 0.4% and 1.0% by weight, cracking is assumed likely to occur within five years, even quicker if carbonation is also present. Where high levels of the chloride are present (above 1% by weight) corrosion of steel can occur, even if the concrete is highly alkaline.

System houses experiencing such problems include:
- Airey
- Cornish Units in South West England
- Reema hollow panel
- Orlit
- Unity.

Thermal problems

Very little consideration was given before the early 1970s to the level of thermal comfort enjoyed by the occupants of dwellings. New houses, including system-built housing, had extremely low levels of insulation incorporated into their construction. Generally, system-built houses constructed with a clad steel or concrete frame suffered more severe problems because their structural form and the lightweight materials used were extremely thermally inefficient. This was a situation that was normally exacerbated by a lack of any central heating system.

SPECIFIC SYSTEMS AND THEIR DEFECTS

Airey houses

Approximately 26,000 of these houses were constructed between 1945 and 1955. They were built with storey height pre-cast concrete columns at 450 centres. These had tubular steel centres and were erected over a cast in situ ground floor slab. They were connected at roof and first floor levels by timber or steel lattice joists.

The external elevations were finished with horizontal pre-cast concrete slabs which were either smooth finished or pebble dashed and secured to the columns with copper wire in shiplap fashion. Internally, the construction was usually of plasterboard at both floor levels, although later houses incorporated clinker blockwork for lining kitchens and bathrooms.

Common problems include:
- cracked columns and first floor beams as a result of poor design and site processes
- corrosion of the tubular steel reinforcement due to water penetration at the joints between the horizontal planks or insufficient concrete cover

- similar problems as a result of condensation due to a lack of any vapour check
- poor thermal insulation made worse by draughts, as the shiplap construction was not windproof
- the original windows were of poor quality steel and have often been poorly maintained
- structural problems caused by the removal of the vertical posts when replacing windows (the original posts acted as mullions which were too closely spaced to suit the new windows).

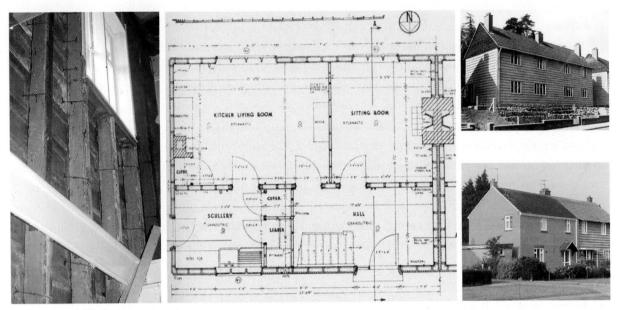

The left-hand photograph shows an Airey house during repair. The external cladding is tied to the columns with twisted copper wire. A timber batten on the inner face of the column will provide a fixing for the plasterboard and fibreboard lining. The top right-hand photograph (1950s) shows a newly constructed Airey house; in the lower photograph, the left-hand 'semi' has had the external walls replaced with loadbearing cavity work. Much of this work was funded through the Housing Defects Act 1984; funds were only available to those who had bought their houses.

British Iron and Steel Federation (BISF) Houses

Some 30,000 of these houses were erected in England and Wales together with over 4,000 in Scotland, out of a total of about 140,000 metal-framed houses, in all, of some 30 different types. There were three different types of BISF house, although the essential feature of each was a frame of either hot-rolled or cold-formed steel. A number of cladding systems were used of which the most common consists of an outer cladding of profiled steel sheeting at first floor level, cement rendering applied to metal lathing at ground floor level, and internal facings of plasterboard on timber battens. Steel trusses supported roof coverings of asbestos or metal sheeting.

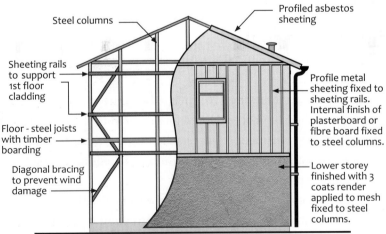

Steel columns

Profiled asbestos sheeting

Sheeting rails to support 1st floor cladding

Profile metal sheeting fixed to sheeting rails. Internal finish of plasterboard or fibre board fixed to steel columns.

Floor - steel joists with timber boarding

Diagonal bracing to prevent wind damage

Lower storey finished with 3 coats render applied to mesh fixed to steel columns.

In situ floor (traditional founds or raft)

These houses have suffered from a number of problems but their overall performance is better than many other types of system-built housing. Some common problems are set out below.

- Corrosion of the profiled sheeting at first floor level and, in some cases, the horizontal (or sheeting) rails that support the cladding. This can be as a result of condensation (the steel sheet is impervious to moisture, thus trapping vapour within the structure) or the failure of the weather stripping on the gable ends.
- Corrosion of the vertical steel stanchions, especially corner members where the cladding of steel or render may be less effective.
- Impact damage to the ground floor render can lead to rusting of the metal lathing and progressive failure of the render.
- The asbestos roofing tends to become brittle and crack with age. There is probably no health hazard unless it starts to shed fibres, particularly into the building. Its removal is strictly regulated. *(Refer to the section on asbestos slates in Chapter 7: Pitched Roofs).*
- The large void between the internal wall lining and the external cladding is now recognised to be a potential fire risk, allowing an easy spread of fire throughout the house. This can be exacerbated where internal linings are of fibreboard or similar combustible material.

Wates houses

These were based on prefabricated, loadbearing, reinforced, tray-shaped concrete panels erected around a special jig. The horizontal joints between the panels interlocked into a mortar bed, while the vertical joints were hollow and required filling with fine concrete after erection. Steel reinforcement was sometimes inserted into the horizontal and vertical joints.

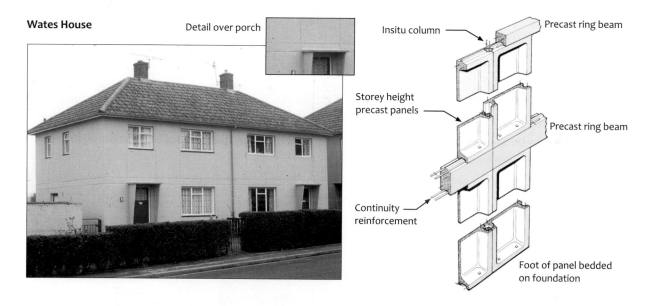

Wates House

Detail over porch

Insitu column — Precast ring beam

Storey height precast panels

Precast ring beam

Continuity reinforcement

Foot of panel bedded on foundation

Some 30,000 Wates houses and low-rise flats (two and three storeys) were constructed in the UK in the 10 year period up to 1955. This panel system was subsequently developed for use in high-rise blocks of flats of the 1960s and 1970s, although the panels were not loadbearing for buildings above six storeys.

Initially, the panels were lined internally with wood-wool slabs, although these were superseded by lightweight concrete blocks and plaster. Similarly, the original prefabricated sectional pitched roof was replaced by 'cut' rafters in later roofs.

Some common problems are set out below.

- Corrosion of the in situ steelwork in the joints and above windows, leading to cracking and spalling of the concrete. This is often due to lack of concrete cover.

- Corrosion of the reinforcement in the flange of the panel may affect either the front face of the concrete or the internal lining.
- Cracking along the vertical joints between the panels, as well as the panel corners, both being defects which may lead to damp penetration.
- The use of clinker blocks for party walls and the inner lining of the external walls was common in the north where Wates loadbearing panel system housing estates were developed by the National Coal Board. In the long term, these may become a problem because of chemical expansion.

Unity houses

There were approximately 15,000 Unity houses built, most of them between 1953 and 1955. The structure comprises a series of pre-cast columns, usually at 3ft centres with pre-cast (but not reinforced) panels forming internal and external skins. In early forms the inner and outer panels were tied together with metal ties but later models had the external panels tied independently to the columns. The inner panels were of lightweight concrete to provide some measure of thermal insulation. The outer panels were profiled along their top and bottom edges to provide weather protection. In addition, they were separated from the columns by a vertical DPC. The panels helped stiffen the frame although additional bracing was provided by steel straps running diagonally between the columns. On completion the outer panel joints were pointed. At first floor and roof level steel beams spanned across the building and also joined the ground and first floor columns.

Most Unity houses were built as two storey houses. However, there were also a number of bungalows and some medium-rise flats. The party walls were formed in cavity construction and comprised two 75mm leaves with a 50mm cavity. Ground floors were usually of solid construction with splayed timber bearers bedded into the screed or slab and supporting softwood boards. Internal partitions were formed from storey-height 50mm reinforced plaster panels or from 50mm breeze blocks. Windows were mostly metal. During the late 1940s and early 1950s there were a number of changes aimed mostly at speeding up construction times.

Some common problems are set out below.

- Corrosion through carbonation and impact damage can result in steel expanding and columns spalling.
- Where claddings contain high levels of chlorides, it is possible that the ties which connect these panels to the columns or internal leaf can corrode.
- Tenants have complained about poor sound insulation between units.
- In some areas, there has been damage to the external cladding panels, probably through frost action.

Unity houses are similar in principle to Airey houses. They both comprise pre-cast concrete columns with a pre-cast concrete cladding. Unity houses also have steel braces to prevent 'racking'.

'Pre-fabs'

At the end of the Second World War, the Government introduced a temporary housing programme in an attempt to meet immediate emergency housing needs. This involved the erection between 1945 and 1948 of temporary dwellings, nicknamed 'pre-fabs', with only a very short designed life span of 10 years. These buildings, of which some 157,000 were constructed, including 32,000 in Scotland, performed better than expected and, although most have subsequently been demolished, there are several thousand still in use.

The temporary houses were normally single storey buildings which were site-assembled in two to three days from a number of prefabricated components, including those manufactured by former aeroplane factories using war-surplus materials, such as aluminium and steel. A number of other readily available materials, such as asbestos-cement sheeting and plasterboard were also used.

The two outer photographs show the simplicity of the 'pre fab' design. The left-hand prefab was of aluminium construction; that on the right was of steel. The middle photo shows the fitted bathroom of the aluminium pre fab.

Eleven different types of temporary house were produced, of which six involved numbers of less than 2,500 units each. Of the remaining five, the following three were produced in the highest numbers.

- Arcon Houses – 39,000 of steel frame, asbestos-cement cladding and plasterboard/building board linings. The Arcon Mark V was the most common of all 'pre-fab models'.
- Aluminium Bungalows – 54,500 houses of aluminium alloy frame, cladding and roof, lined with plasterboard.
- Uni-Seco Structures – 29,000 houses with timber frame, asbestos-cement cladding and wood wool linings. It had a flat roof.

Problems include:

- Poor insulation leading to low comfort levels.
- Condensation affecting the frame, claddings and linings due to the low thermal insulation of the original construction.
- The aluminium sheeting was often formed from re-salvaged scrap from shot down and obsolete planes. There was often no attempt to separate different metals and this has resulted in corrosion of the metal sheeting and frame.
- Poor detailing, leading to water penetration of joints and gradual deterioration of the materials used, has resulted in damp problems of the frames, claddings and linings.
- Deterioration of the asbestos-cement sheeting can lead to a dangerous health hazard if individual fibres are released into the environment (*refer to Chapter 7: Pitched Roofs – asbestos-cement slates, for a more detailed discussion*).

DEFECTS IN HOUSES CONSTRUCTED AFTER 1955

Many of the defects associated with earlier system buildings are found in dwellings of this later period. Problems caused by carbonation, chloride attack, lack of thermal insulation, damp penetration and inadequate sound insulation are all too common across the broad range of industrialised (or system) buildings constructed – low, medium and high rise. Over 150 different systems were authorised in England and Wales alone, with the individual numbers erected ranging from single figures to over 128,000 for Wimpey No-Fines.

SPECIFIC SYSTEMS AND THEIR DEFECTS

Cross wall construction

This form of construction became popular in the 1960s and 1970s as it enables standardisation and repetition of the construction processes for both the structural and non-structural elements of a building. It is based upon the premise that all structural loads are carried on regularly spaced parallel walls; all other walls, either in-between or at right angles, merely being non-loadbearing partitions or cladding panels. The cross walls are constructed at approximately 3.5 – 4.5m intervals in brickwork, blockwork or in situ concrete and form the gable and party walls of semi-detached or terraced housing. In medium height blocks of flats, where reinforced concrete cross walls form the gable and intermediate walls, the process is known as 'box-frame construction'.

Cross wall construction

Tiling on trussed rafters

Weather boarding or tile hanging

Gable walls - concrete with brick facing or brick and block cavity construction.

Non loadbearing studwork infill panels with render or tile hanging.

Return on gable wall to provide resistance to longitudinal deformation.

Trussed purlin carries load of first floor joists.

The walls not only provide the structural element of the building but also offer fire resistance and sound insulation between separate dwellings. The loads from upper floors and roofs are transferred onto the cross walls either directly, where the floor/roof is built into the wall, or indirectly, where the floor/roof is supported on transverse beams or lintols that are built into the cross walls. The cross wall requires lateral bracing against wind pressure at right angles to its length and this is achieved by a number of means; in situ reinforced concrete floors; longitudinal beams or walls; returning the ends of the cross wall in the shape of an 'L' or 'T'. A typical system is shown above.

Problems commonly experienced include the following.

• Water penetration at the vertical junction between the end of a cross wall and the adjoining cladding/infill panel. This detail is often poorly thought out and merely finished with mastic/sealant and, possibly, paint.

• Movement as a result of a lack of lateral restraint may occur where the floor or roof joists run parallel to the cross walls. In this case, no intermediate restraint is given to the walls, which are dependent for stability only upon the end beams receiving and transferring the floor/roof leads to the walls.

• On the other hand, lateral movement may be caused by the expansion of joists that are at right angles to the cross walls. This may be as much as 40mm and be due to either thermal or moisture movement.

• Condensation on poorly insulated and exposed dense concrete walls – this may occur on flank/gable walls and at the outer ends of party walls, ie a form of cold bridging.

Typical 1960s/1970s concrete cross wall construction

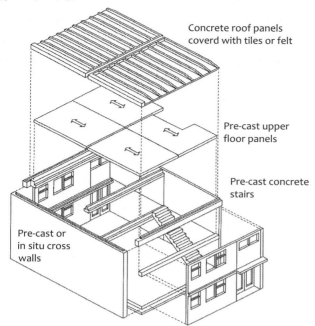

Concrete roof panels coverd with tiles or felt

Pre-cast upper floor panels

Pre-cast concrete stairs

Pre-cast or in situ cross walls

Timber frame cladding to front and rear elevations. Tile hanging between floors.

Wimpey No-Fines houses

This form of construction involves the use of no-fines concrete, ie concrete containing cement and graded aggregate (often 12 or 19mm in size), but no sand nor fine aggregate. The lack of fines results in a wall that can be cast more economically than a normal concrete mix and which contains many air gaps, giving some improved insulating capability. It is also more resistant to damp penetration as its wide pore structure reduces capillary action *(refer to Chapter 14: Damp)*.

The first recorded no-fines house was constructed on the Isle of Wight in 1852 and several hundred were built in England and Scotland between the wars. Over 72,000 of these low and medium rise houses were constructed by Wimpey in England and Wales from 1945 to the early 1960s and a further 128,000 between 1964 and 1979. Several thousand were erected by a number of other contractors, while in Scotland, the Scottish Special Housing Association, Wimpey and other contractors commissioned or constructed several thousand more. The majority of the dwellings were low/medium rise, although a certain number of high rise buildings were constructed.

The Wimpey version had external walls formed with 200–300mm thick no-fines concrete poured on site into prepared shuttering containing continuous reinforcement. External elevations were finished with a wide range of materials, although cement render was very commonly applied. Internal walls were of breeze blocks or timber studwork. Very little prefabrication was involved except for the shuttering panels which were re-usable.

Common problems include the following:

• Rotting windows and external doors caused by poor joinery design and/or inadequately sealed gaps between joinery and structure.

• Excessive heating loss due to poor thermal

An early 1970s No-Fines house above. These particular units had long term problems of rain penetration at roof level and defective render. They were demolished in 2001.

On the right, two residential tower blocks of No-Fines construction are being erected (1950s photo).

insulation. This is often combined with an inadequate or expensive-to-run heating system.

- Surface condensation and mould as a result of poor ventilation, cold bridging, and inadequate thermal insulation/heating.
- The wall construction should reduce capillary action but there have been rainwater penetration problems.
- Cracking and spalling of render on external wall faces often caused by lack of suction in the no-fines concrete, or unsuitable mortar mixes, eg strong over weak.

Pre-cast concrete houses and flats

A number of pre-cast concrete houses were manufactured in the 1970s. These were often adaptations of high-rise buildings. The problems experienced, ie damp penetration, condensation and high heating costs, have led to many units being demolished.

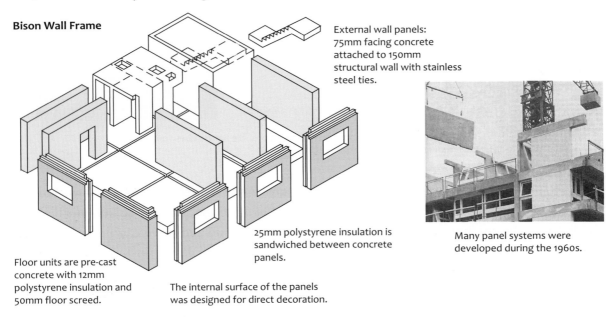

Bison Wall Frame

External wall panels: 75mm facing concrete attached to 150mm structural wall with stainless steel ties.

25mm polystyrene insulation is sandwiched between concrete panels.

Many panel systems were developed during the 1960s.

Floor units are pre-cast concrete with 12mm polystyrene insulation and 50mm floor screed.

The internal surface of the panels was designed for direct decoration.

The Bison system was originally intended for high-rise construction. Later developments included medium-rise flats and two-storey houses.

Some manufacturers concentrated their efforts producing pre-cast concrete low-rise houses. A number of systems were available in the 1960s and 1970s. The system shown below is based on panel construction.

Apart from the problems noted above many of the buildings were incorrectly assembled; panels were sometimes insecurely fixed, joint gaskets were omitted and the properties suffered raised levels of dampness. Problems of penetrating dampness were therefore common.

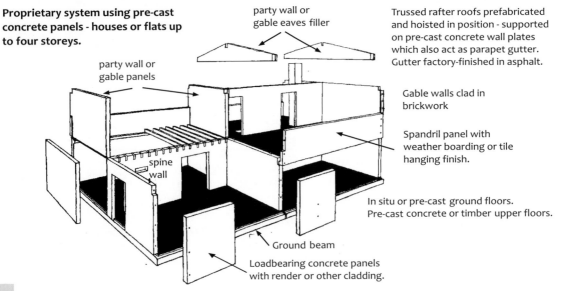

Proprietary system using pre-cast concrete panels - houses or flats up to four storeys.

party wall or gable eaves filler

party wall or gable panels

Trussed rafter roofs prefabricated and hoisted in position - supported on pre-cast concrete wall plates which also act as parapet gutter. Gutter factory-finished in asphalt.

Gable walls clad in brickwork

Spandril panel with weather boarding or tile hanging finish.

spine wall

In situ or pre-cast ground floors. Pre-cast concrete or timber upper floors.

Ground beam

Loadbearing concrete panels with render or other cladding.

CONCLUSION

Historically, prefabricated reinforced concrete and steel systems have promised rather more than they have delivered. The present stock of such houses contains many that have under-performed. However, a number of building companies are currently experimenting with system buildings. Some are investigating steel-framed housing, while other organisations are looking into buildings formed with in situ aerated concrete or timber framing. The underlying factor is no longer simply about increasing the speed of erection but is also about seeking prefabrication methods that reduce dependency on skilled site trades.

SECTION 3

TIMBER FRAME HOUSING

INTRODUCTION

In this country, a modern house is still usually thought of as having a cavity wall with an outer leaf of brick and an inner leaf of blockwork. However, an alternative approach is to construct the house from a framework of timber protected by weather boarding, tile hanging or brickwork. This section briefly describes the development of timber frame housing and explains those defects that are associated with this approach.

Before the ascendancy of brick and stone, most houses of substance would have been constructed from timber. These buildings were generally constructed in one of two structural forms: cruck or box frame. By the end of the 16th century timber buildings, particularly box frames, had evolved into a very sophisticated form of construction.

A cruck frame and a later box frame (17th century).

In these traditional timber framed buildings the structural members were infilled with panels made of wattle and daub. The wattle consisted of oak staves, hazel wattle, and a daub of lime, mud, straw and cow dung. Later, the infill panels were often replaced with brickwork. In later timber framed buildings bricks were sometimes used as the original infill for the panel.

The left-hand and middle photos show wattle and daub in progress. The right hand photo shows brick infill – this is not usually original but a replacement for earlier wattle and daub.

The 20th century saw a revival of timber frame construction, albeit using different methods to those employed traditionally. The timber frame houses of today have evolved from the approach pioneered in Scandinavia and North America where there has been a long tradition of timber construction. Both Scandinavia and North America had plentiful supplies of renewable timber, and, in the latter case, the carpentry skills that the early English settlers took with them. Mechanised sawmills, which enabled speedy production of uniform cross-sections of timber, together with industrial techniques which produced low cost nails, both played their part in developing this form of construction.

Modern timber frame housing

The system built timber frame houses constructed today have evolved from a number of earlier forms including the following.

• Platform – where walls are constructed as storey height elements.

• Balloon frame – where walls are constructed as complete entities that are continuous from sole plate to eaves.

• Post and beam – where, like traditional timber framed buildings, the loads are carried by columns and beams.

• Volumetric box – where the units are delivered and placed on site as a series of pre-constructed boxes.

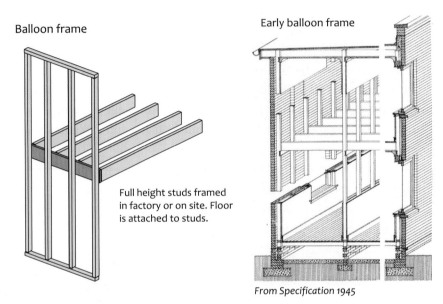

An example of a balloon frame used on a 1940's house. Note the flat roof (right-hand diagram).

Balloon framed construction was common until it was replaced by the platform method. This is usually more economic and better suited to the demands of factory based prefabrication. An example of a platform frame is shown on the right.

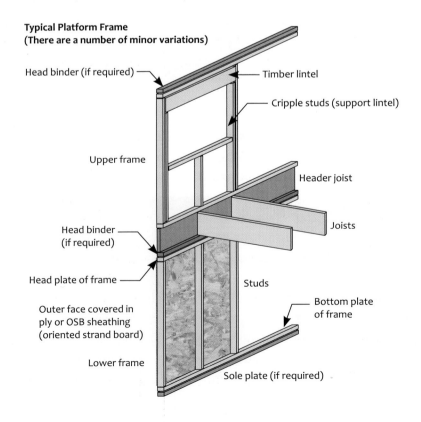

HISTORY OF MODERN TIMBER FRAME HOUSING IN THE UK

Although not significant as a percentage of total houses built, there were a number of timber frame houses constructed in the period between the two World Wars. Many of these were constructed from timber studwork although some were constructed from what were effectively solid timber walls, usually clad on the outside with timber boarding.

In these solid wall structures there was usually a protective breather membrane sandwiched between the cladding and the frame. Its purpose was to prevent rain penetration while, at the same time, allowing moisture in the frame to evaporate. Vapour control layers (vapour checks), which limit interstitial condensation, were not usual at this time.

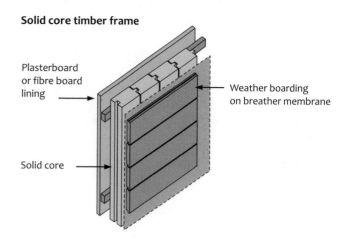

Solid core timber frame

Plasterboard or fibre board lining

Weather boarding on breather membrane

Solid core

Following the Second World War the increasing interest in industrialised systems revitalised timber frame systems. Apart from the reasons for embracing system building that have already been mentioned earlier in the chapter, timber frame became a focus of attention because the experience of several severe winters raised concerns about levels of heating and insulation in both traditional and PRC domestic buildings. An export drive by the Canadian Government in the early 1960s helped to sell the idea that timber frame construction produced warm buildings; this coincided with the UK Government's policy to further develop industrialised systems of building. The picture below shows a Canadian Trend House first exhibited at the Ideal Home Exhibition in 1957.

From The Builder June 21st, 1957

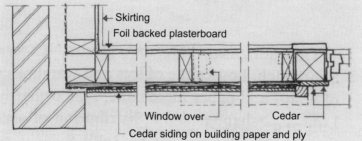

Skirting
Foil backed plasterboard

Window over — Cedar
Cedar siding on building paper and ply

Timber frame housing is by no means new. However, systems used for the last 50 years or so having nothing in common with the traditional timber framing of Medieval England.

In 1957 A Canadian house was exhibited at the Ideal Home Exhibition. For the first time there was a quick and cost effective alternative to traditional housing and existing system building (mostly pre-cast concrete). However, not until the 1960s did it become popular and then in cross wall form rather than true timber frame.

Nowadays most timber frame houses are not clad in Cedar (as this example is) but brickwork. So, there is a loadbearing internal timber frame and an external single leaf of brickwork. Many of the defects in modern timber frame housing are, in fact, caused by the problems of building in these two very different materials.

In the immediate post-war period most timber frame housing was largely based on timber stud walls which were covered with a variety of claddings including; timber boarding, tiles/slates, render on a plywood backing, brick, and asbestos cement sheets. These lightweight claddings would normally be fixed directly to the timber frame, or indirectly using timber battens. In these designs there was no cavity to protect the inner load bearing frame against water penetration. There were also hybrid versions where, for example, there would be a brick cladding at ground floor level and one of the directly fixed claddings on the upper floor. Some of the systems were thermally insulated but not to a very high level. A breather member was normally incorporated between the cladding and the studs although vapour checks were not normally included (foil-backed plasterboard was used from the mid-1960s). Some houses had an external cladding of 100mm brickwork; a cavity separated the brick and timber studding. This approach later became the norm.

In the 1960s (left) timber framed houses were often clad in tiling, slating or weatherboarding. Modern houses (right) are usually clad in brick.

From the mid-1960s to the early 1980s there was a steady increase in the number of timber frame houses. Most of these were in the public sector although many volume house builders were developing their own timber-framing systems to cope with the increased demand for private housing.

Some cross wall houses (see previous section) were constructed with brick party walls but with timber frame panels to the front and rear. These were usually clad with timber boarding and/or tiles (left).

Advantages of timber frame houses

The main advantages of timber frame houses, when measured against brick and block construction, are generally argued to be related to cost savings that can accrue from industrialised processes, including standardisation and speed of on-site construction. Speed of construction is related to the possibility of quick erection times and the ease of making connections between uniform materials. Additionally, there can be a significant reduction in time spent waiting for the finished structure to dry out, because fewer 'wet trades' are involved.

Other advantages of timber frame are generally perceived to be:

- the ability to, relatively easily, increase the amount of thermal insulation incorporated in the walls
- a relatively light-weight structure that allows some savings in foundation size
- the ability to accommodate settlement or other foundation movement more readily than brick/block construction.

The usual method of constructing timber frame in the UK is, as mentioned, the platform frame, ie. the assembly of the ground floor wall panels on the substructure, followed by the first floor structure, which, in turn, provides a platform to erect the next storey, and so on. The timber frame is the main loadbearing element in this type of construction; the brick cladding's primary purpose is weather protection, although it may also have been chosen for its appearance. The brickwork is separated from the timber frame by a cavity and carries none of the building load apart from its own weight. In order to provide stability the brickwork is tied to the frame at regular intervals with metal ties. The other key elements of the timber frame walls include the following.

Breather membrane This is intended to prevent wind driven rain reaching the timber frame. It should also allow any water trapped in the frame to escape – ie it should be 'breathable'.

Sheathing The purpose of this is to add structural stability to the frame by helping to prevent 'racking'.

Thermal insulation This is usually placed within the timber frame (rather than in the cavity).

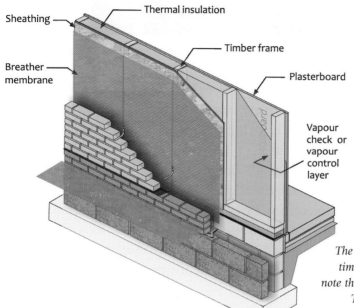

The left-hand graphic shows the construction of a modern timber frame wall. The photo above shows a party wall – note there are two timber frames separated by a small cavity. The strand boarding provides stiffness against racking.

From the mid-1970s Building Regulations required external walls to meet certain levels of thermal insulation. In timber frame housing this could be achieved by insulating the spaces between the vertical studs. A vapour barrier, now usually referred to as a vapour check or vapour control layer, is placed on the inner surface of the insulation (its warm side) in order to limit the amount of vapour moving towards the cavity where it might condense on the inner face of the sheathing. The vapour check usually takes the form of a polythene sheet although foil-backed plasterboard has been used *(see photo bottom right)*.

The breather membrane, which is fixed to the external face of the sheathing, has the function of protecting against any wind driven rain which may pass through the outer cladding. The membrane also protects the timber frame during the construction period. Therefore to be effective the membrane has to prevent water penetration but, at the same time, allow the passage of water vapour.

This photograph shows the brick skin being built as a cladding to a modern timber frame. The frame itself is hidden by the protective breather membrane. Vertical fire barriers can be identified here by their red polythene sheathing.

The breather paper (membrane) can be seen in this photograph prior to the brickwork skin being built. The membrane is usually applied in the factory. If this is not the case it should be fixed in position as soon as the frame is erected in order to prevent the sheathing getting wet. Breather membranes vary in colour depending on the manufacturer.

Identifying brick-clad timber frame houses

It can be difficult to differentiate brick-clad timber frame houses from brick/block cavity walled houses. However, there are a number of features which, while they may not confirm timber frame (because the same features could be on brick/block construction), will provide cumulative indicators. Examples include:

- Windows fixed to the inner rather than the outer leaf.
- Weepholes incorporated just above or just below damp proof course level. These were provided to allow drainage of moisture away from the vulnerable timber frame.
- Movement joints formed around windows, verge and eaves (or signs of distress if not incorporated). The movement joints accommodate the different rates of movement of brick and timber.
- Weepholes at first floor level to indicate the presence of a cavity tray situated over horizontal fire stops.
- Cavity vents – by no means universal but cavity vents in the form of airbricks were sometimes incorporated to help disperse moisture vapour.

An internal inspection might confirm whether or not timber frame has been used by reference to:

- Separating walls in the roof space. If the party walls are finished in plasterboard this will be indicative of timber frame. (The plasterboard provides the necessary fire/sound insulation to the timber studs. A masonry wall does not normally require any surface finish. It should be noted however that some early timber frame houses were of hybrid construction and used brick or block for the party walls.)
- Gable walls – the inner gable leaf will probably be built in timber.
- Eaves – the cavity closer may be timber or mineral wool and the wall plates supporting trussed rafters may be planed (in the factory) rather than 'sawn'.
- The timber studding might be revealed by removing electrical sockets or by lifting floor boards or skirting boards.
- Dry lining – although dry lining is often used on blockwork, with timber frame the position of the fixing nails may indicate the standard centres of the vertical timber members.

DEFECTS IN TIMBER FRAME HOUSING

Timber frame became so popular that by the early 1980s nearly a quarter of house building starts used this method of construction. However around this time reports began to emerge of a number of defects in relatively new timber frame houses. These reports, which included a damaging television broadcast, led to a decline (since reversed) in the construction of timber frame houses. Volume house builders withdrew from timber frame construction on the assumption that adverse publicity would affect sales. It appears however, that public reaction was muted: either because people were not aware of the adverse publicity, or perhaps because occupiers of timber frame did not realise how their houses were constructed.

Many of the reports highlighted defects that arose due to either mistakes and/or ignorance, at both the design and construction stage. Other reports raised more fundamental questions about inherent problems in the use of timber frame. It is clear however that many of these reports, at least in the press, were related to faults that could have been found in any house type. Because they were also found in timber frame houses the faults were erroneously linked to this construction method. For example, reference to rusting wall ties, missing thermal insulation, poorly supported water tanks and defective brick garden walls could all be found in contemporary articles about timber frame, written as though such defects were unique to the method of building. Nevertheless, some criticisms were valid and although the reports made reference to defects that were not specifically related to the construction type (such as mortar droppings on wall ties), they were pertinent because, for example, they could result in the timber frame becoming wet. More specific criticisms were directly related to the design principles of timber frame housing.

With hindsight it is now clear that the incidence of failures from the timber frame 'boom' of the 1970s and 80s has been less significant than predicted. However, theoretically there are some potential problems inherent in timber frame houses from this period – and indeed in those timber frame houses constructed more recently if these issues have not been addressed.

The problems identified in the reports were caused by a number of factors acting both singly and jointly. These include:

- poor design and specification
- poor site practices
- a lack of familiarity with the function of key elements in timber frame design
- a lack of experience about how theories of timber frame design would apply in practice.

Problems also sometimes arose because of the difficulties inherent in constructing factory made timber frames on top of a substructure formed on site. Specific concerns centred on a number of defects which may also be found in current timber frame construction of poor quality. Potential problems include the following.

Excess dampness in the timber frame The normal risks associated with rising damp, penetrating damp and condensation are increased because of the vulnerability of the timber frame and sheathing to fungal attack. Excess dampness allows fungal attack to occur, and the risk is increased if the timber is not treated with 'preservative' and/or the timber is of low natural durability (for example where it contains a high proportion of sapwood). In some cases timber treatment is targeted on those areas, such as sole plates, that are considered most at risk. Where timbers are cut on site (perhaps in the fixing of sole plates, or to overcome assembly problems), any treatment becomes less effective.

Fire Potential defects in design or construction can lead to rapid spread of fire.

Movement Differential movement of the timber studding in relation to the brick cladding can produce stresses which may lead to damage of the wall and/or associated elements.

Condensation

A key element in the design of most timber frame houses is the vapour check (or vapour control layer). Its purpose is to prevent the passage of moisture vapour from the warm side of the insulation to the cold side where there is a danger that it might condense within the wall structure. This phenomenon is known as interstitial condensation *(see Chapter 13)*. Where timber frame houses are not insulated the risks of interstitial condensation are much lower as heat losses through the frame keep the timber structure above dew point. If a frame is insulated the temperature of the studding remains high but there is a rapid drop towards the outer face of the insulation The cold sheathing layer, therefore, is the element most at risk from interstitial condensation.

Where there is a risk of interstitial condensation the performance of the breather membrane is crucial if it is to successfully allow trapped moisture to escape. Unfortunately there have been reports of impermeable materials such as polythene being used for breather membranes.

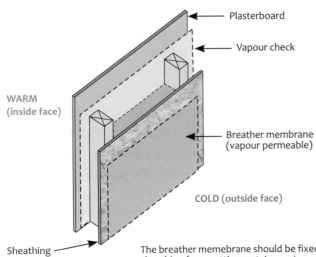

Plasterboard

Vapour check

WARM
(inside face)

Breather membrane
(vapour permeable)

COLD (outside face)

Sheathing

The breather memebrane should be fixed as soon as the frame is erected. It prevents the ply sheathing from getting wet. In most modern systems it is applied in the factory.

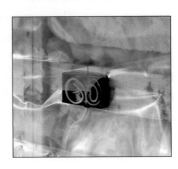

The vapour check should not be punctured in the course of fitting electrical sockets or central heating fittings.

There is a danger of condensation occurring if the vapour check fails to perform adequately. This may occur if it is:

- missing – either not included as part of the design or omitted on site
- placed on the cold, rather than the warm, side of the insulation
- incorrectly lapped to adjoining sheets or with other elements, such as damp proof courses
- damaged during (or before) the construction process
- placed in position while the timbers' moisture content is high.

Additionally, with aluminium-foilbacked plasterboard vapour checks there are two potential problems:

- the gap at the junction between the boards allows water vapour penetration
- the aluminium foil can corrode.

Condensation can also affect the timber frame where it occurs on say, single glazed windows, and the water run-off penetrates through poorly sealed joints.

Rising damp

Although incorrectly specified, incorrectly located or damaged damp proof courses can result in damp problems in any type of house, it is the vulnerability of timber which is of particular concern with timber frame. The bottom section of the timber frame is most at risk, particularly the sole plate or bottom rail. As the sole plate is often fixed in position by nailing through the DPC this is a potential area of damp penetration.

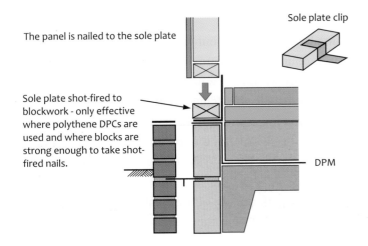

The panel is nailed to the sole plate

Sole plate clip

Sole plate shot-fired to blockwork - only effective where polythene DPCs are used and where blocks are strong enough to take shot-fired nails.

DPM

Many prefer galvanised or stainless steel sole plate clips because they do not puncture DPC.

Rain penetration

In addition to the problems described in the chapter on damp the likelihood of rain penetration occurring in timber frame housing is exacerbated by the following factors.

- Reduced cavity width; site adjustments may be required while trying to marry a factory made wall to an in situ base or foundation.
- Breather membrane that is defective or missing. (Note: in early designs bitumen impregnated fibre board may have been used instead of plywood sheathing with a separate breather membrane.)
- Distortion of wall ties caused by differential moisture movement of the timber studding and brick cladding. Shrinkage of the frame may result in ties sloping towards the inner leaf, which will allow water to be transported towards the timber frame.
- Misplaced insulation. When thermally insulating a timber frame the intention is to leave a clear cavity, and to place the insulation only between the vertical timber studwork. There have been reports of the cavity being filled. This poses the danger that the insulation may provide a path across the cavity to the timber frame.
- Lack of weepholes. These are required above doors and windows and immediately below DPC level; their purpose is to allow moisture, from whatever source, to drain away. In properties constructed before the early 1990s weepholes were also required at upper floor level just above the cavity fire barriers.

The breather membrane is normally stapled to the sheathing in the factory. Laps or hems are left to provide an overlap at adjacent panels and over the substructure. These hems should be re-stapled as soon as the panels are erected. Membranes are also easily torn on site. Repairs should be carried out immediately.

External timber claddings may, of course, also be vulnerable to fungal attack. In those designs where the cladding is fixed directly to the frame the danger may be increased because there is no ventilated cavity. For those systems with no cavity the performance of the cladding and the breather membrane is, of course, critical.

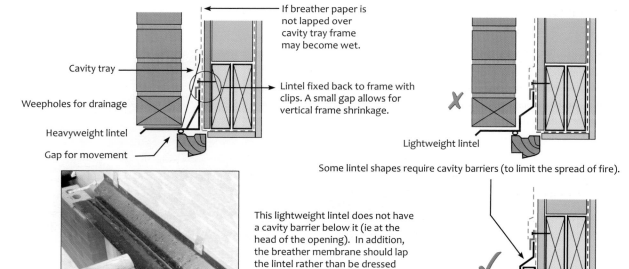

If breather paper is not lapped over cavity tray frame may become wet.

Cavity tray

Weepholes for drainage

Heavyweight lintel

Gap for movement

Lintel fixed back to frame with clips. A small gap allows for vertical frame shrinkage.

Lightweight lintel

Some lintel shapes require cavity barriers (to limit the spread of fire).

This lightweight lintel does not have a cavity barrier below it (ie at the head of the opening). In addition, the breather membrane should lap the lintel rather than be dressed behind it as shown here (otherwise water can run down between lintel and membrane).

Movement

Timber frames are likely to shrink, particularly upon initial occupation when heating systems are first used. Failure to provide movement joints at positions where this shrinkage occurs may cause damage to the structure. Timber shrinks across the grain, rather than along it. It is therefore the horizontal members that mainly cause the problem.

Storing panels in the open will allow their moisture content to increase. The subsequent expansion may cause cracking and problems of condensation as the panels slowly dry out in the completed building. Careless stacking can also tear the breather paper.

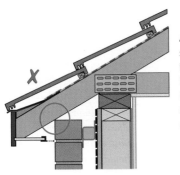

A two storey frame will shrink about 15mm or so in overall height. A gap between rafter and brickwork is essential to accommodate differential movement.

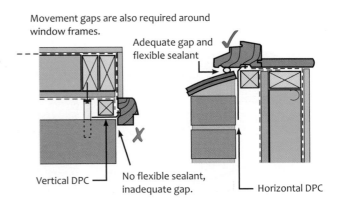

Movement gaps are also required around window frames.

Adequate gap and flexible sealant

Vertical DPC

No flexible sealant, inadequate gap.

Horizontal DPC

The most vulnerable areas are those:

- where there is a direct connection between the brick and the timber elements, eg wall ties, which if not flexible may snap
- elements attached to the timber frame which pass into or across the brickwork, eg windows, roof structure, balanced flue boilers
- elements such as flashings and sealants, as they may be disturbed or damaged. This in turn may lead to rainwater penetration.

As the panels dry they will shrink slightly. The brickwork, on the other hand, may expand slightly. In the left-hand photo shrinkage of the panel will result in the tie sloping backwards. The right-hand image shows better construction.

Poor structural connection

One of the advantages of timber framing is the ease of connecting the various timber elements. As long as the timbers are nailed in accordance with the appropriate specification, structural integrity will be assured. In some instances however connections such as simple nailing are ineffective because they have been carried out carelessly. Mistakes on site such as wall tie fixing nails passing through the sheathing layer but missing the stud framing have been observed. Reports have also revealed that similar poor connections have occurred in the factory.

Factory nailing should ensure this problem does not occur. The nails are not located in the studs.

There have also been instances where the omission of head binders has resulted in distortion of the panels.

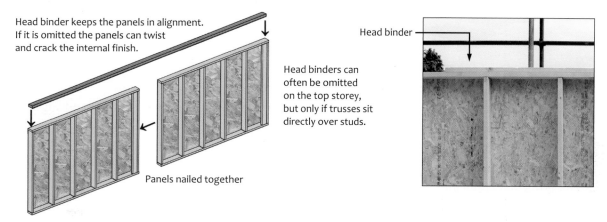

Head binder keeps the panels in alignment. If it is omitted the panels can twist and crack the internal finish.

Panels nailed together

Head binders can often be omitted on the top storey, but only if trusses sit directly over studs.

Head binder

FIRE

In timber frame buildings with an external cladding of brickwork there is a danger of fire spreading in the cavity. The fire can start within the building but spread into the cavity through, for example, an incorrectly fixed internal plasterboard lining. Fires can also result from careless soldering of pipes adjacent to the cavity. The breather paper in older timber frame houses was flammable. In more modern timber frame houses, where flame retardant breather membranes are specified and used, this particular fire risk should not be a problem.

Cavity barriers are required to limit the spread of fire within the cavity itself. In modern construction they are required at the positions shown in the drawing below.

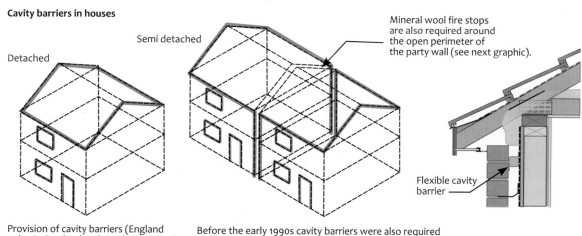

Cavity barriers in houses

Detached

Semi detached

Mineral wool fire stops are also required around the open perimeter of the party wall (see next graphic).

Flexible cavity barrier

Provision of cavity barriers (England only - In Scotland and Northern Ireland the requirements are more onerous)

Before the early 1990s cavity barriers were also required vertically at 8 metres intervals and horizontally at first floor level (look for weep holes over the cavity tray).

305

In addition fire stopping is required at party walls. This usually takes the form of a mineral quilt.

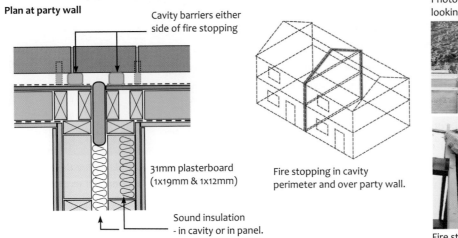

Plan at party wall

Cavity barriers either side of fire stopping

31mm plasterboard (1x19mm & 1x12mm)

Sound insulation - in cavity or in panel.

Fire stopping in cavity perimeter and over party wall.

Photo taken from first floor looking down at the party wall.

Fire stopping at party wall

There is a danger of fire spreading within a cavity if:
- fire barriers are missing or placed in the wrong position
- fire barriers do not completely span the cavity, thereby allowing the passage of fire and smoke
- fire barriers have deteriorated.

Some early types of barrier were constructed from sheet metal or timber battens. With both of these materials it is relatively easy to fix them to the plywood sheathing but it is likely that some gaps will occur at the junction with the relatively irregular inner face of the brick cladding.

To overcome this problem some fire barriers are made from a resilient material as shown below.

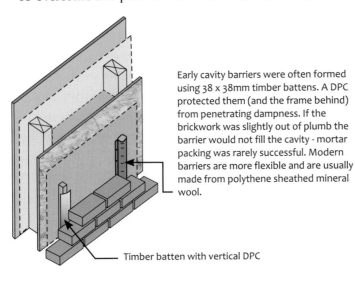

Early cavity barriers were often formed using 38 x 38mm timber battens. A DPC protected them (and the frame behind) from penetrating dampness. If the brickwork was slightly out of plumb the barrier would not fill the cavity - mortar packing was rarely successful. Modern barriers are more flexible and are usually made from polythene sheathed mineral wool.

Timber batten with vertical DPC

This small block of four flats has vertical and horizontal cavity barriers to separate dwellings.

Sound insulation

In timber frame houses, incorrectly constructed party walls can lead to problems of sound insulation. The party wall normally comprises two leaves of timber studding (some early houses have a hybrid form of construction with dense concrete block party walls). A sound deadening quilt and extra layers of plasterboard help reduce transfer of airborne sound. If this quilt or additional plasterboard is missing, or if heavy metal ties bind the two leaves together, the sound insulation will be reduced.

SUMMARY OF POTENTIAL DEFECTS DURING CONSTRUCTION PROCESS

Potential Problem Areas	Examples	Implications/Manifestation
Fabrication errors	Careless nailing of studs and sheathing	Loss of strength
Incorrect storage or handling on site	Storage in the open will increase moisture content of panels	Panels will shrink, twist and may affect plasterboard finish
	Incorrect stacking can damage breather paper	Torn breather paper will allow moisture penetration
Base out of square or out of level	If the base is out of square, panels will not fit properly without additional work on site	Panels which are a bad fit are more likely to deform
	If the base is not level the top of panels will not line up	Window openings may distort, floors may not be level
Construction errors	Incorrect fixing of sole plate and DPC	Rising dampness
	Inadequate nailing	Racking of finished building
	Failure to provide bracing on internal partitions	
	If panels are not vertical it may affect the brick cladding	Maintaining correct cavity may be difficult. Brickwork may not be plumb
	Omission of head binders	Panels may be out of alignment
Finishing the panels	Breather paper is non-continuous	Damp penetration
	Insulation must be a good fit to avoid risks of cold bridging	Condensation
	Vapour check must be on warm side of insulation and must not be torn	Interstitial condensation
	Brick cladding not tied back to studs with flexible ties, or ties incorrectly positioned (ie no back fall on ties)	Damp penetration, lack of restraint for cladding
	Lintels not clipped back to frames in a way which permits differential movement	Cracking and distortion of window heads
	Cladding fixed without allowance for differential movement at eaves and around windows	Cracking and distortion
Fire protection	Header joists omitted between joist ends	These three problems can all lead to a reduction in fire protection and a subsequent risk to life
	Cavity barriers non-continuous and/or in incorrect position	
	Cavity fire stopping in party walls non-continuous – does not run up to roof apex	

CONCLUSION

The timber frame houses built in the 1960s, 1970s and 1980s were subject to a significant amount of adverse publicity from the media. In retrospect the number of houses affected by serious defects is probably less than was implied by much of the rhetoric of the time. Nevertheless, when dealing with timber frame housing one should be aware of the potential problems explained in this text. Because of the risks of rot and fire, the need for good site practice and thorough site supervision cannot be over emphasised.

CHAPTER 16

Water and Heating

COLD WATER SUPPLY

Incoming mains

In modern houses the incoming water main is made from plastic; in older properties it is more likely to be lead or copper. Lead pipes, like copper, are easily deformed, thus restricting the flow of water in them. Some soils can cause corrosion in metal pipes and they should, therefore, be protected with a plastic coating or some form of bituminous paint. Modern incoming mains should be at least 750mm below ground level to avoid the effect of frost but in older properties the main is often very close to the surface. If the rising main within the property is situated on the inner face of an external wall there is the consequential risk of freezing. Where the main enters the property there should be a stop tap to isolate the distribution pipework. In older properties these are often missing or they may have corroded to such an extent that they are no longer operational. Temporarily freezing the pipe (not as difficult as it sounds) may be the only option where internal maintenance or isolation is required.

Direct cold water systems

These systems are the most cost effective, providing there is a reliable and continuous water supply and the building has no need of a stored water supply. As water enters the property directly from the mains it is at a high pressure and the fittings throughout the system must be able to cope with this high pressure. This applies especially to the ball valves feeding cisterns and tanks for heating, hot water and toilet facilities. Water hammer is a problem experienced with variations in water pressure and this can be caused by incorrectly specified valves and fittings. Worn fittings can also start vibrating and this noise may be amplified in other parts of the building.

Where pipework changes direction using bends, there is a tendency for the high water pressure to exert forces on the pipework in its efforts to straighten the run in the same way it will in a garden hose.

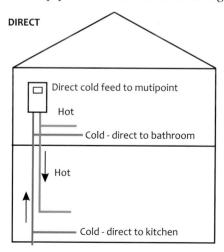

DIRECT

Direct cold feed to mutipoint

Hot

Cold - direct to bathroom

Hot

Cold - direct to kitchen

Direct supply is often used where mains pressure is constant. No water is stored so there is no need for a tank in the roof space.

However, where hot water is supplied through a cylinder rather than a multipoint the cylinder will require an indirect feed and a storage tank. Alternatively, a special pressurised cylinder can be used. In practice, many houses like this use a combination boiler to supply the heating and the hot water; this does away with the need to store hot water.

The brackets must be fixed at appropriate distances apart to eliminate this tendency and reduce the forces acting upon the bends. (On the water company's mains the forces can be so high as to warrant special concrete securing blocks known as 'thrust blocks' to prevent the water pressure lifting the water mains out of the ground.)

In a direct supply, double check valves are required where there is the risk of contaminated water being sucked back into the mains. A typical example is where one end of a hose is connected to the mains and the other end is inadvertently dropped into a drainage gully. Any sudden drop in the mains pressure will suck the drain contents into the main, resulting in contamination.

Indirect cold water

This system is necessary where the mains water supply is unreliable for any reason. The internal supply becomes indirect when it is fed from a storage facility within the building. It is often necessary in a very densely populated area where, if a direct system were provided, the many people calling for water at any one time would warrant the size of larger, uneconomical main.

Drinking water should never be consumed from a tank supply unless the tank has been manufactured specifically for the purpose. Most houses with an indirect supply have a direct supply to a downstairs tap (in the past it was not uncommon to see a tap specifically labelled 'drinking water'). Modern water bye-laws require all tap water to be potable.

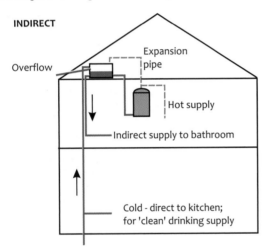

INDIRECT

Overflow

Expansion pipe

Hot supply

Indirect supply to bathroom

Cold - direct to kitchen; for 'clean' drinking supply

Stored water, ie an indirect supply, allows for intermittent supply. A storage tank in the roof supplies all taps except the kitchen and also feeds the hot water cylinder. The cylinder is vented to allow for expansion of heated water and in case of possible boiling due to thermostat failure.

The pressure in the hot water circuit is determined by the height between the taps and the cold water storage tank. The distribution pipes from the tank are normally of slightly larger diameter than the rising main due to the decrease in pressure.

Problems common to cisterns, tanks and pipework etc

Lead pipework was a very common material until the 1950s, but it has been found to be detrimental to health as lead may be absorbed into the water; a particular problem in soft water areas. The skills involved in making lead joints are also no longer readily available.

Forming bends in copper pipes inevitably stretches the copper so fittings of capillary and compression types are often preferred. The thickness of copper pipe has reduced in recent years and bending copper pipes can result in failure of the pipework.

Plastic pipework is now becoming more common but the support distances need to be matched to the particular material specification. Some water has been found to absorb the taste of plastic from certain grades (usually earlier ones). The expansion rates of plastics are very high and the surrounding temperature will also affect the composition of the pipes. Earlier pipes could become very brittle, developing hairline cracks if subjected to shock loads when very cold, and becoming quite soft and malleable when hot. Some grades can even acquire a sticky surface under adverse conditions and certain colours are affected by ultra-violet light.

The materials used in the manufacture of cisterns and tanks has changed with time, and, while old tanks can still be found in use, made from timber, slate, steel, ceramics and copper, the most widely used materials now are either plastic or fibre-glass. While the older materials were quite rigid and could safely be mounted on beams or joists, it is now essential that plastic is uniformly supported on suitable load-bearing boarding, capable of withstanding possible water penetration.

All tanks and cisterns should have tight fitting lids which prevent dust, dirt, bird droppings and even rodent and bird carcasses from polluting the water.

The correct specification of ball valve is also essential to prevent back siphonage, noise and water hammer, and should be matched to the required inlet water pressure.

Well-supported overflow pipes should run to their point of discharge without sagging, which could become a problem in freezing conditions. These overflows are essential as a safety precaution should valve washers, seals or floats cease to function properly and water overfills the storage vessel. The point of discharge should be in a very prominent position so it will be noticed quickly and steps taken to avoid water wastage.

Any leaking overflows should be fixed promptly to prevent waste, unsightly staining and to resolve any float and valve problems before they become serious. Lead overflows, unfortunately, have a high scrap value. Storage cistern (right) should be insulated.

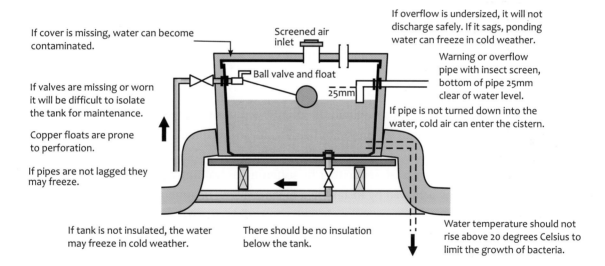

If cover is missing, water can become contaminated.

Screened air inlet

Ball valve and float

25mm

If overflow is undersized, it will not discharge safely. If it sags, ponding water can freeze in cold weather.

Warning or overflow pipe with insect screen, bottom of pipe 25mm clear of water level.

If pipe is not turned down into the water, cold air can enter the cistern.

If valves are missing or worn it will be difficult to isolate the tank for maintenance.

Copper floats are prone to perforation.

If pipes are not lagged they may freeze.

If tank is not insulated, the water may freeze in cold weather.

There should be no insulation below the tank.

Water temperature should not rise above 20 degrees Celsius to limit the growth of bacteria.

Nowadays a wide range of outlet taps is available – some are more complex than others. Generally tap washers will need more attention on high pressure supplies; even the ceramic types have a limited life expectancy. Dripping taps and ball valves will ultimately suffer internal damage if worn washers are not replaced at the first sign of seepage. It is much cheaper to replace the washer than the whole tap or valve. It is also cost effective to have isolation valves provided to cover sections of the installation and avoid total system drain down with its resulting water wastage or excessive time involved when washer replacement is necessary.

Many modern taps (usually shower mixers and monobloc taps) will not work properly if the water pressure is low – they may not be suitable, therefore, for indirect water supplies. Both the hot and cold water pressures must normally be above 1.0 bar if they are to supply a high pressure tap. Pressure can be calculated by measuring the vertical distance in metres between the bottom of the cold water storage tank and the tap (or shower) and dividing the answer by 10 – thus 1 metre of Head = 0.1 bar.

HOT WATER

Direct hot water systems

The main problems which occur in direct hot water systems are water discoloration, possible scale formation, poor temperature control and noise. Older systems are usually heated with coal or gas fired back boilers; more modern assemblies are single/multi-point gas and electrical water heaters of both storage and non-storage types. A typical coal fired back boiler system is shown below.

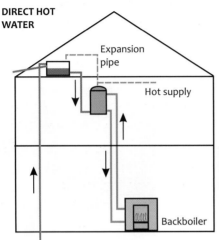

DIRECT HOT WATER

Expansion pipe

Hot supply

Backboiler

Direct hot water supply was quite common before the 1960s. The water from the cylinder was heated by a cast-iron heat exchanger situated behind the open fire. As the water heated up, it gravitated up the pipework and heated the water in the cylinder. Thus, all water heated ran through the heat exchanger behind the fire.

'Furred' pipe

Rust in the heat exchanger could contaminate the hot water and eventually come out of the taps. In addition, in hard water areas, the pipework coming from the back boiler could quickly get blocked by limescale deposits.

Early back boilers were made from either copper or cast iron. Providing pipes were made of the same materials, the only problems experienced with copper were in hard water areas, resulting in scale build-up. Cast iron, however, could oxidise and the resulting rust deposits would cause water discoloration; even more of a problem in soft water areas. The lack of thermal control, however, sometimes leads to the water boiling, creating noise and vibration, heat wastage and in extreme cases, scale build up with a risk of explosion.

Modern systems

Scale formation remains the major problem with modern systems as the alkaline deposits, through which ground water passes, are drawn into solution. On heating, these precipitate onto the higher temperature surfaces forming scale – the hardness depends on temperature. The only way to eliminate the problem is to provide some form of water treatment either by softening or holding the troublesome salts in suspension; the latter can be achieved by installing a magnetic-type treatment unit.

Most modern direct hot water systems are made with compatible materials so that electrolytic action and water discoloration problems have been eliminated. However, with many of the modern heating boilers, the numerous electronic controls, safety devices, fans, heating elements and burners, still require regular maintenance, together with care to ensure adequate ventilation rates are fully provided. The more refined and efficient our technology becomes, the more there is potentially to go wrong and regular servicing has become essential.

Indirect hot water systems

This has been the most popular means of providing hot water domestically, because it is simple and eliminates most of the potential problems associated with direct systems. In an indirect system there are two discrete volumes of water. The primary system circulates through the boiler and through a heat exchanger in the hot water cylinder. It will also circulate through radiators, where they are fitted. A secondary system feeds the cylinder and serves the hot water taps. The two systems must be kept separate to prevent contamination of the water which comes out of the taps.

Scale formation is still seen as a problem, but at a much reduced level. This is because the primary hot water medium, which is the means of indirectly heating the domestic hot water, does so with much lower temperatures on the secondary heat exchanger surfaces (inside the indirect cylinder) than are experienced within a boiler or on the surface of an electrical element.

The primary water, which is first heated in the boiler, is in what is described as a sealed system, so that the water circulating is being continually recycled without replacement, giving off its heat to the secondary system and returning to the boiler for reheating. Once the salts in the original supply water are in this primary circuit they will be precipitated out at their first heating and, with no further fresh water being introduced, the primary circuit stays free of further scale.

The secondary water flowing into the cylinder and on to the tap outlets is from a different supply and is heated by surfaces at temperatures not usually more than 85°C. This lower temperature means that only some of the salts precipitate out on heating, the remainder staying in solution, passing out of the taps without coating the surfaces.

In small housing developments the pipework lengths and sizes are usually quite small and so the amount of heat which is lost from these pipes is minimal. Single run pipes in which the hot water can cool down when the water is stationary are called 'dead legs'. These must be kept to a minimum to avoid heat loss and wastage of water (when cold water runs to waste awaiting the arrival of the hot through the tap). On larger installations a water circulation system is incorporated into the system to ensure the hot water is continually pumped around to avoid this wastage.

Certain bacteria are known to multiply in water based systems and can cause problems of ill health to building occupants. Legionnaires Disease is one of these, and the most straightforward way of preventing these problems is to ensure water services run either hot or cold and never at tepid or luke warm temperatures. Hot water should always be available at above 65°C and the cold water below 15°C. Should an intermediate temperature be required the hot and cold should be mixed at the point of use. Mixing valves are available for this purpose if required. Thermal insulation is used to prevent heat loss or gain, and pipework carrying water at different temperatures should take separate routes through the building to avoid problems.

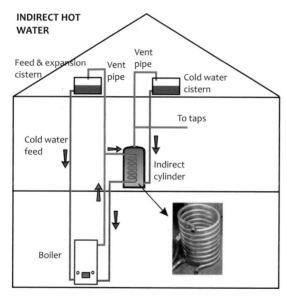

INDIRECT HOT WATER

Feed & expansion cistern

Vent pipe

Vent pipe

Cold water cistern

To taps

Cold water feed

Indirect cylinder

Boiler

Nowadays, most water is heated indirectly. The water which passes through the boiler does not come into direct contact with the water in the cylinder. Instead, it passes through a coiled tube in the cylinder and then returns to the boiler. The boiler usually has the additional function of providing hot water for radiators.

There are, therefore, two separate volumes of water in this system, each with its own feed and expansion tank. The risk of contamination and limescale is reduced in the indirect system.

It is not always possible to incorporate indirect systems within our buildings, especially those which have low floor to floor dimensions or lack of roof spaces for location of storage tanks. Some manufacturers will supply combined tank and cylinder components for mounting within cupboards, but the lack of height above the likes of a shower outlet will then make the indirect system less attractive. For very small hot water users the indirect system may also be expensive.

A rather alarming problem can occur with hot water cylinders if the roof storage tank and ball valve become frozen. If a coating of ice seals the feed tank and the user draws water from the cylinder a potential vacuum can occur within the cylinder. This in extreme cases may cause the cylinder to implode in a similar fashion to an empty drinks can. Adequate thermal insulation is obviously essential around the roof mounted tanks.

Primatic cylinders

During the 1960s and 1970s efforts were made to reduce capital costs for indirect systems by introducing an alternative type of indirect cylinder. These were called 'primatic' cylinders and were used extensively in domestic property. The design of the heat exchanger within the cylinder enabled the cold feed to the boiler and central heating circuit to be fed from the domestic hot water side of the installation, so eliminating the need to provide the separate feed and expansion tank and pipework necessary in the more conventional system. These worked quite well but there is a possibility of the boiler water (primary circuit) mixing with the domestic hot water (secondary circuit) and, as some of the more recent additives now used in central heating systems to reduce corrosion are toxic, it is no longer acceptable to allow this to happen. The result is that primatic cylinders are usually replaced by conventional cylinders and the additional tanks and pipework are incorporated during building or system refurbishment.

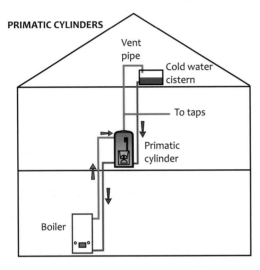

PRIMATIC CYLINDERS

Vent pipe

Cold water cistern

To taps

Primatic cylinder

Boiler

A Primatic cylinder is installed as a direct cylinder but functions as an indirect one. When the cylinder is first filled, the primary circuit is charged with water. When the cylinder is full, air seals separate the primary and secondary circuits. When the water in the primary circuit is heated, expansion forces some of the air from the upper chamber into the lower chamber. The primary system vents air into the secondary system which also tops-up the primary system as, and when, necessary.

NB In most of the above systems an immersion heater can provide a secondary form of water heating. A thermostat prevents the water from over heating. There have been examples of thermostats failing, causing scalding hot water to flow into the feed and expansion tank. If the overflow is blocked or if the tank is old and made from plastic it can split. Thermostats should be of the fail-safe type – if one fails the power is cut off.

Pressurised cylinders

In a pressurised cylinder the expanding water is contained within a pressurised expansion vessel. This is unlike a vented system where the water expands back into the cistern and up the vent pipe. Pressurised cylinders have to be stronger than vented ones and are often made from stainless steel or glass enamelled steel. There are usually three main 'functional' controls to protect mains water quality and prevent damage to the cylinder.

- A pressure reducing valve reduces the mains pressure to the safe working pressures of the cylinder.
- A single check valve prevents contamination of the main (ie prevents backflow).
- An expansion valve protects the cylinder from over pressure caused by failure of the pressure reducing valve.

In addition there are normally three safety controls to protect the householder.

- A control thermostat usually set between 60–65°C – to close a motorised valve on the primary flow or switch off the immersion heater.
- One or two energy cut out devices with manual reset, usually set between 85–89°C – to close down the primary flow and to switch off an immersion.
- A temperature and pressure relief valve set to 90°C or so – to allow water to escape if pressure or temperature get too high.

These installations require local authority approval and should be serviced every year.

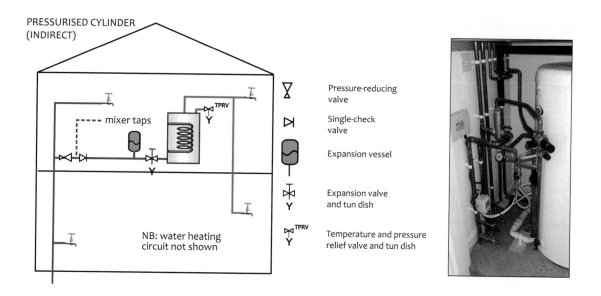

PRESSURISED CYLINDER
(INDIRECT)

mixer taps

TPRV

NB: water heating
circuit not shown

⋈ Pressure-reducing valve

▷ Single-check valve

Expansion vessel

Expansion valve and tun dish

TPRV Temperature and pressure relief valve and tun dish

CENTRAL HEATING SYSTEMS

Boiler types

Most modern boilers are of three types *(see below)*. However, they work on broadly similar principles; in other words, water is heated in a heat exchanger and is then circulated around a series of radiators. In a 'regular' boiler and a 'system' boiler the water from the heat exchanger is also fed through a coil (another heat exchanger) in a separate cylinder where it heats up water for washing. In a 'combi' boiler water for washing is heated by a small heat exchanger located inside the boiler.

Regular (or vented) boilers These 'traditional' vented boilers have been around, in one form or another for nearly a hundred years. Modern ones are much more efficient than their predecessors. Regular boilers are designed to provide hot water for radiators and to heat water in a separate cylinder (usually located in an airing cupboard). The boiler is fed by a storage (feed and expansion) cistern usually located in the roof void. This cistern accommodates the expansion of hot water (its level rises slightly) and a pipe from the boiler vents back over, and into, the cistern if the boiler controls go wrong and the water boils. Nowadays most regular boilers are wall-mounted; early ones were floor-standing.

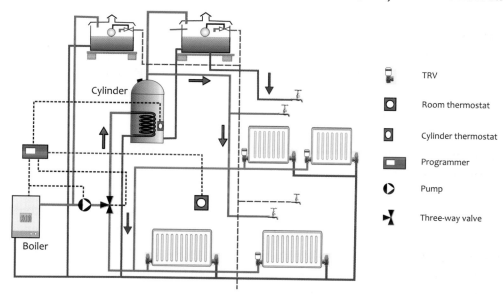

Cylinder

Boiler

TRV

Room thermostat

Cylinder thermostat

Programmer

Pump

Three-way valve

System boilers These are designed along similar lines to a regular boiler but for use with an expansion vessel rather than a feed and expansion cistern. In some boilers the expanding water is accommodated in an expansion vessel fitted inside the boiler. In others it is not part of the boiler but is usually located near to it. To provide hot water for washing a system boiler heats water in a cylinder. This can be either a vented (traditional) cylinder or a modern pressurised one. **NB** Boiler controls not shown in drawing.

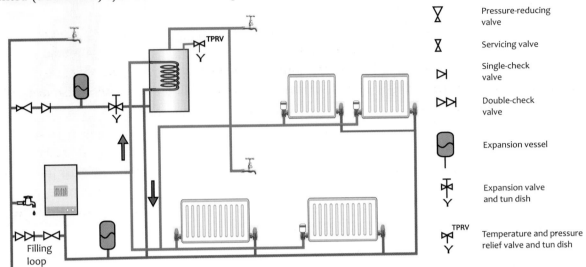

	Pressure-reducing valve
	Servicing valve
	Single-check valve
	Double-check valve
	Expansion vessel
	Expansion valve and tun dish
TPRV	Temperature and pressure relief valve and tun dish

Combination boilers These boilers (so called because they deliver hot water for washing and heating) now account for over 60% of all the new domestic boilers installed in the UK every year. They were originally developed for flats to save having lots of tanks in the roof space. A combi eliminates the need to store hot water – a hot water cylinder is not required. Hot water for washing is heated and delivered via a water-to-water 'plate' exchanger as and when required – not always at very high flow rates but at mains pressure. The savings in installation cost (no cylinder and no tank) are probably offset by the slightly higher cost of the boiler. These boilers do have minimum requirements in terms of mains water pressure. They also need a good gas supply – at times these boilers work quite hard.

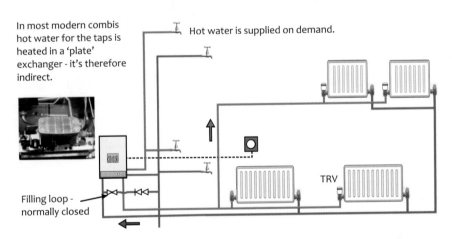

In most modern combis hot water for the taps is heated in a 'plate' exchanger - it's therefore indirect.

Hot water is supplied on demand.

Filling loop - normally closed

TRV

Some older combi systems are likely to have TRVs on every radiator but no room thermostat. It is better practice to have a thermostat in the lounge, for example, one radiator in the lounge without a TRV, and TRVs elsewhere.

Condensing boilers (regular, system or combination) These are available in regular, system and combi versions. In a modern (non-condensing) oil or gas fired boiler about 70–80% of the fuel is converted into useful heat. The remainder is lost through the flue – the flue gases can be as hot as 250°C. In condensing boilers more heat is extracted from the fuel; their efficiency can exceed 90%. In fact, so much heat is extracted that the flue gases are only about 50–60°C. In a typical condensing boiler there are two heat exchangers. After crossing the first one the combustion gases are directed over the second heat exchanger to pre-heat the water returning from the radiators. If the water in the radiator return pipe is cool enough the exhaust gases will condense on the heat exchanger releasing all their latent heat. The secondary heat exchanger must, therefore, be corrosion resistant and a drain pipe must be fitted to remove the condensate. The exhaust gases of condensing boilers 'plume' – this is quite normal and reflects their lower temperature. Nearly all new boilers are condensing boilers.

General boiler problems

Some boiler problems are unique to specific models; others can occur in all boilers. This section describes some of the more general problems.

Most modern boilers are designed with sealed combustion chambers so that the poisonous gases produced during combustion of the fuel cannot pass into the building. As part of the regular servicing arrangements this must be checked. Adequate ventilation of the room housing the boiler may be required for certain types of boiler in case there should be a failure of the combustion chamber seals. Most boilers will not operate if they sense a leak from the combustion chamber.

Some of the most common failures within boilers relate to the controls and any fans or pumps within the unit. High limit thermostats (which switch off the boiler when overheating occurs) are usually incorporated to prevent the water boiling should the boiler temperature thermostat fail. If the fans or pumps stop working, the unit should 'fail safe'.

The air to fuel mixture controls the efficiency of the boiler and the flame colour gives a very good indication of its efficient operation. Blue/purple flames indicate correct settings but any indication of yellow means the fuel is not being burnt properly. This may be caused by contaminated fuel, dirt particles on the burner jets causing some form of blockage or a restriction of the air supply/flue gas discharge. Unburnt fuel results in a carbon/soot build-up which will occur rapidly, requiring additional cleaning and maintenance and causing possible damage to other components.

During summer shutdown periods when the boiler has totally cooled down, it is often found that a water leak occurs from within the boiler casings. This is often cured as soon as the boiler is returned to its normal working temperature as the gaskets and boiler sections expand to fill the gaps which occur during the cooling down processes. It is often more advisable to leave the pilot lights on (where they exist) or retain the boiler for heating the domestic hot water, unless total shutdown is required for maintenance purposes.

Shutdown periods are also the most common times of pump and other component failure as many will continue to work when quite badly worn out, up to the time they are switched off.

There are a number of safety systems associated with fuel burning appliances, and these are provided to ensure fail safe or warning is given to the user. Safety and vacuum release valves protect the structural integrity of appliances and reduce risks of explosion. Core plugs are also sometimes fitted for the same reason. Pilot lights (common in early boilers) impinge on flame failure devices which automatically switch off the gas supply if the pilot light is extinguished. Other controls sense when the oil has not ignited in a burner. Sensors are available connected to alarms measuring excess quantities of carbon monoxide gas, which can cause loss of consciousness without smell or other warning. Automatic fire fighting systems may also be required where risks warrant them.

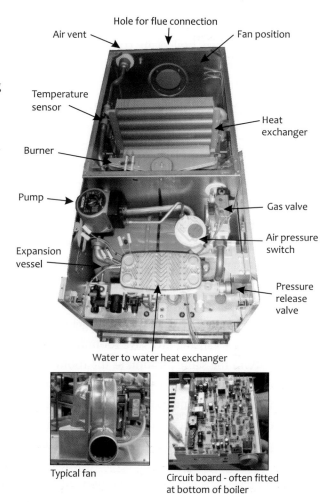

The boiler shown here dates from about 2000 – it's a combination boiler (non-condensing). The arrows identify the main components (many of which can be found in other boiler types).

According to *Which?* magazine the most common causes of boiler breakdown are failure of (in descending order):

1. circuit board
2. automatic air vent
3. pump
4. overheat thermostat
5. heat exchanger
6. expansion vessel (not present in a regular boiler)
7. gas valve
8. burner.

Flue fans can also fail quite often – sometimes within a couple of years. Most boilers will not start if the fan is not working properly.

In older boilers thermocouple failure is quite common. It is the sensor that sits in the pilot light flame and delivers an electrical current when hot, holding a solenoid valve open, allowing gas to flow. If the pilot goes out the thermocouple cools, the electricity stops and the gas gets shut off.

Flues

There are a number of regulations regarding the siting of boiler flue terminals. These are quite detailed and well beyond the scope of this chapter. A fan assisted balanced flue gas boiler, for example, should have its terminal sited at least:

- 300mm away from an opening
- 200mm below the eaves
- 150mm away from a vertical soil or drainpipe
- 600mm away from a boundary facing the terminal (1200mm if it's opposite another terminal)
- 300mm away horizontally from a terminal on the same wall (1500mm vertically).

Problems with whole heating installations

Although most new boilers are combis, the majority of existing central heating systems are vented. In other words, the boiler is fed by a storage tank and the water in the boiler can discharge back into this tank if the water overheats and reaches boiling point. Because of the risk of this discharge the storage tank for the boiler must be separate from the cold water storage tank which serves the cylinder.

Most modern central systems also heat the domestic hot water via a heat exchanger in the indirect storage cylinder. Modern control systems allow the user to select heating, hot water or both, switching on and off the systems at variable times according to need. A summertime problem is sometimes experienced when the radiators on upper floors continue to get hot even when the heating is apparently switched off. This is due to what is called gravity circulation. The only way to cure this problem is to insert additional isolation valves into the upper floor circuits, making sure that the vent pipes are not affected. Other problems with older systems include:

- gravity circulation (no pumping) to the hot water cylinder, which results in stored water being slow to re-heat
- lack of cylinder thermostat, resulting in excessive stored water temperature
- lack of room thermostat – rooms get too hot
- no TRVs – causing excessive room temperatures and poor system balancing
- a TRV in the same location as the room thermostat
- no boiler interlock, causing the boiler to stay hot and to cycle unnecessarily during programmed periods
- incorrect positioning of the pump in the system can cause a number of problems leading to water 'venting' into the feed tanks or air being drawn into the system through the bleed valves of the radiators.

Sometimes the float valve in the feed and expansion tank can seize (usually in the closed position). Over a long period the water level in the tank and heating pipework drops. If, due to minor leaks etc, the water level falls too much the pump will only operate intermittently – if at all. Sometimes, this will be accompanied by strange (and loud) banging noises from the boiler. The boiler can also stop working because the boiler thermostat, sensing a rise in water temperature, turns the boiler off. Another problem can be caused by a build-up of deposits in the cold feed to the boiler. These deposits can slowly block the pipe and starve the system of fresh water. It can be difficult to diagnose because the supply cistern shows the correct water level.

Deposits or sediment in the pipework or a badly sited pump can cause 'over pumping'. If this occurs water will discharge into the feed and expansion tank. If this is not addressed promptly the constant replenishing of the water will add oxygen to the system which will hasten rusting of the radiators. It can also split a plastic tank or lead to its collapse.

Specific problems of sealed installations (combi and system boilers)

Most combi and sealed system manufacturers recommend a pressure of about 1.0 to 1.5 bar (when the boiler is cold). Over a period of months some heating systems steadily lose water; this is usually from tiny leaks which can be very difficult to trace. Water, and therefore, pressure, will also be lost if the radiators are bled. If left unattended the pressure will drop to zero (many boilers have a dial on the front of the casing showing the pressure). At low pressure air can enter the system and speed up internal corrosion of the radiators. Most boilers have an internal sensor which prevents the boiler firing if the pressure is too low – this can be mistaken for boiler failure.

An expansion vessel accommodates the increased volume of expanding water. If it fails, possibly because of a perished or torn diaphragm, or due to leakage of its air charge, the increased pressure should trigger an automatic pressure relief valve; this valve usually opens at about 3 bar and ejects water to the outside. The ejection of water will lower the pressure in the system and, after the boiler has been switched off for a while (with a further lowering of pressure), the boiler will not restart because the pressure switch senses the low pressure.

In combi and system boilers the water pressure is adjusted by opening the filler loop valve. If this valve is not turned off properly the pressure in the system will slowly build up until the pressure relief valve opens. If this problem is not rectified the slow but constant supply of fresh water will lead to corrosion (and a steady stream of water being discharged from the pressure relief valve).

If mains pressure water leaks into the heating pipework from a faulty combi water-to-water plate exchanger the heating circuit pressure will rise and cause similar problems.

Radiators, pipes and thermostats

The majority of existing installations utilise copper for the pipework and steel for the radiators. Using these dissimilar metals can cause problems as an electrolytic action takes place due to the water acting as an electrolyte between them. Over a period of time, varying between two and 30 years, pin holes may appear in the radiators requiring their replacement. This problem is generally regarded as acceptable because of the initial high capital cost of alternatively supplying radiators manufactured in copper. The problem is even more severe when the steel pipework has been galvanised; the pin holes can appear quickly and the steel looks as if it is infected by woodworm.

Concealing pipe runs within walls (ie within the actual structure – not just behind plasterboard) and solid floors was considered acceptable a few years ago. Damage can occur by nail penetration when securing future fixtures, or chemical reaction may occur over time with the in situ building materials. A nail driven through a water pipe may not leak immediately and so lead to an unexpected problem months or even years later. Some copper pipework embedded in floors in the early 1970s was of a specification inferior to the normal British Standard, being imported from overseas during a time of severe shortage. This was expected to cause problems of interaction with normal in situ building materials unless it was suitably wrapped by protective tape at the time of installation. Further problems have occurred with some cheaper pipework manufactured in the UK in the early 1980s, resulting in pitting and premature leaks.

The route chosen for the pipework may, in some buildings, have to rise to a higher level, eg to pass over a doorway. The resulting high point is where air within the system becomes trapped and will eventually build up, stopping the circulation of the water. Air-release valves need to be incorporated but are often hidden away and forgotten. Should the system not be working, this possibility should be investigated after faults with boiler operation and pumps have been checked. Note that where pipes sit in notches in floor joists etc they should be wrapped in felt to prevent them from 'knocking'.

Air trapped in heating systems reduces water flow and, if lodged in radiators, will reduce their heating efficiency. After a system has been refilled it may take the air some days to arrive at radiators and other high points. Regular checks may need to be carried out in order that the air can be located and removed through the air-release valves. If the problem persists, then the position in the system of the pump (as mentioned earlier) should be examined together with possibly system leaks, where air could be sucked in.

Some heating systems may have had inexpensive valves installed especially on radiator connections. The process of opening and closing the valves may have the effect of loosening the gland, resulting in the fibrous washers or packing becoming loose. This means water can leak out and damage ceiling and floor coverings, especially when the system has cooled down. Repacking and tightening the glands of the valves should be carried out on a regular basis, or it may be considered more economical in the long term to replace the valves with those of a higher specification. Most modern systems tend to have thermostatic radiator valves controlling individual room temperatures. Some early valves only function correctly when they are installed on the flow pipe to the radiator (not the return pipe).

Older heating systems are controlled by a room thermostat switching the pump on and off, so regulating the heat source to all the radiators. The positioning of the thermostat is important and some improvement in comfort levels can be made by moving the thermostat away from ancillary heat sources such as the back of a television or other heat generating appliances. The room in which the thermostat is installed may also affect the overall economy of the heating system and the quantity of fuel burnt.

Micro-bore pipework systems utilise very small diameter tubes of a soft copper specification. If bumped by the likes of a vacuum cleaner during room cleaning, the resulting damage can reduce water flow or, in the worst case, cause a major leak. Some form of protection may be necessary where pipework is exposed.

In modern installations plastic pipework has become common. It's easy to cut and join and ideal for use where the floors are made from metal web joists (in coil form it's flexible enough to be fed through the webs). If not properly supported it tends to sag; in some cases sagging pipes, together with plastic's high thermal expansion and contraction, can actually pull joints apart. Sagging pipes are also difficult to drain down and they look unsightly – one reason why, in modern houses, the pipework is often hidden behind plasterboard. On the other hand, unlike copper, it does not need to be bonded (earthed) and its scrap value is nil, so there is no risk of the pipes being removed while you are away on holiday.

Radiator problems

If the radiators are hot or warm even though the heating is turned off (hot water on) there may be a problem with the two or three port valves.

If the radiator is cold at the top and warm at the bottom, it probably needs 'bleeding' to remove air.

If the radiator is hot at the top and cold/warm at the bottom, it could be blocked with sludge.

Thermostatic valves sometimes stick if heating is left 'off' for long periods.

If the radiators near the boiler are hot and the others are cool the radiators may not be balanced properly.

If the radiators upstairs only are cold it could mean that the feed and expansion tank in the roof is empty or, on a combi system, that the boiler pressure is too low.

If the radiators downstairs are cold there could be a problem with the pump.

RETURN

FLOW

On a single pipe system the radiators at the end of the circuit are likely to be cooler.

In micro-bore systems the pipes are easily damaged and small 'blobs' of solder are more likely to block the pipework.

The first source of leaks is likely to be the joint between the copper pipe and the steel flange of the radiator. Electrolytic action occurs here (where different metals are in contact).

Drainage

ABOVE GROUND DRAINAGE

In the Georgian period sanitation was, when compared to modern standards, primitive. Better quality houses had rainwater pipes, made from lead, or in some cases timber. A crude system of brick sewers, often built around natural water courses, carried rainwater away to streams and rivers. Collection and disposal of foul matter, a much more hazardous material in terms of health, was dealt with in a variety of ways. Perhaps the most common was the cesspool. This comprised a pit dug in the back yard, possibly lined with brick, which was emptied periodically by nightmen (so called because they were compelled by law to operate during the early hours before dawn). Some cesspools were fitted with overflows; surplus liquid matter would percolate into the surrounding ground where it often contaminated local water courses and water supplies. A privy was often sited directly over the cesspool; few houses had flushing water closets until the early 19th century. Although servants would be expected to use the privy the house owners would have used commodes, often discreetly hidden in elegant furniture. Foul matter from commodes and waste water from washing would have been carried down to the cesspool in buckets.

Following the Industrial Revolution there were a number of technological developments which improved sanitation; for the wealthy at least. The first was the steam engine which enabled water to be pumped to reasonable heights. The second was the introduction of machine made lead piping to carry the mains system. This, together with the development of relatively cheap cast iron drainage pipes, led to the introduction of water closets and, eventually wash basins, on upper floors.

By the middle of the Victorian period most prosperous town houses had a system of mains water supply and a system of mains drainage. Although a sewer would contain foul, waste and rainwater, the above ground drainage system was usually a two pipe system. The foul pipe served the WCs; the waste pipe served the sinks, baths and rainwater guttering.

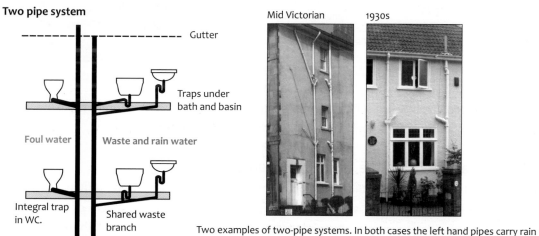

Two pipe system

Gutter

Traps under bath and basin

Foul water   Waste and rain water

Integral trap in WC.

Shared waste branch

Mid Victorian   1930s

Two examples of two-pipe systems. In both cases the left hand pipes carry rain water and waste water. The right hand pipes carry foul matter from the WC. The foul pipe extends beyond the eaves to act as a vent pipe for the drain and sewer.

Traps

To prevent foul smells from the sewer entering the house, traps were fitted below each sanitary appliance. A trap contains a small volume of water 'trapped' in the 'U' bend. Unfortunately, there are a number of ways in which traps can lose their seals. This can be caused by siphonage, or by back pressure created when a large volume of water running down the discharge pipe slows at a tight bend. Failure of traps (or no traps at all) is probably the most significant defect in early above ground drainage systems.

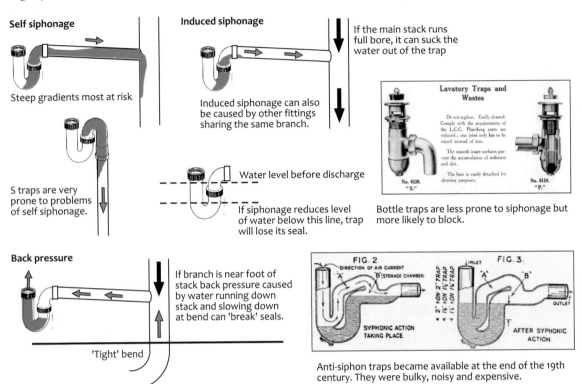

Self siphonage

Steep gradients most at risk

S traps are very prone to problems of self siphonage.

Induced siphonage

Induced siphonage can also be caused by other fittings sharing the same branch.

If the main stack runs full bore, it can suck the water out of the trap

Water level before discharge

If siphonage reduces level of water below this line, trap will lose its seal.

Lavatory Traps and Wastes

Bottle traps are less prone to siphonage but more likely to block.

Back pressure

If branch is near foot of stack back pressure caused by water running down stack and slowing down at bend can 'break' seals.

'Tight' bend

FIG. 2 — DIRECTION OF AIR CURRENT — 'B' (STORAGE CHAMBER) — SYPHONIC ACTION TAKING PLACE.

FIG. 3. — INLET — OUTLET — AFTER SYPHONIC ACTION.

Anti-siphon traps became available at the end of the 19th century. They were bulky, noisy and expensive.

The risk of siphonage, however, could be reduced by ventilating each branch pipe. This, although effective, created a complex system of pipework which was both expensive to install and expensive to maintain.

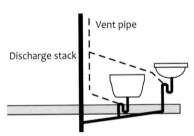

Vent pipe

Discharge stack

As water runs down the branch, any siphonage action pulls air down the vent pipe, thus maintaining the seal in the trap.

Traps can also become blocked. The three main 'culprits' are: disposable nappies blocking WCs, grease blocking kitchen sinks, and hair blocking wash hand basins.

Typical installations of the early 20th century

During the first 50 years of the 20th century millions of low rise houses were built as towns expanded and the suburbs were created. Most of these houses adopted the two pipe system of above ground drainage. Discharge pipes were usually made from cast iron, with yarn and lead joints. From the 1940s asbestos cement pipes with cement mortar joints became common. The discharge stacks were connected to the mains drainage by salt glazed clay pipes (salt glazing provides a smooth surface and improves impermeability). An interceptor, or rat trap was often required by the local authority to prevent smells (and rats) from entering the house drainage system. There were a number of proprietary interceptor traps, although they all worked on the same principle, ie a 'U' bend in the trap held a small volume of effluent. A short length of pipe fixed with a cover allowed access for rodding down the drain should blockages occur. To prevent build up of gases, a one way vent was normally provided just above ground level. This vent, when operating correctly, allowed air to flow into the chamber to ventilate the house drainage system but prevented foul air from escaping into the atmosphere. Unfortunately, these vents, known as Mica valves, often became blocked with paint, dust and debris, which affected their intended function. Interceptor traps, in most cases, were unnecessary and in modern construction they are rarely found.

Some common problems found in these early twentieth century installations are shown in the following drawing.

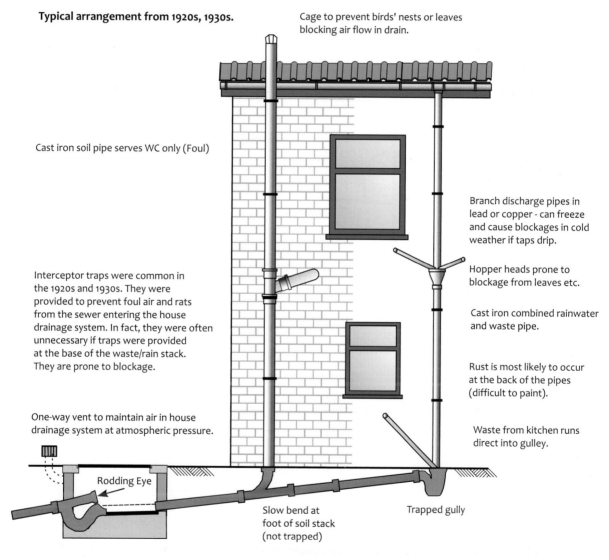

Typical arrangement from 1920s, 1930s.

Cage to prevent birds' nests or leaves blocking air flow in drain.

Cast iron soil pipe serves WC only (Foul)

Branch discharge pipes in lead or copper - can freeze and cause blockages in cold weather if taps drip.

Interceptor traps were common in the 1920s and 1930s. They were provided to prevent foul air and rats from the sewer entering the house drainage system. In fact, they were often unnecessary if traps were provided at the base of the waste/rain stack. They are prone to blockage.

Hopper heads prone to blockage from leaves etc.

Cast iron combined rainwater and waste pipe.

Rust is most likely to occur at the back of the pipes (difficult to paint).

One-way vent to maintain air in house drainage system at atmospheric pressure.

Waste from kitchen runs direct into gulley.

Rodding Eye

Slow bend at foot of soil stack (not trapped)

Trapped gully

In the above example the waste stack runs into a trapped gully. The trap is required to prevent foul smells escaping from the open hopper head. Cast iron pipes, which were the norm until the 1940s, are very durable. However, they can leak at defective joints, sometimes caused by settlement of the property over many years. The back of the pipes can also develop leaks; because this is the surface that is hardest to paint, and therefore the first part to rust. In better quality installations the stack is supported on spacer blocks to ensure it is well clear of the wall. This makes it easier to paint.

In the post-war period the two pipe system of drainage was abandoned and replaced with the single stack system. In this system foul water and waste water are collected in the same discharge stack (rain-water is collected separately), which usually ventilates the system as well. The Building Regulations specify a number of rules to ensure that these systems work effectively. These, in the main, control the depth of traps, the size and gradients of pipes, and the number of fittings which can be served by branches.

Modern single stack systems in low rise housing are invariably constructed with plastic pipes. The discharge stacks have flexible 'O' ring joints; discharge branches can be jointed with 'O' ring seals or solvent welds. In the 1960s it was common to fix the discharge stacks externally; nowadays, they are normally fitted internally and boxed in with plasterboard. Although internal stacks may cause more problems if leaks occur, they are less prone to degradation by ultra-violet light and less prone to thermal movement.

Typical problems with single stack systems are shown in the drawing at right.

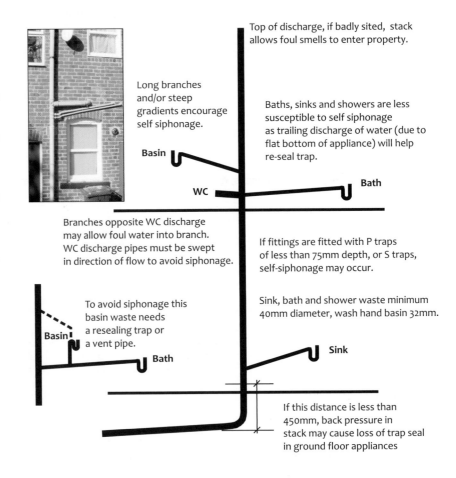

Top of discharge, if badly sited, stack allows foul smells to enter property.

Long branches and/or steep gradients encourage self siphonage.

Baths, sinks and showers are less susceptible to self siphonage as trailing discharge of water (due to flat bottom of appliance) will help re-seal trap.

Basin

WC

Bath

Branches opposite WC discharge may allow foul water into branch. WC discharge pipes must be swept in direction of flow to avoid siphonage.

If fittings are fitted with P traps of less than 75mm depth, or S traps, self-siphonage may occur.

To avoid siphonage this basin waste needs a resealing trap or a vent pipe.

Sink, bath and shower waste minimum 40mm diameter, wash hand basin 32mm.

Basin

Bath

Sink

If this distance is less than 450mm, back pressure in stack may cause loss of trap seal in ground floor appliances

Appliances at ground floor level are often connected direct to the drainage system. This may be because the branch length exceeds the rules set out in the Building Regulations. It may also be necessary because of problems of back pressure.

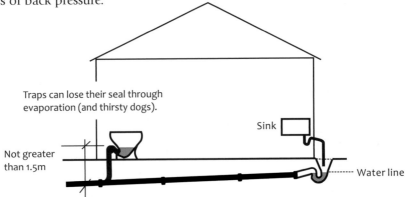

Traps can lose their seal through evaporation (and thirsty dogs).

Sink

Not greater than 1.5m

Water line

If trap is lost while flushing, it is probably caused by self-siphonage. This is a risk if distance from crown of trap to invert exceeds 1.5m.

If waste runs into water in gully trap, self siphonage can occur. The waste pipe should terminate below the grating but above the water line.

The diagrams directly above highlighted a number of design criteria which affect the correct operation of single stack systems. There are also a number of practical problems related to good construction practice.

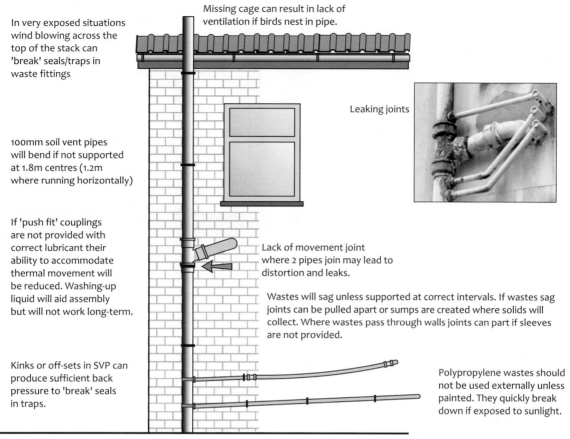

In very exposed situations wind blowing across the top of the stack can 'break' seals/traps in waste fittings

Missing cage can result in lack of ventilation if birds nest in pipe.

Leaking joints

100mm soil vent pipes will bend if not supported at 1.8m centres (1.2m where running horizontally)

If 'push fit' couplings are not provided with correct lubricant their ability to accommodate thermal movement will be reduced. Washing-up liquid will aid assembly but will not work long-term.

Lack of movement joint where 2 pipes join may lead to distortion and leaks.

Wastes will sag unless supported at correct intervals. If wastes sag joints can be pulled apart or sumps are created where solids will collect. Where wastes pass through walls joints can part if sleeves are not provided.

Kinks or off-sets in SVP can produce sufficient back pressure to 'break' seals in traps.

Polypropylene wastes should not be used externally unless painted. They quickly break down if exposed to sunlight.

Wastes with solvent joints cannot be used in long runs. Provision for thermal movement must be provided with additional 'push fit' couplings at 1.8m intervals.

Rainwater systems

Until the 1950s most rainwater goods were formed from cast iron, asbestos cement and even timber. Since then plastic has been the norm. It is available in a range of colours and profiles. The Building Regulations contain guidance on the size of guttering, and the size and number of downpipes for given roof areas. There are a number of common defects, most of which are shown in the drawings below:

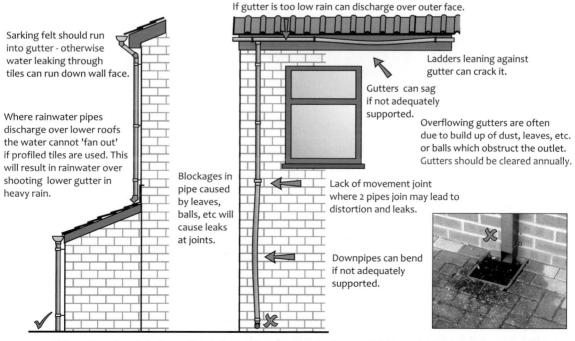

Sarking felt should run into gutter - otherwise water leaking through tiles can run down wall face.

If gutter is too low rain can discharge over outer face.

Ladders leaning against gutter can crack it.

Gutters can sag if not adequately supported.

Overflowing gutters are often due to build up of dust, leaves, etc. or balls which obstruct the outlet. Gutters should be cleared annually.

Where rainwater pipes discharge over lower roofs the water cannot 'fan out' if profiled tiles are used. This will result in rainwater over shooting lower gutter in heavy rain.

Blockages in pipe caused by leaves, balls, etc will cause leaks at joints.

Lack of movement joint where 2 pipes join may lead to distortion and leaks.

Downpipes can bend if not adequately supported.

If downpipes do not discharge directly into gullies or rainwater shoes, splashing can bridge DPC and wet pavings. Splashing can also lead to problems of ice (and injury) in freezing weather conditions.

Gutters may overflow because they are full of debris, undersized or the outlet is blocked.

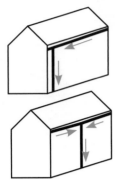

It can also occur where downpipes are incorrectly positioned. If downpipes are sited at the end of the gutter, the lower parts of the gutter may be carrying excess water. Siting the downpipe in the middle avoids this problem.

This gutter falls the wrong way. It also has to carry water from another roof above.

The photos below show a number of common problems with rainwater pipework. Most of them are self explanatory. Two are worthy of specific mention. The bottom right-hand image shows a concrete gutter with a recently applied bitumen lining (just visible). These gutters were made from pre-cast concrete sections. They tend to leak at the joints – hence the development of a whole service industry devoted to lining them. The image to the left of the concrete gutter shows brick damage caused by long term wall saturation – in this case due to a split at the back of the pipe.

BELOW GROUND DRAINAGE

Combined system

Mains drainage is comparatively recent. Although a number of Georgian buildings had a rudimentary form of drainage, it was not until the end of the 19th century that it became an automatic requirement to provide a comprehensive system of mains drainage for all houses in urban areas. Early systems were usually 'combined'; in other words rainwater and foul water (from WCs and wastes) shared the same drains.

A typical terraced house from about 1900 is shown in the diagram opposite. At this date bathrooms, as we know them today, did not exist in most modest houses. People washed themselves over a sink in the scullery. Hot water would have to be heated on a coal fired stove; galvanised steel or tin baths were filled (and emptied) by hand.

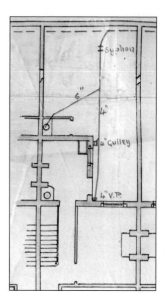

The drawings above and right show a typical drainage installation from the 1900s. The pipes above ground level were usually cast iron, those below, salt glazed clay. Textbooks of the time refer to unscrupulous builders using reject pipes – 'seconds'.

To prevent smells from the drainage system rising into the air, traps were provided just below ground level *(see earlier drawing)*. Some local authorities required a trap at the junction between drain and sewer (interceptor traps).

Most early pipework was made from clay; usually glazed with salt to provide a smooth surface. Pipes were laid in trenches, often in the bare earth, although some local authorities required concrete beds below the pipes. Drainage pipes had a socket at one end and a spigot at the other. Joints were formed in mortar alone or by ramming a tarred rope (caulking) around the junction of the socket and spigot; the joint was finished with a mortar fillet. Any minor ground movement could quickly crack the joint and allow effluent to leak into the surrounding ground.

Rigid joists are not necessarily unreliable. Where pipes were laid (and sometimes surrounded) on a concrete bed (common practice in the 1920s to 1950s) there was little movement in the pipework and therefore limited stress on the joint. However, where pipes were laid straight onto the floor of an excavated trench or on a thin layer of stone, rigid joints often failed quite quickly due to minor movement in the pipe line. In the 1960s the design of drainage pipework was 're-visited' and it was concluded that it was not necessary to bed pipes on concrete as long as the pipes were strong enough to resist compression forces, and as long as flexible joints could be provided to minimise the risks of leakage in case of minor ground movement.

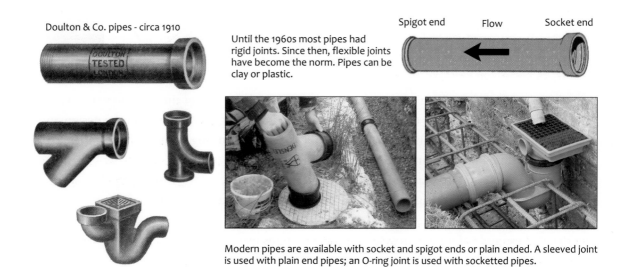

Doulton & Co. pipes - circa 1910

Until the 1960s most pipes had rigid joints. Since then, flexible joints have become the norm. Pipes can be clay or plastic.

Spigot end   Flow   Socket end

Modern pipes are available with socket and spigot ends or plain ended. A sleeved joint is used with plain end pipes; an O-ring joint is used with socketted pipes.

Nowadays clay and plastic drain pipes nearly always have flexible joints. The only exceptions are likely to be where small runs of pipework are being repaired. There are two types of mechanical push-fit joint applied to modern clay pipes; the sleeve joint and the O-ring joint. The sleeve joint is used with plain end pipes, and the O-ring joint is used with socketted pipes.

For O-ring jointing the pipes have polyester fairings cast round the outside of the spigot and the inside of the socket. In the O-ring joint, the sealing action is achieved by a rubber ring, located in the groove in the spigot moulding, and compressed between the spigot and socket fairings.

The coupling for the sleeve joint is made from polypropylene with two rubber gaskets incorporated into the sleeve to give the sealing action between the pipe and the sleeve. The coupling is fitted onto one end, before lowering the pipe into the trench. The pipe is placed on a clean, firm base; lubricant is applied to the chamfered pipe end and the coupling is pressed onto the lubricated end.

With the O-ring type joint, the ring is placed into the groove on the spigot moulding and the socket of the preceding pipe is lubricated. The spigot is then placed up to the socket of the mating pipe and pushed in with a slight side-to-side action. The rubber ring is thereby compressed to make a flexible water-tight seal. Note: drains are normally laid uphill, in other words starting as the lower level and laid with the socket ends at the higher end. With both types of flexible joint it is essential that the mating surfaces are clean and that the lubricant supplied by the manufacturer is used. Any other type of lubricant could have a detrimental effect on the long-term performance of the joints; washing-up liquid should never be used.

Separate system

During the last 60 years or so, there has been a move towards 'separate' drainage systems. This requires extra pipework but does reduce the load on sewage treatment works; the storm, or surface, water being directed into rivers or straight out to sea.

A modern separate drainage system

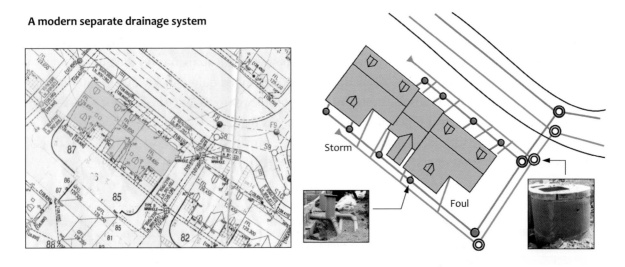

The principles of modern drainage systems have not changed much over the last 100 years; what has changed is the rules regarding their installation. The Building Regulations control the design and installation of drainage systems. They contain a number of rules regarding (among other things):

- pipe size, bedding material and gradients
- protection to pipes under roads and buildings
- provision for clearing blockages, siting of access points, size of access points
- ventilation.

A more detailed example of a modern system is shown opposite.

Separate system

Shallow access point suitable for garden.

SVPs do not require traps as all fittings in dwelling are trapped

Back inlet gully

Inspection chamber

Rodding eye allows rodding down drain

Foul Sewer

Back inlet gully

RE

RWP

SVP

Road gully

WC contains integral trap

WC

RWP

Drive

Gully

Storm Sewer

RWP

RWP

RE

Gully

Manhole

In a combined system, traps are required for RW pipes - not necessary in a separate system

SVP	soil vent pipe
WC	water closet
RE	rodding eye
RWP	rainwater pipe

Manholes (with working space at drain level) can be formed in brick or pre-cast concrete. Iron covers cope with the loads of traffic.

Trenches should not be excavated too far in advance of pipe laying and should always be kept to the specified design width. Excessively wide trenches increase the amount of excavation, bedding and backfilling, and may impose a load on the pipe in excess of the design load. The trench width should not be less than the pipe diameter plus 300mm.

Recommended pipe gradients vary depending on pipe diameter, number of dwellings served, and the pipe contents. 100mm foul pipes are usually laid somewhere between 1:40 and 1:80. 100mm storm pipes can be laid as flat as 1:100. There are no maximum gradients although it should be remembered that steep gradients (ie quickly dropping trenches) are expensive. In practice, branch drains (ie from dwellings to the main drain run) are likely to be quite steep; the main runs will normally as shallow as possible to minimise excavation.

Narrow trenches may impede the proper placing and consolidation of bedding material around the pipe and restrict working conditions in the trench. Where a specific trench width is defined for the narrow trench design condition, this must be maintained vertically to a height of at least 300mm above the crown of the pipe. The load-bearing capacity of an installed pipeline depends on the construction of a suitable bedding and surround. The bedding should level out any irregularities in the trench formation and ensure uniform support along the pipe barrel. Clay pipes are high strength rigid units which have been designed to carry applied loads with no deformation. Additional bedding and side-fill can, if necessary, enhance the pipes' load carrying capacity. Correctly sized granular material should be placed to the required level and extend the full width of the trench. Socket holes (where they exist) should be formed to prevent the weight of the pipe bearing onto the socket. Bricks or blocks must never be placed in the bedding material for setting the pipes to level; at the very least any bricks or blocks must be removed before backfill.

Plastics pipes are flexible and have little resistance to load without the support of granular bedding and sidewall material, which should be correctly sized and compacted. The bedding should normally be placed to at least the full height of the pipe. The side-support provided is essential in keeping deformation to the acceptable maximum of 6% of the pipe diameter. However, although this bedding and surround is expensive it must be accepted as the trade-off for using pipes which are lighter and easier to handle than their clay counterparts.

The Building Regulations with regard to drainage are straightforward but beyond the scope of this book. Most drainage systems were installed a long time before the present Regulations were written and, of course, the Regulations are not retrospective. The drawing on a later page shows typical problems in a modern installation. Its two main differences from the late Victorian example shown earlier are the increased use of access points (inspection chambers and manholes) and the separation of foul and storm water.

SUMMARY OF DRAINAGE DEFECTS

Many existing houses will have drainage systems which appear to work but do not meet with the demands of current practice. Early systems, for example, had few access points so blockages were difficult to clear. Fortunately, modern technology means that drain jetting is a relatively straightforward practice. This short section concentrates, not on some of the general problems highlighted earlier in the text, but on defects or poor practice which can cause blockages and/or leaks.

There are a number of ways in which drains can become blocked. Identifying the cause can sometimes take a surveyor minutes, although it may require specialist input and specialist equipment. Nowadays, there are a number of companies who can inspect drains internally using remote controlled cameras. These are relatively cheap when the alternative cost of digging up several sections of road or gardens is considered. Blocked drains can be caused by:
- inappropriate materials finding their way into the drainage system, eg disposable nappies, fat which solidifies as it cools
- in new developments, blockages may be caused by building materials finding their way into the drainage system during construction
- pipes crushing (see next section on leaks)
- pipes laid at incorrect gradients (excess or inadequate fall, pipes backfalling)
- undersized pipes
- sharp bends
- debris, earth etc falling into gulleys
- ground movement, damage caused by tree and shrub roots
- crushed inspection chambers.

Leaks (but not necessarily blockages) can be caused by:
- broken pipes generally – damaged by traffic, incorrect bedding, surround or backfill, levelling on bricks during construction, ground movement, tree roots
- fractures caused by lack of movement joint where pipes enter buildings, concrete or brick inspection chambers
- leaking – as above plus damage to sleeves or 'O' rings during construction.

Typical Drainage Problems

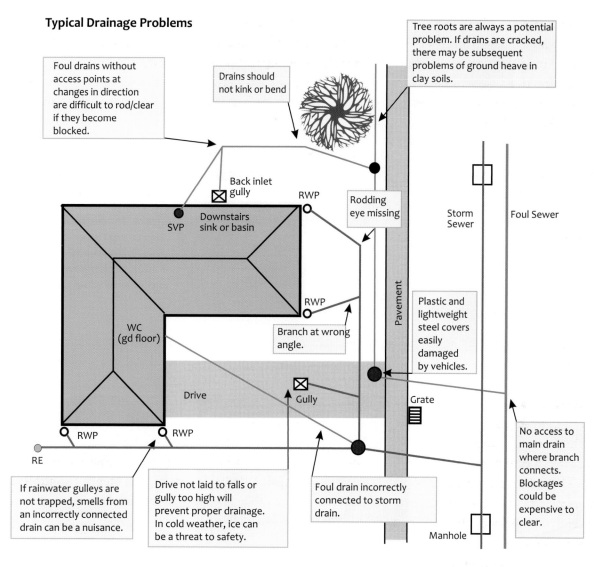

Foul drains without access points at changes in direction are difficult to rod/clear if they become blocked.

Drains should not kink or bend

Tree roots are always a potential problem. If drains are cracked, there may be subsequent problems of ground heave in clay soils.

Back inlet gully

SVP

Downstairs sink or basin

RWP

Rodding eye missing

Storm Sewer

Foul Sewer

WC (gd floor)

RWP

Branch at wrong angle.

Pavement

Plastic and lightweight steel covers easily damaged by vehicles.

Drive

Gully

Grate

RWP

RWP

RE

No access to main drain where branch connects. Blockages could be expensive to clear.

Manhole

If rainwater gulleys are not trapped, smells from an incorrectly connected drain can be a nuisance.

Drive not laid to falls or gully too high will prevent proper drainage. In cold weather, ice can be a threat to safety.

Foul drain incorrectly connected to storm drain.

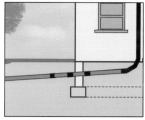

Where pipes run through walls they are likely to fracture unless there is provision for movement. Joints should be made either side of the wall. The pipes should run through a clear opening with a lintel over the top.

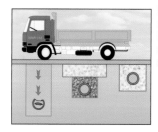

In roads or under drives deep pipes may not need protection. Shallower pipes can be crushed unless they are covered with a reinforced concrete slab or bedded and surrounded in concrete.

In deep trenches below, and close to, foundations, pipes must be protected by concrete.

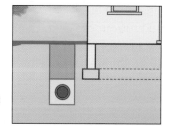

Other problems:
- *plastic inspection chambers in roads may collapse if not surrounded in concrete*
- *deep but narrow inspection chambers will be difficult to rod/clear*
- *drain pipes laid too shallow will not run properly; drain pipes laid too steeply will cause unnecessary excavation and are more prone to blockage – water flows but solids get left behind.*

Gulleys can block quite easily – especially if they are taking the 'run-off' from patios etc.

Two photos taken with a drain camera. The one on the left shows a small sewer – the pipe is clean. The one on the right shows a sewer which is slowly becoming blocked by solidified fat.

Until the 1980s shallow inspection chambers were normally formed in brickwork. If they exceed 900mm in depth they should be one brick thick. The middle example (mid 1970s) has never been finished (there is no benching). Nowadays inspection chambers are usually glass fibre or plastic. They are easily damaged, both during, and after, construction.

CHAPTER 18

Electrical Installations

INTRODUCTION

The electrical supply to most domestic buildings is known as a single phase supply and is currently standardised within the European Community at 230V 50Hz. The electrical supply to large buildings is usually supplied with three phases, each with its own colour cable. These are generated out of phase with each other and can be arranged to supply large electrical motors which then operate more effectively and with greater efficiency.

The majority of our local electrical supply is made up of these three phase distribution systems running in the pavements and roads from the local supply sub-station. To receive a single phase supply in a house a connection is taken from one of the colour coded phase supplies, balancing the load across all three phases by alternating the colour to different houses. Most domestic properties only have single phase supplies because the motors on domestic appliances are small, many below a rating of 1kW and, while some more powerful cleaners may be rated up to 1.5kW, this is considered acceptable for a limited period of use. In large buildings designers will usually provide a three phase supply for any permanently wired motor over the 1kW rating.

Over the years the electrical supply has been moved from overhead to underground, but the alternatives can still be seen. Generally, exposed electrical supplies are more likely to suffer damage during inclement weather from wind, ice build-up or overgrown and falling trees. Underground supplies may still suffer from road and building alterations and, occasionally, subsidence.

Electrical meters need to be fitted to measure and charge for the energy used, but there are many different 'tariffs' available and it is important to arrange for the most economic one to be selected depending on the nature of use. For example, a domestic user may be able to justify a night-time tariff even if there is no permanent electrical heating system, if a washing machine, dishwasher, tumble dryer and hot water immersion heater can be arranged to operate regularly overnight. Commercially, there are many more choices available.

The electrical wiring within the property comprises three wires, live, neutral and earth. The live wire supplies the power when required. The neutral completes the circuit allowing the current to flow when switched on. The earth wire is a protective wire diverting fault current away to achieve circuit isolation.

The distribution of electricity within properties is arranged in circuits for differing needs. These circuits start their distribution from a control panel called a consumer unit. The electrical supply is received by the consumer unit from the meter and can be fully isolated before splitting into circuits.

Lighting circuits are limited to a circuit load of 5A and the number of circuits is dictated by the number of lighting fittings, but all properties nowadays have a minimum of two. These are generally arranged to cover upstairs and downstairs, or day time/night time use in a bungalow. Socket outlets are supplied by a ring main which loops into all fixed outlets and returns to the consumer unit. Again there is a provision of two circuits usually, more as a property increases in size. However the number of socket outlets is usually chosen for convenience to give flexibility of room layout and not always for exact number of appliances required. The maximum rating of socket outlet is usually 13A to avoid overload. Immersion heaters and cookers both have their own individual radial circuit.

A typical wiring layout from the 1970s is shown below.

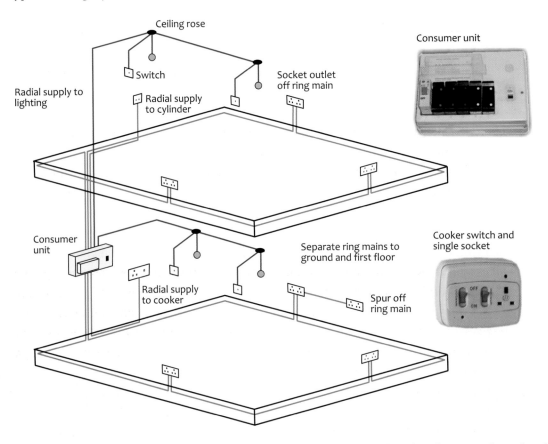

In modern houses there are usually additional radial feeds for showers, burglar alarms and smoke alarms.

Cables and flexes

Cables and flexes (the latter originally known as flexible cable) are quite different and should never be confused. Cables are usually hidden in the structure and carry the power to the sockets and lights. Flexes connect portable appliances and lamps to the power sockets. They are also used for light pendants and connecting items like storage radiators and immersion heaters to fused connections.

Cables and flexes are available in a range of sizes – they are given a rating based on the area of the conductor's cross section. The cable size depends on the amount of current likely to be carried in the circuit. A cable for a ring circuit, for example, is normally described as 2.5mm². In other words it contains two conductors each of 2.5mm². The third conductor, the earth, is usually slightly smaller – it only carries current when there is a fault.

Until recently the live and neutral conductor insulation (PVC) were coloured red and black respectively. The earth conductor is not insulated although it is normally covered with a green and yellow sleeve where it is exposed (ie in a socket). This type of cable is known as two core and earth. In 2004 colour coding changed (European harmonisation). Live is now brown and neutral blue – these colours have, in fact been used in flexes for several years.

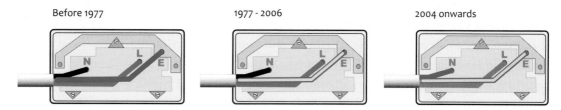

Before 1977   1977 - 2006   2004 onwards

Flexes are made up of multiple strands. Unlike cable the earth conductor in a flex (if there is one) is covered with green and yellow PVC insulation. Some appliances only have a two-core flex – these are items where there is no risk of the casing becoming live.

Unlike flexes, cables do not have to move during their working life – the expensive multi-stranded thin wire of flexes is not required and they can therefore be made from single strand conductors. However, thick cables, serving a cooker or electric shower for example, would be impossible to bend if they were made from single strands and so cables over 2.5mm^2 tend to made from five or seven strands.

Fuses

Protective devices are inserted in the consumer unit to prevent overload and these are usually fuses in older installations, with miniature circuit breakers (MCB) in more modern applications. Fuses also protect circuits at the point of connection of the appliances within either the plug or the socket outlet. Re-wirable fuses have a piece of special fuse wire running between the two terminals. The fuse wire generates heat when a fault current flows through it and it melts when the heat exceeds the acceptable level. This can happen if too many appliances are operated at once or if the current runs to earth due to a faulty connection. The melted fuse breaks the circuit and stops the electricity supply. Fuse wire is available in 5, 15 and 30amp ratings. Cartridge fuses work along similar principles but once this type of fuse blows the colour coded cartridge has to be renewed.

Circuit breakers are automatic protection devices fitted in the consumer unit which switch off a circuit if there is a fault. Circuit breakers are similar in size to fuse holders, but give more precise protection than fuses. If an MCB 'trips', it can be re-set using the switch on its face. However, it would be wise to find and correct the fault first.

Modern consumer units also include a residual current device (RCD) which disconnects a circuit whenever it detects that the flow of current is not balanced between the live and neutral conductors. Such an imbalance can be caused by current leakage through someone's body if it is 'grounded' and touching a live part of the circuit. To comply with the IEE Wiring Regulations, all sockets which could have outdoor equipment plugged into them must be protected with an RCD.

Older installations

Diagnosing many electrical problems requires specialist help. However, if nothing else, a competent surveyor should be able to recognise evidence of old installations. The cables will give an immediate visual clue (a socket or switch plate may have to removed to see the cable).

1920s – 1940s: Rubber insulated, two core (multi strand) covered with an outer lead, or India rubber and cotton braided, sheath If there is any of this very old wiring still installed, it should be isolated and replaced as soon as possible. The inner and outer rubber insulation will almost inevitably have degraded; in lead covered cable any break down in the insulation can cause the outer lead covering to become live – and an old installation still using this type of wiring is unlikely to have modern trips fitted in the consumer unit or an RCD. In addition, as the rubber deteriorates and starts to flake, sparks between the exposed wires can cause a fire. Lighting pendants and bed switches were made from a rubber insulated conductor covered with cotton or silk. The conductors were then twin-twisted. For portable appliances the best flex was circular braided, in other words the two conductors (three where an earth was included) were separately insulated but within one braiding. This was less liable to kinking than the twisted type.

Very occasionally, old fuse boxes are still found – they usually have a wooden back and cast iron switches.

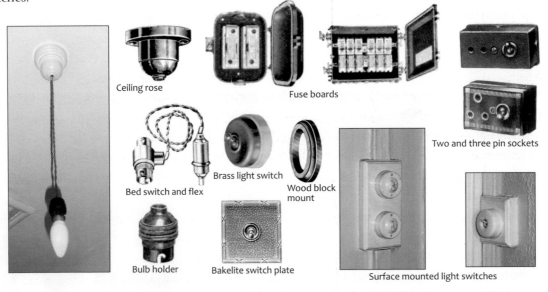

Ceiling rose

Fuse boards

Two and three pin sockets

Bed switch and flex

Brass light switch

Wood block mount

Bulb holder

Bakelite switch plate

Surface mounted light switches

Late 1940s – early 1960s: Rubber insulated, two or three core (multi strand) covered with an outer rubber sheath The danger with this type of cable is the degrading of the rubber especially where a terminal has become overheated.

Some houses (mostly from the 1960s) were wired with PVC insulated, three core (multi strand), with an outer PVC sheath. This may still be in good condition although broken strands sometimes occur at the terminals which reduces current carrying capacity of the cable and can cause local overheating.

Some common electrical defects

Plugs become unduly hot, scorch marks on sockets or plugs	Fuse of incorrect rating Loose cable terminals Poor connection between plug and socket Prolonged high loading (3kW or above) Poor quality (non BS) plugs
Sparks at socket outlet	Loose wires or faulty socket
Repeated blowing of fuses	Incorrect fuse rating Cable or accessory fault Overload on circuits
Smells of burning	Old wiring, loose connections, cables overheating
Bulbs blow repeatedly	Cheap bulbs Bulbs of high wattage in enclosed light fittings Mains voltage fluctuations
General loss of power	Main fuse blown, trip switches over sensitive, mains failure
Loss of power at individual socket	Fuse blown Loose connection

Care is required in selecting the routes of cabling within buildings to avoid possible physical damage or overheating. Where embedded in walls the routes should always be vertical so the socket outlet or light position is known to indicate the position of wiring. Nasty electrical shocks have been experienced by DIY enthusiasts when fixing pins into walls. It is also not recommended that cables are tightly bunched together, nor should they be tightly enclosed beneath thermal insulation where heat cannot dissipate.

A poor electrical connection or cable joint can be dangerous as the electrical energy attempts to jump the gap and any plug or fitting which shows signs of discoloration or a smell of burning should be attended to immediately. All electrical installations should be tested regularly to ensure their continued safe operation.

Outbreaks of fire are often linked to electrical faults or misuse, and the attack of systems by rodents can be devastating. Where mice or rats are detected it is essential that they are not only removed as a health hazard, but also from an electrical safety point of view. Finally, existing cabling may overheat of buried below thick roofing insulation – if possible the insulation should be laid below the cable.

More complex problems – periodic inspection
The Institution of Electrical Engineers recommend that, for domestic buildings, an inspection is carried out every 10 years (by a 'competent' person) to identify any deficiencies in an electrical installation. A periodic inspection will:

- reveal if any of the electrical circuits or equipment is overloaded
- find any potential electrical shock risks and fire hazards in the electrical installation
- highlight any lack of earthing or bonding
- identify any defective DIY electrical work.

Tests are also carried out on wiring and associated fixed electrical equipment. A schedule of circuits is also provided. The electrical installation is assessed against the requirements of BS7671:2008 – Requirements for Electrical Installations (IEE Wiring Regulations) and will include (among other things):
- adequacy of earthing and bonding
- suitability (or other wise) of the switchgear and control gear
- serviceability of equipment, eg switches, socket-outlets and light fittings

- type of wiring system and its condition
- provision of residual current devices for socket-outlets that may be used to plug in electrical equipment used outdoors
- presence of adequate identification and notices
- extent of any wear and tear, damage or other deterioration.

The overall condition will be assessed as either 'satisfactory', in which case no immediate remedial work is required, or 'unsatisfactory' which means remedial work is required to make the installation safe to use. The Inspection Report includes a section, *Observations and Recommendations for Actions to be Taken*, which highlights any departures from BS 7671. Each 'action' is coded.

- Code 1: Requires urgent attention.
- Code 2: Requires improvement.
- Code 3: Requires further investigation.
- Code 4: Does not comply with BS 7671 (but not necessarily unsafe).

A few examples form each category are set out below. Most of these are items that can only be identified by a qualified electrician or electrical engineer. The examples are taken from *Good Practice Guide No.4*, published by the Electrical Safety Council.

Code 1
- Exposed live parts that are accessible to touch.
- Conductive parts that have become live as the result of a fault.
- Absence of an effective means of earthing for the installation.
- Evidence of excessive heat (such as charring) from electrical equipment causing damage to the installation or its surroundings.
- Incorrect polarity, or protective device in neutral conductor only.
- Circuits with ineffective overcurrent protection (due, for example, to oversized fuse wire in rewireable fuses) Absence of RCD protection for socket-outlets in bathrooms or shower rooms, other than SELV (Separated extra-low voltage) or shaver socket-outlets.
- Absence of earthing at a socket-outlet.

Code 2
- A ring final circuit discontinuous or cross-connected with another circuit.
- A public utility water pipe being used as the means of earthing for the installation.
- A gas or oil pipe being used as the means of earthing for the installation.
- A 'borrowed neutral', for example where a single final circuit neutral is shared by two final circuits (such as an upstairs lighting circuit and a separately-protected downstairs lighting circuit).
- Absence of a warning notice indicating the presence of a second source of electricity, such as a microgenerator.
- Fire risk from incorrectly installed electrical equipment, including incorrectly installed downlighters.
- Undersized main bonding conductors, where the conductor is less than 6mm² or where there is evidence of thermal damage.
- Immersion heater does not comply with BS EN 60335-2-73 (that is, it does not have a built in cut-out that will operate if the stored water temperature reaches 98°C if the thermostat fails), and the cold water storage tank is plastic.
- Absence of RCD protection for portable or mobile equipment that may reasonably be expected to be used outdoors.

Code 3

- Unable to trace final circuits.
- Unable to access equipment or connections needing to be inspected that are known to exist but have been boxed in such as by panels or boards that cannot be easily removed without causing damage to decorations.

Code 4

- Circuit protective conductors or final circuit conductors in a consumer unit not arranged or marked so that they can be identified for inspection, testing or alteration of the installation.
- Undersized main bonding conductors (subject to a minimum size of 6mm^2), if there is no evidence of thermal damage.
- Absence of circuit identification details.
- Bare protective conductor of an insulated and sheathed cable not sleeved with insulation, colour coded to indicate its function.
- Installation not divided into an adequate number of circuits to minimize inconvenience for safe operation, fault clearance, inspection, testing and maintenance.
- Fixed equipment does not have a means of switching off for mechanical maintenance, where such maintenance involves a risk of burns, or injury from mechanical movement.
- The use of rewireable fuses (where they provide adequate circuit protection).

ELECTRICAL THERMAL STORAGE HEATING

During the 1950s the power supply industries were trying to find means of levelling out the peak demands for electricity and one of the ways introduced was to offer electrical power at a cheaper rate during the times of low demand. The most significant period was overnight, and the introduction of electrical thermal storage heaters proved very effective. The economies to the user however, were questionable under certain circumstances and led to a generally poor reputation for this type of heating. The main problem lies in the difficulty of anticipating how much heat is required for the following day and initially the systems introduced were unable to account in any way for this. Many buildings were overheated some days but other buildings cold by mid afternoon. The size of building would also dictate how much electrical power was consumed and the fact that the capital cost was very attractive hid the revenue implications of high energy costs despite the discounted prices.

The principle of stored heat relies on a mass of material such as clay or concrete blocks being heated overnight by electrical elements inserted between them, utilising the off-peak electrical supply. The thermal store is then dissipated during the day to keep the building warm. In cold weather, following complaints of insufficient heat to last the day, the supply authorities would allow a late afternoon boost but the cost to the customer then became even more expensive. Many different designs of storage heater were tried, including water as the thermal store, and a number of control systems incorporated to make the systems more effective.

Nowadays, thermal storage can be effective providing certain criteria are met.

- These systems should not be installed in large poorly insulated buildings. The low capital cost cannot justify the high energy costs.
- Ideally the building should be of a low air volume and the construction should be with materials of high thermal insulation values.

- Ideally the heaters should be capable of dissipating the heat under fan control and some form of external optimising controls, regulating energy input relative to weather conditions, should be incorporated.
- If these parameters are adhered to, electrical thermal storage systems can have distinct advantages over gas-fired boiler alternatives, especially in timber-framed designed housing as long as the power supply remains reliable.

Storage radiators can be very reliable and durable; their obsolescence is often due to changing tastes rather than mechanical failure.

Index